MW00611734

System
Synthesis

System Synthesis

Product and Process Design

Jeffrey O. Grady
JOG System Engineering, Inc.
San Diego, California, USA

CRC Press
Taylor & Francis Group
Boca Raton London New York

CRC Press is an imprint of the
Taylor & Francis Group, an **informa** business

CRC Press
Taylor & Francis Group
6000 Broken Sound Parkway NW, Suite 300
Boca Raton, FL 33487-2742

© 2010 by Taylor and Francis Group, LLC
CRC Press is an imprint of Taylor & Francis Group, an Informa business

No claim to original U.S. Government works

International Standard Book Number: 978-1-4398-1961-6 (Hardback)

This book contains information obtained from authentic and highly regarded sources. Reasonable efforts have been made to publish reliable data and information, but the author and publisher cannot assume responsibility for the validity of all materials or the consequences of their use. The authors and publishers have attempted to trace the copyright holders of all material reproduced in this publication and apologize to copyright holders if permission to publish in this form has not been obtained. If any copyright material has not been acknowledged please write and let us know so we may rectify in any future reprint.

Except as permitted under U.S. Copyright Law, no part of this book may be reprinted, reproduced, transmitted, or utilized in any form by any electronic, mechanical, or other means, now known or hereafter invented, including photocopying, microfilming, and recording, or in any information storage or retrieval system, without written permission from the publishers.

For permission to photocopy or use material electronically from this work, please access www.copyright.com (http://www.copyright.com/) or contact the Copyright Clearance Center, Inc. (CCC), 222 Rosewood Drive, Danvers, MA 01923, 978-750-8400. CCC is a not-for-profit organization that provides licenses and registration for a variety of users. For organizations that have been granted a photocopy license by the CCC, a separate system of payment has been arranged.

Trademark Notice: Product or corporate names may be trademarks or registered trademarks, and are used only for identification and explanation without intent to infringe.

Library of Congress Cataloging-in-Publication Data

Grady, Jeffrey O.
    System synthesis : product and process design / Jeffrey O. Grady.
        p. cm.
    Includes bibliographical references and index.
    ISBN 978-1-4398-1961-6 (alk. paper)
    1. Systems engineering. 2. Industrial design. 3. Concurrent engineering. I. Title.

TA168.G654 2010
620.001'171--dc22   2009050498

Visit the Taylor & Francis Web site at
http://www.taylorandfrancis.com

and the CRC Press Web site at
http://www.crcpress.com

CV 09.25.2020 1530

Contents

Part one: The fundamentals

List of figures

List of tables

Preface

This book was originally conceived as a companion to an earlier work titled *System Requirements Analysis* (SRA) (Grady, 1993, 2005) and was titled *System Integration* (Grady, 1994). These two activities, requirements analysis and integration, form twin buttresses for the system engineering process connected through the creative design process for the many items that comprise a system. Yet, there were few books available on these subjects particularly as they were being influenced by the concurrent development notion. I was stimulated to start work on the predecessor book after a conversation with a colleague, G.D.H. (Dan) Cunha, who had volunteered to teach a course on this subject at the University of California San Diego Extension systems engineering certificate program while we were both employed by General Dynamics. Dan could find no adequate textbook and we had some difficulty determining the boundaries and focus of the subject.

That I could not help Dan determine a sound approach in teaching this course after getting him to agree to do it was an especially bitter pill for me to swallow because one of the areas in the department I managed at the time was titled "System Integration." Dan selected what turned out to be a very effective teaching technique composed of requiring each student to research two or three sources on integration in papers, books, and magazines and report upon them in class. At the end of the class, he published all of these collected sources and distributed them to the students. It turned out to be a very helpful collection of information in steering my own attitudes about integration, but there was still something missing in my own mind relative to teaching others how to do this work well.

Later, work on a generic system development process diagram at General Dynamics Space Systems Division yielded the insight that several very different kinds of integration work were needed in a concurrent engineering environment. Some integration work had to be done in a cross-functional and coproduct fashion while other work aligned with a cofunctional and cross-product orientation. By applying a common system engineering thought process to these insights (decomposition and a search for completeness), it evolved that one could use a simple three-space coordinate system to express all integration situations in terms of points, lines, planes, and spaces.

With this aid, one could be sure that all of the possibilities would be covered. It also appeared that it would be possible to apply this theoretical construct in a practical way to help engineers understand the full sweep of integration situations. That is, it would be possible to teach people how to do system integration work in a classroom. Thus was born a book to fill the void that Dan and I had earlier uncovered. (Please visit http://www.crcpress.com/product/isbn/9781439819616 for more details.)

Throughout the writing of this book I tried very hard to detune the content to recognize the widest possible range of customers from the Department of Defense (DoD) to commercial sectors in all product fields. I confess this was difficult because my principal experience had been in DoD programs and systems. There is an increasing interest I know in applying the systems approach in the non-DoD government and commercial sectors, and several colleagues in the International Council on Systems Engineering (INCOSE) were helpful in sensitizing me to these efforts. Doubtless, I am not completely cured and the reader interested in commercial applications may have to tolerate an appeal to what some would consider unnecessary rigor in some parts of the book. There is, however, another side to this question and that is that the DoD-derived process has a great deal to offer everyone. No organization has devoted more energy to encouraging an effective system development process than DoD because no other organization has had to deal with such complex problems with more adverse consequences of failure.

Acknowledgments

The original version of this book would not have been written without the encouragement of my wife, Jane. Shortly after finishing a previous book, I expressed to her that it would take too long to write a book on integration, the other part of the system engineering story. She reminded me how fast things seem to come together once you start. And so it is. Thank you, Jane.

Many thanks to Bernard Morais, president of Synergistic Applications Inc., and Jim Lacy, president of Jim Lacy Consulting, for their helpful suggestions as reviewers of an earlier version of this manuscript. I also benefited from conversations with Gene Northup, executive VP of Leading Edge Engineering, Inc., who shined a light in the direction of the future of the system engineering profession unconstrained by the boundaries of our facilities through advances in computer technology.

I most appreciate the discussion in 1991 with Dan Cunha, then of General Dynamics Convair Division, about the difficulty in developing a strategy for teaching this subject, for it was this stimulus that started me thinking thoughts that now appear here.

Subsequent to the publication of *System Integration* (Grady, 1994) Dan Cunha was tragically killed in an auto accident and the world lost a man who would have matured into a tremendous system engineer and manager. He had already achieved much as a naval officer, father, professional member of our engineering community, and human being.

Several parts of *System Requirements Analysis* (SRA) (Grady, 1993) have been used in this book with minor changes. The content of *System Integration* (Grady, 1994) is also included with changes combined with considerable additional information to broaden the scope from integration to synthesis.

We are the sum of every story we have ever heard or read. I have heard many, many system engineering stories, very few of them of a humorous nature but all of them leading to new understandings, from my colleagues in the field of system engineering over more than a few years, especially in the past several since the founding of the INCOSE. After a while one cannot associate the sources with the specific stories nor even discriminate a story from its aggregate effect on our current knowledge base. Paraphrasing Willie Nelson, let me say, "To all the system engineers I have ever known, thanks for the effect you have had on me, no matter your intent at the time." I hope the content of this book, some of which could have come from you at some time in the past, will have a positive effect on you and your future.

Jeffrey O. Grady
San Diego, California

Author

Jeffrey O. Grady has been the president of JOG System Engineering, Inc., San Diego, California, a system engineering consulting and training company since 1993, working with a wide array of companies with a broad cross section of product lines. Prior to this, he worked for 30 years in aerospace companies as a customer-training instructor at Librascope on antisubmarine rocket (ASROC) and submarine rocket (SUBROC) underwater fire control systems; field engineer, project engineer, and system engineer at Teledyne Ryan Aeronautical on unmanned day and night as well as strategic and tactical photo reconnaissance, SIGINT, strike, EW, and target aircraft launched from ground launchers, aircraft, and shipboard; system engineer at General Dynamics Convair on advanced cruise missile; and system development manager at General Dynamics Space Systems Divisions on space transport (Atlas) and energy systems.

Jeff has authored nine published books and numerous papers in the systems engineering field. He holds a bachelor's degree in mathematics from San Diego State University, a master of science in system management with a certificate in information systems from the University of Southern California, and a systems engineering certificate from the University of California San Diego. Jeff teaches system engineering courses around the country onsite at companies and universities including the University of California San Diego. He is an International Council on Systems Engineering (INCOSE) fellow, founder, and certified system engineering professional (CESP).

Acronyms

ABS	Automatic Breaking System
AFB	Air Force Base
AFSCM	Air Force Systems Command Manual
ALS	Advanced Launch System
ASIC	Application-specific integrated circuit
BIT	Built-in test
CAD	Computer-aided design
CCB	Configuration control board
CDRL	Contract data requirements list
CEO	Chief Executive Officer
CG	Center of gravity
CMM	Capability maturity model
CMMI	Capability maturity model integrated
COTS	Commercial off the shelf
C/SCS	Cost/schedule control system
DBS	Drawing breakdown structure
DCSM	Design constraints scoping matrix
DET	Design evaluation testing
DID	Data item description
DIG	Development information grid
DoD	Department of Defense
DoDAF	Department of Defense Architecture Framework
DSMC	Defense Systems Management College
DTC	Design to cost
DT&E	Development Test and Evaluation
ECP	Engineering change proposal
EFFBD	Enhanced functional flow block diagram
EIA	Electronic Industries Alliance
EIT	Enterprise integration team
ERB	Engineering review board
EVS	Earned value system
FAA	Federal Aviation Administration
FCA	Functional configuration audit

FFBD	Functional flow block diagram
FMEA	Failure modes effects analysis
FMECA	Failure modes effects and criticality analysis
GD	General Dynamics
GPS	Global positioning system
HMO	Health maintenance organization
HW	Hardware
ICBM	Intercontinental ballistic missile
ICD	Interface control document
ICDB	Interim common database
ICWG	Interface control working group
IDEF	Integrated computer-aided manufacturing definition
IEEE	Institute of Electrical and Electronics Engineers
ILS	Integrated logistics support
IMP	Integrated master plan
IMS	Integrated master schedule
IOC	Initial operating capability
IPPT	Integrated product and process team
IRFNA	Inhibited red fuming nitric acid
ISO	International Standards Organization
IV&V	Independent verification and validation
LCC	Life cycle cost
LCI	Logistics critical item
LSA	Logistics support analysis
M	Mean active maintenance time
MBA	Manufacturing breakdown structure
MDT	Maintenance downtime
MID	Modeling identification
MSA	Modern structured analysis
MTBF	Mean time between failures
MTBM	Mean time between maintenance
MTTR	Mean time to repair
NASA	National Aeronautics and Space Administration
NDI	Nondevelopmental item
NFL	National Football League
OA	Operations analysis
OEM	Original equipment manufacturer
OMG	Object management group
OOA	Object-oriented analysis
OT&E	Operational test and evaluation
PIT	Program integration team
PMP	Parts, materials, and processes
PSARE	Process for system architecture and requirements engineering

QA	Quality assurance
RAM	Reliability, availability, and maintainability
RAMCAD	Reliability, availability, and maintainability computer-aided design
RAS	Requirements analysis sheet
RDD	Requirements driven design
RFP	Request for proposal
SAC	Strategic Air Command
SBD	Schematic block diagram
SDRL	Supplier data requirements list
SEM	System engineering manual
SEMP	System engineering management plan
SOW	Statement of work
SW	Software
SysML	System modeling language
TAC	Tactical Air Command
TBD	To be determined
TPM	Technical performance measurement
TQM	Total quality management
TRW	Thompson Ramo Wooldridge Inc.
TSA	Traditional structured analysis
UFDM	UML Framework for DoDAF Modeling
UML	Unified modeling language
USAF	United States Air Force
WBS	Work breakdown structure

part one

The fundamentals

chapter one

Introduction

1.1   Systems and their birth

A system is a collection of things that interact to achieve a common purpose. Natural systems exist that consist of organic and inorganic materials. These systems come to be through natural processes unaided by man. It is true that man is increasingly influencing these natural processes but, to date, the forces man is able to control are minuscule compared to those of nature. Even nuclear weapons pale compared to the forces applied through hurricanes, volcanic eruptions, and potential meteor strikes. In addition to natural systems on this Earth and beyond, man, over a period of thousands of years, has evolved a capability to create things, very complicated things, that satisfy our definition of a system.

Man-made systems come into being through a process that will be referred to in this book as system development, a process for solving a complex problem initially framed in a customer need statement verbally presented or in writing uttered or stated by someone who concludes, "There is something I must do that I cannot do with what I have." The system development process may be applied to a new, unprecedented problem never before solved resulting in a clean-sheet-of-paper development but can also be applied to major modifications or reengineering of existing systems. The system development process is accomplished by programs organized, staffed, and run by enterprises for organizations often called acquisition agents to be used by organizations often called users.

These man-made systems may be of two kinds but most often include both aspects. Product systems consist of material things commonly manufactured specifically for a particular purpose. These things in the system are organized into a hierarchical product entity structure forming a physical solution-oriented model of the real system. Process systems are also possible where things are accomplished in some fashion generally by people performing in prescribed ways and most often including machines as well. Man-made systems dominated by new or modified machines should be created in combination with the development of the corresponding processes in which they will be used.

In the past, some program managers have encouraged integration and optimization of the product system, but that alone is not adequate. The product system cannot be developed in isolation from the corresponding process without considerable risk of product design changes late in the program or unfulfilled customer expectations. Thus, an integrated product and process development process is encouraged in this book where programs employ teams organized around the elements in the product that are responsible for concurrently developing product and related processes. This is a much more difficult integration and optimization task than simply optimizing the product system, but success in so doing will yield a superior system of greater value to the customer in terms of cost, schedule, and performance.

Still, a more difficult integration job even than concurrent development of product and process is encouraged in this book. Programs that develop product and process systems as well as the enterprises within which they function are man-made systems themselves. These are primarily process systems and the ultimate integration and optimization capability includes a coordinated development of product and process systems for delivery to customers and continuous improvement of the enterprise system based on lessons learned from the product and process systems developed by the enterprise programs. This is what the author refers to as grand systems development where the enterprise is integrating and optimizing about the enterprise level and insisting on its programs doing the same for their products and processes relative to program requirements.

1.2   Synthesis, optimization, and integration: What are they?

Synthesis, according to the dictionary, is an act of combining constituent elements of separate material or abstract entities into a single or unified entity. This is not unlike the dictionary definition of the word integration. The author uses the word synthesis (system development process step 2) to refer to the design process responsive to the predefined requirements (system development process step 1) addressed in *System Requirements Analysis* (Grady, 2005). This design process encompasses three significant components: (1) engineering design of the product, (2) design of the material and procurement processes and operation of these process to make material available to the manufacturing process, and (3) manufacturing engineering design of the manufacturing process coordinated with the quality engineering design of the corresponding quality assurance process and operation of these processes to produce a quality product.

There are other process design activities that must occur and be coordinated with the product design process certainly such as logistics or sustainment and verification (system development process step 3). The former is treated in this book as a specialty engineering activity, to the extent that it deals with product and process development, in Part five, and the latter as the third major system development process step in the development activity trio that is covered in *System Verification* (Grady, 2007). The author does recognize a major program life cycle phase referred to as sustainment within which the user operates and maintains the developed system over its life, but this is outside the area of interest in this book and the author has not published in this field. The process of developing a system to have good sustainability (operability, maintainability, and supportability among other things) features is an important part of the system engineering process discussed in all of the author's published work, however.

The author includes not only the design of the manufacturing, quality, and material and procurement processes, but also their operation under the title "synthesis." Thus synthesis, in this book, is the complete transform from a set of requirements that define the problem and prescribe the boundary conditions for the design solution into a physical product and accompanying process description that will satisfy the customer need and solve the problem defined in the requirements. There are, therefore, three major transformations involved in the synthesis process:

1. Transformation between a clear definition of the problem space in a set of specifications into an integrated set of product design solutions expressed in engineering drawings, parts lists, and a software architecture
2. Transformation of the engineering design into a set of procurement sources, plans, and contracts together with plans for processing the arriving material between the receiving and the manufacturing locations
3. Transformation of the available material into a finished product ready for delivery through a manufacturing process where the steps in this process have been under the watchful eye of a quality assurance influence throughout

The design engineers who accomplish synthesis work are putting many ideas together that have been defined in the item specification. These designers are synthesizing a design solution from the many performance requirements and constraints or boundary conditions such that the resultant design is compliant with all of those requirements. Persons doing the other forms of synthesis are doing the same thing relative to other sets of constraints. Both the product and process designers must work together exchanging and debating alternative solution ideas in search of the optimum solution to the true system problem composed of both the product and process solutions. The principal card in the hand of the design engineer in accomplishing this work is creativity. However, since every engineer working on the solution is a specialist, there must be another force applied to the solution process. That force is optimization and integration. This reality is manifest in so many ways in the development process. Whenever we decompose or partition anything it is necessary to integrate and optimize as well. We are forced to partition man's knowledge base into the heads of many people—all specialists—because no person's mind is sufficiently grand to master the whole of man's vast knowledge base. Each of these minds involved in a development effort will be doing its best to solve the design problem, but no one mind can completely understand the whole problem space. Thus, teams are formed that contain the right mental capabilities to solve some part of the problem provided they can communicate among themselves to create the equivalent of one great mind that really does contain all of the knowledge needed to solve the problem. Man's method of performing this combinatorial process is called integration and optimization in this book.

One of the most often used words, yet most neglected notions, in the application of the system engineering process in industry is the word integration. It is a word like system. It has so many meanings and shades of gray that the listener or reader is never quite sure exactly how another person is using the word. Whatever integration is, however, it is a universally accepted necessity in the development of systems that attempt to solve very complex problems. It is at the downstream end of an organized process of decomposing complex problems into many smaller related problems, solution of the smaller problems by teams of specialists, and integration of the results into a solution to the original larger problem.

The system integration process is the art and science of facilitating the market place of ideas that connects the many separate solutions into a system solution. System integration is a component of the system engineering process that unifies the product components and the process components into a whole. It ensures that the hardware, software, facilities, and human system components will interact to achieve the system purpose by satisfying the customer's need. It is the machinery for what some call concurrent development.

The general approach in system engineering involves decomposing a customer need into actions or functions that the system must satisfy in order to satisfy that need. These lower tier functional requirements are then allocated or assigned to specific things or components in the system. These allocated functions are translated into performance requirements and combined with design constraints to form a set of minimum attributes (requirements) that a design team must satisfy in order for the component to fit into the overall system solution.

These requirements are synthesized by a team of engineers into one or more design concepts and, if more than one, the alternatives are traded one against the other in accordance with a predefined value system to select a preferred solution. The design concept is expanded into a preliminary and detailed design interspersed with reviews to assure the team is on a sound path toward success and the work of several other teams is in agreement. Specialists from many disciplines work with the principal designer or design team to assure that the solution accounts for specialty requirements with which the designer may not be fully familiar.

Designs are translated into manufacturing planning and procurement decisions, and the production and procurement works begun. Special test article versions of the product elements may be subjected to special testing to build confidence that a solution of the customer's need is possible.

1.3   Toward a more effective process

But in this process, is there not some way to characterize the activity called integration so that we may all agree on its meaning and upon which we can base a more effective and foolproof deployment of its power into our own system development programs? In this book, we establish a premise that integration has a finite number of component parts, each uniquely definable, and follow the consequences of this premise to their ultimate conclusions. We will decompose integration work into its fundamental parts and integrate the results. Yes, we are going to apply the system engineering process to improve our understanding of this word and concept.

This book was originally conceived as a companion work to *System Requirements Analysis* (Grady, 1993, 2005). These books cover problem space decomposition, solution space product entity identification, and the requirements definition process occurring on the front end or down stroke of development programs. Integration work is necessary across the development effort but principally occurs in the middle of the program during the product design effort as a part of the synthesis process, and is followed on the tail end or up stroke of development programs by verification where we prove to what extent to design solution satisfies the requirements that were the basis for the design. These two system engineering activities, together with verification by testing and analysis, comprise the heart and soul of the systems approach to the development of systems to solve complex problems.

Some observers of the process expressed in several books by the author have concluded that it is only appropriate to apply it to grand systems with a Department of Defense (DoD) customer known for its deep pockets. The author believes the system engineering process is perfectly applicable to smaller scale developments as well as large ones and those oriented toward commercial customers as well as DoD and National Aeronautics and Space Administration (NASA) customers. The Mars mission space transport system will include within it an onboard computer (many actually, but certainly one) that can be treated as a system by its supplier development team. That computer will include a power supply that could be treated as a system by its vendor design team. A company that develops and manufactures video recorders for the commercial market can profitably apply these same principles as could a company whose product line is farm tractors.

Some of the component parts of the process may be overkill for some combinations of product lines (like bean bags), customer bases, and degrees of difficulty in the development process or maturity of the corresponding technology needed. The skillful development team will select, or the experienced customer will insist on, the components that make sense for the particular program.

It is true that the process, in the past, has been primarily applied by the aerospace industry for large government customers interested in solving very complex problems. Much of the language used to express system engineering ideas and references for parts of the process are derived from this historical reality. The author has made a conscious effort to present the system integration process from a generic perspective but is himself a captive of the same history and may not have succeeded completely. There is another side to this story, however. The author believes that people in firms in the commercial market place are little different from those in the DoD industry in their resistance to integrated and structured development.

Individual creative engineers are commonly not fond of the systems approach because they feel it deprives them of an opportunity to more fully express their creative talents. Many of the design engineers the author has worked with would have preferred to have been left alone to try to understand what kind of design would best satisfy the problem solution even if they knew from past project failures that an organized approach to a solution to the problem would provide a better chance of success.

It is true that there are aerospace firms with very fine system engineering capabilities that they effectively apply with good results. The systems approach to problem solving has been available for many years yet there are also many managers, directors, and vice presidents in engineering organizations that support DoD and NASA customers who distrust it and avoid its successful application. The stories are many and in agreement with some of the author's own experiences in industry of frustrated attempts to implement an effective system engineering process. The story often goes something like this:

1. The customer requires in a request for proposal some evidence that a company has a system engineering capability. Typically, this may take the form of a requirement that a system engineering management plan ·(SEMP) or system engineering plan (SEP) be submitted

with the proposal. The company will write a credible SEMP/SEP with no intent to organize or work as described therein once the contract is awarded.

2. When the contract is awarded to this firm, work begins using autonomous design groups interacting on an ad hoc basis. The customer becomes concerned that the company has neither a system engineering organization nor process, and encourages the company to comply with their process requirements stated in the statement of work and the SEMP prepared by the contractor.

3. Finally, to get the customer off their backs after the design solution has been essentially determined, the company creates a system engineering function and assigns personnel to the program to accomplish a parallel system engineering process along the lines described in the SEMP. This team may be isolated from the design people and even told not to interact with them as it would be a distraction from their important work. They produce the standard set of system engineering products that may entail differences from the predetermined design solution, and these products are essentially ignored.

4. At the earliest possible moment when the customer has been minimally satisfied that a systems approach has been applied, this group is disposed of.

This is called paying lip service to the systems approach. It occurs in commercial firms as well, where everyone knows that to be certified at level 5 for software and system engineering as well as certified against ISO 9001 for quality can be good for business. One of the fundamental requirements for a high ranking in these assessment criteria is possession of a written practice description and compliance with that documentation while performing program work. In one company the author did some work for, he asked to review their system engineering manual as a precursor to the work knowing they were recently ISO 9001 certified. He was told that they had no such manual. The author asked how that could be, and was told that since they had no system engineering organization they would have been in conflict with certification criteria if they had a manual for a nonexistent organization.

Despite this history, even in some of our finest companies, some tremendous products have been developed that have served their customers (DoD and the general public) well. In some cases, not all, we will never know how much less these products would have cost or what other features or capabilities they may have had if a sound systems approach had been applied in their development. There appears to be a natural resistance to the systems approach driven by a human urge for power and control on the part of those in functional management and by a human urge to belong to a group by the members of these organizations. The resulting autonomy of the functional organizations is the enemy of the systems approach on programs, and it is alive and well in more places doing DoD work than one would think possible after decades of DoD insistence on an effective systems capability and the emergence anew of concurrent development ideas.

The point is that this same resistance, inspired by the same human motives, exists in commercial firms as well and prevents them from crossing the bridge sincerely looking for an effective organized product development approach that DoD has inspired but not always been fully successful in obtaining from its contractors. Implementing an effective systems capability requires leadership from the top to impress everyone with the understanding that product success and a happy customer take priority over employee or management attitudes and company organizational dynamics. Product success results from customer value and that can be encouraged through an organized approach to problem solving called system engineering.

The author believes that the systems approach and happy employees are not mutually exclusive. The organizational structure and techniques encouraged in this book will provide an environment within which people, design engineers, as well as system engineers and specialists, can excel in their profession. It will result in the imposition of the minimum set of constraints on the solution space within which the design community must seek a solution. It will result in the lowest cost and greatest value to the customer in the product they receive. Commercial companies that introduce appropriate elements of this process and educate their employees to perform them well will have an advantage over their competitors in terms of product cost, time to market, and product performance.

As we depart from a phase of history characterized by intense global military competition requiring increasingly sophisticated military systems, we will find that complex systems will be associated with a broader range of needs than in the past and the pace of development will be retarded from that of the past driven by a survival instinct. There will be a greater need to reengineer old systems to upgrade their capabilities than to develop totally new systems pushing the state of the art in several directions at once. There will be a greater appeal to precedented systems than unprecedented ones involving a completely clean sheet of paper. Regardless of these changes, however, we are unlikely to discover a method for developing these new and modified systems that is more effective than the methods discussed in this and similar books for they are based on the way we humans function.

1.4   Scope of integration work

Integration is necessary in any case where we partition work into subsets coordinated with specific knowledge domains, Since we accomplish every part of the system development process following this pattern including requirements, design, manufacturing planning, testing, and analysis, it appears we will have to do integration work throughout the development period. The author believes this to be true. But some companies prefer to break this integration work into parts and refer to some of these parts as something other than integration. For example, a company may accomplish integration work in requirements identification but refer to that as simply part of requirements management. Similarly, the integration work done during the design period may be thought of as an integral part of the design process. This same company may link the integration and verification words together where integration is the work that connects all of the isolated test and evaluation work together into hierarchical planes of work, each higher level integrating the lower tier elements together. This process begins with the availability of a physical product and a code but is certainly not the complete integration task. Later in this part, we will explore the intellectual integration task that is accomplished early in the program prior to the physical integration activity that occurs later in the program.

1.5   Book organization

The author treats this book as the third book of four books covering the system engineering work over the system development life cycle time frame in an enterprise. At the time this book was being written, *System Requirements Analysis* (Grady, 2005) had been released, *System Verification* (Grady, 2007) had also been released, and *System Management: Planning, Enterprise Identity, and Deployment* (Grady, in press) was being published. The *System Management: Planning, Enterprise Identity, and Deployment* book combined two earlier books by the author *System Engineering Planning and Enterprise Identity* (Grady, 1995) and *System Engineering Deployment* (Grady, 2000), and this book has been based on *System Integration* (Grady, 1994). Figure 1.1 illustrates the common process the author applies throughout his published work. The circled numbers coordinate major life cycle tasks to the four books noted above.

This book is composed of seven parts. Part one (Chapters one through seven) explores the fundamentals of integration and offers a suggested organizational structure used throughout the book in other discussions. Part two (Chapters eight through eleven) focuses on generic process integration for the organization accomplishing a development program. The enterprise should have a generic process that satisfies its ultimate

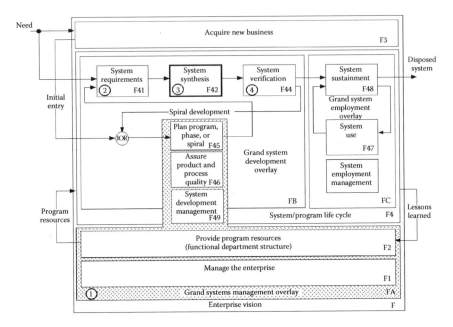

Figure 1.1 Enterprise vision decomposition.

enterprise requirement referred to in this book as a vision statement (note the top level function in Figure 1.1) and it should be possible to show traceability from this process description to external standards of excellence such as the following:

Quality	ISO 9001
System engineering	ANSI/EIA 632 or IEEE 1220
Software development	IEEE/EIA 12207or EIA J-STD-016

Further, when the organization is audited in accordance with a capability maturity model (CMM) such as the Carnegie Mellon University Software Engineering Institute (SEI) model for software, EIA 731 for system engineering, or CMMI for the whole enterprise, the enterprise should score very well because they have effective practices that they actually follow in performing program work and they have a way to continuously improve their process.

Part three (Chapters twelve through fifteen) covers the program planning process based on the generic model covered in Part two torqued for the peculiarities of a specific program. Part four (Chapters sixteen through twenty-two) discusses the implementation of program planning in four chapters.

Part five (Chapters twenty-three through thirty-one) deals with product and product use process integration. This book maintains that the successful system developer must integrate and optimize not at the product level, not at the combined product and product use process, but jointly the enterprise and multiple program level as well. Commonly, we speak of system integration focused on the product system elements, but this is not sufficient. Part six (Chapters thirty-two through thirty-five) extends integration ideas to this fusion of enterprise and program processes. Part seven (Chapter thirty-six) closes the conversation.

Parts two through four and six are the province of system engineers who look at their profession as a management science the application of which allows them to get complex things done. People who do this work are commonly called project engineers and they are responsible for technically planning programs, setting up budgets, defining planned work in statements of work and program plans, and tracking programs to the planned budget and schedule. Part five is tuned to the system engineers who view their profession as an engineer solving complex product-oriented problems. These engineers work at the product interfaces and seek out inconsistencies between the system requirements and the current concept or design and within the overall work product of several different development teams often focusing on product and development organization interfaces.

System integration embraces both of these perspectives, process, and product integration. While it is not uncommon for people with very different experiences to perform these two different functions on actual programs, the essence of the work they perform has a great deal in common. They work in the same patterns with different subject matter. Also, U.S. Air Force work on a concept called integrated management system has helped to more intimately connect these two component parts of the integration picture than heretofore, and we will take advantage of that work in this book extending it to embrace the generic development process. In addition to isolated product and use process integration, we must accomplish integration between these two entities as well in order to achieve the best possible product and customer value. Our reputation and reward will flow from how well we produce sound value for our customers. Excellence in system integration is a necessary prerequisite to the development of effective systems that customers will find useful, effective, affordable, and delightful.

1.5.1   Part one: The fundamentals

We begin in this part with an overview of the system development process offering a framework within which the technical and programmatic work may proceed. We then discuss the foundation stone upon which all system work rests, man's knowledge base and our knowledge capacity limitations. The need for integration is driven by human limitations and competition because these same forces encourage specialization and decomposition of complex problems. The universal solution to individual knowledge deficit is specialization. Because we decompose knowledge and large problems, we must engage in very effective communication and integration work. Decomposition or specialization is easy to perform. Integration is very difficult to perform.

Chapter four breaks the overall integration problem up into 27 simple integration components, each of which can be easily understood and applied. Why are there 27, and not 25 or 13, and how can we be sure there are no more? We can be sure because we will apply a sound systems approach that ensures completeness in our search for integration components.

Chapter five covers four key roles in the system development process: program managers, functional department managers, system engineers, and domain engineering leaders. Program managers run programs. Functional managers provide programs with resources. System engineers specialize in generalism and complexity. Domain specialists and managers provide the enterprise with the detailed knowledge needed to solve very complex problems represented by systems and their related customer needs. Chapter six offers what the author claims is an optimum organizational structure for performing system development work in an enterprise that simultaneously develops multiple systems. Since integration is nine-tenths human communications, understanding the relationship of integration work to information systems is essential. We will find in Chapter seven that a big part of the communications answer is available in the computer systems that many companies have already paid for. But, how can we use these resources more effectively?

1.5.2   Part two: Enterprise common process integration

In Part two, the book applies a subset of the 27 integration components to the development process, to the system that gives birth to programs that in turn give birth to product and process systems. This is the system of which we humans are a part. The point is made that an enterprise should have an identity formed of its vision statement, process description, and coordinated resources base available for application to all of the programs implemented by the enterprise.

1.5.3   Part three: Product system definition

Part three explains how the U.S. Air Force initiative called the integrated management system can be applied by any company in any product field to satisfy customer needs and improve contractor efficiency through development of sound planning that is clearly related to customer needs. Chapter thirteen extends the U.S. Air Force approach to a generic enterprise and program peculiar planning process resulting in a seamless program planning process based on the generic data. Chapter fourteen then discusses how to deploy this structure into early program work and how this pattern changes over time in the program life cycle.

1.5.4 Part four: Product design synthesis

In Chapter sixteen, we discuss the principal integration techniques for program execution and the greatest challenge in industry today: how do we organize the work of many specialists to be brought to bear on common problems without destroying their creativity? This is the proper realm of management. This chapter also deals with the common situation of program execution under faulted conditions leading to discontinuities that must be healed. We must understand the problem, replan the remaining program, and get back on the plan. Chapter seventeen discusses the use of modeling and simulation during development. Chapter eighteen covers decision-making techniques that can be effective at a time when we have not yet amassed all of the information we would like to have when the decision must be reached. This includes accomplishing trade studies.

Chapter nineteen discusses integration in the context of cross functional teams while Chapters twenty and twenty-one cover two of the three design steps recognized in this book—preliminary and detailed. These are commonly preceded by concept development that must take place in context with the requirements analysis work. Chapter twenty-two completes the integration story by extending it to verification.

1.5.5 Part five: Specialty engineering methods and models

Part five covers specialty engineering integration work. Chapter twenty-three provides an integration to this work and eight other chapters discuss specific disciplines such as reliability and maintainability. These chapters are not intended as offering stand-alone treatises on the work of the disciplines discussed, rather an attempt to disclose to the system engineer what the people in those disciplines can be depended upon to do, how that work is performed, and the value of that work.

1.5.6 Part six: Concurrent post-design process synthesis

Part six extends the development life cycle process beyond the period of primary engineering responsibility into procurement, manufacturing, and quality, as well as into the system employment period. The system engineer should not feel that program experiences outside of the engineering area of responsibility are off limits. Often important knowledge resides there that will help to understand and resolve difficult development issues. While employed at Teledyne Ryan Aeronautical, the author took every opportunity to visit the factory to see what the product actually looked like inside and how the factory actually fabricated parts of it.

1.5.7   Part seven: Closing

Here we will look back over the aggregate process, peek into the future a bit, and attempt to draw some conclusions of merit.

chapter two

System development process overview

2.1 The global situation

Systems are developed by enterprises. Enterprises are invented by people or other enterprises to satisfy the needs of people and other enterprises while obtaining supporting economic benefit from those people and enterprises in compensation for their efforts. There are only a few strict rules for the formation of these enterprises in free societies, and the marketplace determines the extent to which those enterprises will be successful in a free market economy. It is not a simple problem, however, to create and nurture an enterprise so that it enters the set of successful ones. Those enterprises that would develop systems to satisfy complex needs have special problems as do those that must cooperate with other enterprises to codevelop systems to satisfy complex needs.

Enterprises design programs to deal with each customer and related product or collection of products. These programs can appeal to many strategies to move from program entry conditions to completion, but generally a program plan of some kind is useful that clearly tells what must be done, when it must be done, and who shall do it. A cost and schedule budget is established for each task at some level of indenture and those responsible for accomplishing the tasks are required by a manager to complete the task within that budget.

Figure 2.1 illustrates the overall process within which programs are implemented to produce product systems desired by customers. Where the enterprise deals with large acquisition agents like DoD and NASA, the enterprise designs a program based on the content of a request for proposal (RFP) from the customer. In the ideal case, the enterprise has a clear definition of its preferred process that it is able to torque into an effective program plan oriented toward solving the customer's problem. If the enterprise wins the competition for the program, they implement their plan and manage in accordance with it. This book encourages the three-step process introduced in Chapter one in all cases, but all programs are not necessarily alike and an enterprise must be sufficiently agile to respond to the special problems of some of its programs.

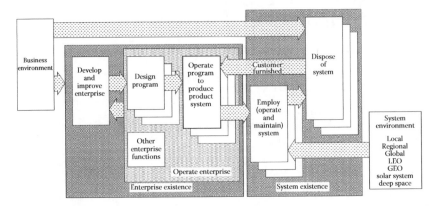

Figure 2.1 The general program situation.

A fairly simple program involving a clearly understood problem can be planned to employ any of the three work sequence models covered in this chapter. Where the problem to be solved is very complex, it will commonly be necessary to employ a spiral model that repeats the three-step process described in Chapter one as necessary. Each spiral involving the three-step process exposed in Figure 1.1 uncovers a clearer view of the problem space and defines a solution at some level of indenture, increasing in depth over time. A subsequent spiral may uncover a rationale for an alternative approach at some higher tier possibly requiring iteration, and this process can occur several times over the run of the development work.

The strength of the cohesion between the organizations responsible for the parts of the development effort will have a lot to do with how much iteration will occur. If all of the teams working on the program are from the same enterprise, that enterprise should be able to manage those teams effectively with minimal iteration. If some of the teams are suppliers, it will introduce a need for very effective supplier management to preclude wasteful iteration. If multiple enterprises are involved with loose contractual coupling between them, it can result in considerable wasteful iteration.

However complex the problem to be solved, the author maintains that the three-step process discussed in Chapter one and followed throughout this book will be effective when applied iteratively for each layer of functionality and related product entities repeated as necessary to move to deeper layers and to correct for nonoptimum selections in a prior spiral.

2.2   Top-down work sequence models

Several models of a structured, top-down system development process have been offered by various authors and practitioners, and three such models have been summarized here. The author used the first one discussed in Section 2.2.1 as his model, but the content of this book can be applied to any of these three models for they are but different views of the same process.

2.2.1 *Traditional waterfall model*

The first model is illustrated in Figure 2.2 by a simple Gantt chart showing the application of the three-step process in time. The name waterfall was obviously selected to reflect water flowing from earlier tasks down to later ones. Many programs that produced good products have been managed using this visualization, but it does fail to recognize some important aspects of program realities. The principal problem is that there is actually some iteration common in program implementation, and the waterfall model does not clearly illustrate this.

While employed at General Dynamics Space Systems Division, the author recalls a new general manager who was disappointed in the scheduling computer tool used because it printed the waterfall flowing down and he wanted it printed with water flowing up, that is, the later tasks on top of the chart. The author cannot recall if the gentleman was still there when the conversion, that did not take a lot of time, was complete.

The pure waterfall model implies that it is possible to accomplish all requirements work in a single sweep and then move on to synthesis without a concern for possible errors or omissions in the requirements work. The reality is that in order to accomplish the requirements work well, we have to follow it in an interactive fashion with concept development work that, one would have to conclude, is part of the synthesis process. We identify each layer of the product entity structure through some form of structured analysis, and it should be validated through a concept development effort with the potential for a need to change the requirements for compatibility with available technologies among other reasons.

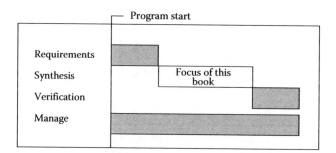

Figure 2.2 The waterfall sequence model.

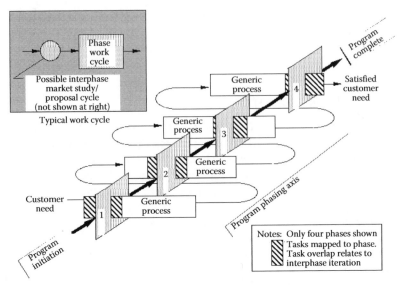

Figure 2.3 Traditional model of development. (From Grady, J., *System Integration*, CRC Press, Boca Raton, FL, 1994. With permission.)

Programs to serve DoD, NASA, and other large acquisition organizations are commonly accomplished in program phases to control the cost risk between major review and decision events. Figure 2.3 illustrates this general process. Goals are established for each phase and work to date is reviewed at the end of each phase against those goals. A subset of the complete generic development process is accomplished in one phase to achieve these goals. In the next program phase, which may be preceded by a competitive proposal cycle, some of these tasks are repeated and some new tasks are accomplished while moving closer to program completion.

Iteration occurs in this process for two reasons. First, while working from the top down, we become much better educated about the higher system levels as we work our way through lower tier requirements and solutions. This often requires some rework of prior conclusions at the higher level possibly arrived at on a previous program phase. Also, we iterate our work in order to move deeper into the system hierarchy. Many of the same tasks required to characterize a subsystem, for example, are then required to characterize the major elements of that subsystem and are repeated for each of these elements, and so this process continues down to the component level. The generic process diagram portrays a comprehensive set of sequential tasks needed to fulfill system engineering goals, while the phasing diagram gives the programmatic and contractual view of the world, a spiral in fact. This is the model used by the author in this book, simplified as a waterfall, but the others can be applied.

2.2.2   The "V" model

The Center for Systems Management in Santa Clara, California, in the persons of Kevin Forsberg and Harold Mooz, has popularized a process diagram formed into a "V" shape with a decomposition and definition downstroke on the left and an integration and verification upstroke on the right. Figure 2.4 shows the general nature of this concept, but it does not illustrate the detailed product entity structure levels uncovered on the downstroke and the correspondingly layered integration and verification process on the upstroke characteristic of that model fully depicted.

The principal strength of this model is that it shows a longitudinal relationship between the requirements that define the problem to be solved in synthesis, taking place at the meeting of the planes, and the verification process where we seek to prove that the design solution satisfies those requirements. This encourages the notion that one should identify the verification requirements while writing the product requirements and use these as a basis for verification plans and procedures to be prepared a little later. Where one follows this encouragement, it can often result in the requirements being better stated because, if the requirements writer has difficulty writing the corresponding verification requirements, he or she may conclude that their difficulty is motivated by having written an unverifiable requirement leading to restatement of the requirement. Unfortunately, we often discover the poor quality of the system

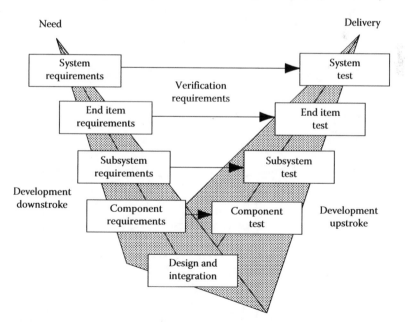

Figure 2.4 The V model of system development.

engineering work during item qualification, and this is much too late to encourage any kind of corrective action for the program in question.

2.2.3   The spiral model

The spiral model, first adapted to software development by Barry W. Boehm of University of Southern California Professor, presents an image of iterative cycling between several fundamental development activities while working toward more detailed knowledge of the preferred solution to the customer's need. Figure 2.5 offers a simplified variation on this model. While the original intent of this model followed a predictable pattern to a final conclusion, some persons with a commercial interest see this as a useful model of reality where they get off the spiral when the product is good enough to go to the market ahead of the competition within the context of a competitive situation. Others take this one step further and think in terms of a continuous spiral over time releasing market-ready products, with increasingly noteworthy features, flowing to market. This is a valid model of reality for many long-serving systems such as automobiles from most manufacturers, B-52 bombers, and many

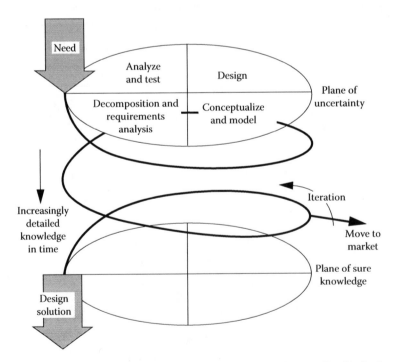

Figure 2.5 The spiral model of system development. (From Grady, J., *System Integration*, CRC Press, Boca Raton, FL, 1994. With permission.)

software products, all of which continue over a very long time to improve and acquire capabilities not envisioned for the product in its initial introduction. The author prefers to draw this spiral with an increasing diameter from top to bottom signifying the increasing product knowledge acquired over time, but it is harder to draw it that way.

It should be noted that by 2006, the DoD model for system development had migrated to the spiral model, not just for software but for everything, reflecting an interest in buying systems with sound chassis, airframes, and hulls that could be maintained in inventory for many years but be upgraded over time with new capabilities through, primarily, software and electronics changes.

Some developers of construction and farming equipment among others evolve a final system through as many as four prototypes where the final prototype is intended to be the first production article. This too is a spiral process that may start with an initial set of requirements based on some improvements on a prior product encouraged by access to new enabling technologies. A manager from one of these companies expressed to the author the intent of the company to move from what it called iron verification through prototyping to simulation verification to the extent possible in order to cut cost and time to market.

During a break in an interface tutorial the author presented at the Daniel Hotel in Herzlia, Israel in September 2005 for the 2005 Israel INCOSE conference, an Israeli Army officer assigned to the Israeli Army tank program explained a variation on the spiral that he referred to as a telescopic spiral to reduce the overall development time span. Normally, the exit criteria for a spiral is the basis for departing one spiral and movement to the next spiral generally based on some kind of program review concluding that the criteria has been satisfied. One does not begin spiral $N+1$ until spiral N is complete. The model the Israeli officer explained permitted an exit into spiral $N+1$ prior to the end of spiral N so long as the exit criteria had been completed. In this case, spiral N would continue to be worked to its conclusion while spiral $N+1$ was also being worked in parallel.

One sees this same effect on programs managed through the waterfall model where the contractor is allowed to proceed with manufacturing the end items needed for the development test and evaluation (DT&E) prior to the completion of the item qualification work for the items comprising that end item in the interest of shortening the schedule path particularly where a series of slips in the past has placed the schedule in some jeopardy. The risk is that the item qualification work will expose problems that require changes in the product being manufactured for DT&E resulting in entry into DT&E with a product that may not be able to be operated at full capability until modifications are performed. Often, DT&E will also unearth a need to make changes so that the system flowing into operational test and evaluation (OT&E) will require changes from qualification and DT&E resulting in a string of field modifications and delayed user availability of the full capability of their system. So, in spiral or either of the other two models, management should carefully weigh the risk before committing to the telescoping effect.

2.3 A common desire

In all of these sequence models of the system development process, we seek to illustrate a multidimensional phenomena in two dimensions on paper, and it is very difficult to picture it clearly. We work to develop unprecedented systems from the top down because it reflects the order in which we humans are most effective in discovering information about something new. We begin with relatively little knowledge of the details and work toward complete knowledge of the solution to the problem expressed initially by the customer need. There is an experimental aspect to this process in that you must make decisions without sure knowledge as you work your way into a condition of more knowledge and some of these decisions will later be shown to have been less than the best of choices. Figure 2.6 illustrates the dilemma of the early decision-making process.

Early development work can also be characterized as a learning process. We start not having mastery of the problem space and it is during this period when it is very inexpensive to make decisions. A one-hour meeting between three experts each costing $100/h in program budget will cost the program only $300 total. If that decision is not a good one due to their ignorance at the time, it may cost $3,000,000 to fix the problem 4 years later when the knowledge level has increased to the level needed to expose the error. At the same time, a program manager who cannot make a decision for fear of the potential for errors cannot manage a program well. We simply have to understand the risks as best we can and keep working toward better knowledge.

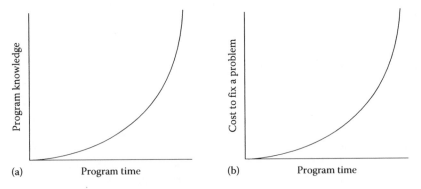

Figure 2.6 The cost of knowledge delay: (a) increasing knowledge and (b) increasing cost.

There is a need, therefore, for iteration in the system development process. At the same time, it is economical to apply essentially the same development process for all elements of the system. Since the system is composed of many things in a hierarchy and we work from the top down, we will be repeating one process over and over again in time. So, the same process occurs in time repeatedly in an expansive pattern during decomposition and definition and in a retracting pattern during integration and verification. This combination of cyclical action and the unidirectional nature of time make it difficult to portray the process clearly. The accompanying illustrations are imperfect examples of attempts to do so.

It is interesting to note that all of the three sequence patterns are really manifestations of the same process. The V model is the waterfall bent in the middle to emphasize the relationship between requirements development and verification. The spiral model is the waterfall model worked at a higher rep rate. Instead of an intended single pass through the process during a program run, we make as many passes as necessary. The spiral model is especially useful where the problem is extremely difficult involving a human interacting with the product in some fashion requiring mental activity to join the stimulus–response actions. The difficulty in these situations is that the internal human interface between stimulus and response is very complex and not necessarily the same for all people unless the stimulus is very carefully crafted. Military systems depend on sufficient training exercises to make the response to a given stimulus routine.

2.4   Variations on a theme

This chapter and book are primarily concerned with the development of unprecedented systems for which the top-down model of development, starting at the system level and working downward, is generally most effective. As reflected in the V model, the definition process in this case runs top down but the design and integration process actually runs bottom up. There are other possibilities. It is conceivable that the enterprise could apply a bottom up definition approach identifying lower tier entities first, but it is hard to imagine this being a systematic approach that would lead to success given an unprecedented problem.

In the unprecedented case, the author prefers to apply the idea of Louis Sullivan, an architect in Chicago, Illinois in the late nineteenth century, who encouraged that form should follow function, that one should first determine the function of a building before designing its form. However, when developing a highly precedented system based on a prior implementation, this process should be reversed. Given that the development program has a prior reality to work with, the developer can start with the reality of an existing physical product entity structure that

hopefully provided good service for some period of time with the operator of the system becoming concerned eventually that it was no longer
serving their needs completely. In this case, one can start with that product entity structure and selectively sort the entities that compose it into
four piles: (1) those items that will continue to provide good service with
no changes, (2) those items that can be modified to provide good service,
(3) those items that must be eliminated from the system, and (4) those
items that will have to be added to the system.

The three-step development process can then be applied to the problems posed by the new and modified entities based on an understanding
of the functionality differential between the desired and prior systems. In
the process of doing this work, one could follow an outside-in or inside-
out sequence of work driven more by an interest in the interfaces between
the entities than any hierarchical relationships between the entities themselves. It often happens even in the unprecedented development case that
customers will require that particular entities that are available in suitable
quantities be used in a new development. These entities are generally available from some prior system being withdrawn from service to be replaced
by the product of this new development effort. In these cases, the existing
interface terminals of these items must generally be respected. We must
keep in mind in any development orientation that systems are composed
of both the entities that comprise them and the interfaces between them as
well as outside entities that are in the system environment.

One enterprise the author dealt with as a consultant decided to apply
an unprecedented approach on a new program despite their history of
developing their product for decades using a precedented approach.
Actually, they made this decision because of their history that was proving
to be a problem in battling with their competition who was beating their
pants off on some key requirements. They created a new system engineering department and began trying to understand the system functionality.
They also applied cross-functional teams rather than using the functional
departments for a program organizational structure for the first time. So,
there may be cases where an unprecedented approach may be helpful
even in a case where the changes that result may not be dramatic.

But, no matter the difficulty of the problem to be solved or the orientation selected to approach its solution, the author maintains that the three-
step process is appropriate applied in total to the development or to a sorted
new and modified collection of elements that in combination with some
existing entities will form a new system. There are, however, some situations where development may not be successful depending solely on a linear, deterministic, planned sequence of work like that offered in this book.

At the time this book was being written, there was a lot of interest in
what is called complex system development. Surely, over time, the systems
that we seek to create will become increasingly complex because we will

become increasingly adept at identifying problems of increasing complexity. When we have such wonderful world utilities as the Internet, Global Positioning System (GPS), and earth satellites we will find other world utilities and imagine wonderfully inventive ways to integrate these capabilities into our earthbound systems. The work that DoD is undertaking now to create a global information grid (GIG) to interconnect sensory systems and weapons systems is monumental, essentially providing a military nervous system of global scope. If the reader thinks this kind of capability will be applied only to military situations, they are mistaken. The Internet is already growing into a powerful commercial marketplace that will provide the backbone for some fantastically complex commercial capabilities in the future.

Systems that appeal to these kinds of capabilities combined with tremendously complex entities within new systems can become a very difficult chore to develop within reasonable cost and schedule constraints, and traditional linear development methods such as the three-step process encouraged in this book may not be completely effective for all parts of those systems. The author has, in the past, preferred to recognize that the problems that we deal with are complex but the resulting systems should be as simple as possible. However, the scope and density of the problems we are starting to confront will result in the development or merging of systems that qualify for the term "complex systems." A complex system has been defined as any system featuring a large number of interacting components (agents, processes, etc.) whose aggregate activity is nonlinear (not derivable from the summations of the activity of individual components) and typically exhibits hierarchical self-organization under selective pressures.

The three-step process applies a deterministic, linear, mechanistic, or reductionist approach to problems that may not fit the model. It produces good results where the problem can be partitioned into a finite number of parts each of which can be understood on their own and their mutual effects can be understood. It tends to fall apart when the number of entities become very large, and especially when the number and complexity of the inter and intra relationships becomes very large.

In the linear approach to system development, we seek to produce a system that will have a predetermined outcome resulting in the solution to a problem we call a customer need. In order for that to occur, we evolve a system design that given particular initial conditions, the system will produce the desired outcome when operated in accordance with a set of rules within an intended environment. There exist natural systems where it is not possible to predict the outcome given a set of initial conditions other than in a probabilistic fashion. Sometimes, it is a lack of information that causes us to think a situation is random where as more information may cause us to conclude that it is actually a deterministically ordered case that can be explained under the rules of chaos. But, in other situations, we may properly conclude that the laws of chance apply. In the former case, given sufficient information about the initial conditions and the context of the intended actions, it would be possible to predict the outcome but it is still difficult.

We understand how to team individuals together, each understanding a part of the problem space to solve a complex problem that none of the individuals can solve by themselves due to knowledge deficiency so long as that problem is linear in nature and permits the application of simple deductive and inductive logic to create an understanding of strings of reasoning explaining the paths between causes and effects. This book attempts to explain that process. But, is it possible to chart an affordable, understandable, and repeatable process for attacking problems that are not linear in nature and cannot be simplified as a series of linear problems?

One approach we did not explore in Section 2.2 was ad hoc. System engineers have traditionally been hostile to the ad hoc approach, but it should be noted that it has a lot in common with the natural methods that shaped our universe and ourselves. The problem with natural processes involving self-selection is that they commonly require a very long time to take effect, often on the order of thousands of years. This kind of schedule would not sit well with a client interested in a fairly near-term solution to their problem.

Experience with the attack on very complex problems in the defense realm as well as biology, economics, and other fields will in time offer some effective techniques for dealing with systems that satisfy the complex definition. In the near term, we can chart two development strategies: (1) a deterministic, linear strategy and (2) one characterized by complexity. In the former case, any of the three sequence models in Section 2.2 can be applied successfully, but in the latter case none of them may result in success though spiral is the closest to providing hope for success in that it recognizes possible failure and additional attempts with alternative approaches.

In a large system, some but not all of the system will be complex. For the time being, that part that is complex may best yield to the concentrated efforts of a small collection of very smart people thinking about the nature of the problem they are dealing with and producing insights that others can carry to fruition or failure in a more deterministic fashion. Yes, this is an appeal to trial and error that has been the underpinnings of natural development for a very long time.

2.5   *Process definition*

Process diagrams are used throughout this book and the other books mentioned previously to provide a specific, documented process description

commonly called for in ISO 9001 for system development process quality, IEEE/EIA 12207 for software, and ANSI/EIA 632 or IEEE 1220 for
system engineering. We make the point repeatedly throughout the book
that an enterprise should have a standard, written process description
telling what it must do and how it intends to accomplish the system
development process. At the top level, this document might be termed
a system development manual but the author prefers the title Enterprise
Definition Document that is explained in *System Management: Planning,
Enterprise Identity, and Deployment* (Grady, in press). Subordinate manuals
may include the enterprise's system engineering manual (SEM), program
and functional management manuals, and generic software development
plan (SDP). Other domain manuals should also be included, of course.
If your company has an effective, documented system engineering process derived over the years from trial and error and conscious thought to
improve it based on program experiences, you should consider yourselves
fortunate and the author salutes you.

If you do not have such a history and documentation, you should start
now on that road while there is still time. The process diagrams included
in this book as they relate to system synthesis may be useful as a basis
for a similar set created by you for your company. *System Engineering
Planning and Enterprise Identity* (Grady, 1995) included a generic SEM/
SEMP embodying the principles discussed in this book.

If you do not have such a written manual or plan, you may wish to use
the one included in that book as a basis for preparation of one. The text of
a manual was on a computer disk provided with that book. One can edit
the text and insert their own graphics describing the special features of
their own situation. At some point in time, a more up-to-date version of
this SEM will be available where anyone will be able to view or download
a set of documents related to each of the four courses.

System Management: Planning, Enterprise Identity, and Deployment
(Grady, in press) includes a system engineering process specification illustrating the complete set of process diagrams used selectively in this book
to discuss the design and integration process as well as the process diagrams used in *System Requirement Analysis* (Grady, 1993, 2005) and *System
Verification* (Grady, 2007). This book also includes a SEM in support of a
four-course certificate program available via distance learning around the
world. All of this supporting data is available through courses offered to
industry by the author's company as well.

Much of the content of this book has been consciously coordinated
with ANSI/EIA 632 because that standard will influence how we think
about this process for a long time to come. This will be true whether you
are defense practitioners or commercial suppliers. In the former case, we
will be directly influenced because we may be called upon through a
contract to comply with it in our process definition. In the latter case, we

will be indirectly influenced by it because no one has done a better job of defining the requirements for an effective system engineering process.

A lot of good thinking by many very talented system engineers and managers has gone into ANSI/EIA 632. The author admittedly is not in complete agreement with all of its content, and such is the case with any grand design. So, the company plan should be coordinated with ANSI/EIA 632 tailored as needed to make it compatible with the company's preferred process. This standard does not actually identify a particular process, rather it contains a set of 33 requirements for a process. You have the obligation to design a process that complies with those 33 requirements.

Commercial practitioners may first view the contents of ANSI/EIA 632 as overkill, excessively costly, damaging to time-to-market needs, and stifling of creativity. Members of the design community in some defense firms where the system engineering process has only been given lip service will have similar feelings. It is very hard to accept that a sound system engineering job on a program will reduce program cost, improve program schedule, and result in greater customer value in the product. The fundamental problem in reaching this conclusion is that panic too often sets in at the beginning when time must be allotted to truly understanding the problem before leaping to design solutions. When we later have to redesign the product a thousand times (several thousand engineering changes after the design is complete is not all that uncommon on large defense programs) based on information derived out of problems exposed in testing and manufacturing, we strangely and foolishly accept that as part of our normal way of doing business referring to it as necessary iteration.

Life cycle cost is one of the several specialty disciplines often applied on a DoD program to control the total cost of a system over its life from cradle to grave. What many practitioners find is that an effective system engineering process may cause the nonrecurring cost of a system to be somewhat increased but the recurring cost substantially reduced leading to a significantly smaller life cycle cost figure. However, the author believes an effective system engineering process should reduce even the nonrecurring cost due to specifications done well as a prerequisite to design and timely risk mitigation throughout the development effort.

Commercial practitioners may wish to steer toward a condition of fewer formal phases than commonly applied on DoD programs and more intense iteration volatility than a DoD practitioner would be comfortable with. But all will realize benefits from an organized process for the coordinated development of product and process described here.

A common concern expressed by design engineers in companies dealing with DoD as well as those in a commercial product line is that "imposition" of a systems approach will stamp out creativity. While this has happened before, where the process has been implemented by zealots with no appreciation for the need for balance between order and creativity as a function of the program position in time and a lack of appreciation for the individual human being, it represents a poorly implemented systems approach. Perhaps the most extreme example of this exists in the imposition of the Communist state on the peasants of Russia for 70 years. State planning was accomplished at the highest level with insufficient information about lower tier need. No closed loop system existed like our market system to encourage adjustments leading to improved performance. Rigid prescriptions and a lack of any reward for new ideas destroyed creativity. In the end, this society could not compete with the less-well-organized societies that worked, with which they were at odds. In this book, we seek to strike the optimum balance between order and creativity, far removed from this highly regimented situation, that protects the development team from bad choices while not screening them from great opportunities.

chapter three

The human basis for integration*

3.1 Human limitations drive integration

The need to integrate the work of two or more people stems from some very fundamental principles of system engineering: human psychology and human existence. Large and complex systems are composed of components that appeal to many different technologies because they must satisfy very complex needs that will not yield to simple solutions. We humans are limited in the amount of information, knowledge, and technology we can master, and the amount of knowledge that we have created and provided access to far exceeds our individual limitations. There is an economic benefit realized by those organizations that can take full advantage of available knowledge since they will be able to solve more complex problems than those who have mastered a smaller knowledge base. As the world turns, mankind finds new and more complex challenges to conquer encouraging more complex combinations of available technology and spurs renewed interest in pushing back the frontiers of science further. Figure 3.1 illustrates the fundamental knowledge space problem.

The universal solution to the mismatch between man's individual knowledge capacity and available knowledge is for individual engineers to specialize deeply in narrow fields and find ways to pool the talents of a team of specialists from engineering, quality, manufacturing, procurement, and other fields to form one equivalent super-engineer. This solution is the basis for the birth of system engineering as a process, career, and, in some companies, an organization. It is also the basis for the need for requirements analysis and synthesis, two of the three foundation stones of the system engineering process.

The specialization solution to this problem is thousands of years old. The time is long past, by several thousand years, when every human could master all available knowledge. In very early societies, men specialized in hunting or gathering probably based on their individual physical abilities. Men with excellent eyesight, instincts, and physical ability were much better at hunting game then and have excelled in hunting their fellow man in jet fighters more recently. In medieval Europe, blacksmiths and tinsmiths had very different techniques. The industrial revolution accelerated this

---

* The material in this chapter is from Grady, J., *System Integration*, CRC Press, Boca Raton, FL, Chapter 2, 1994. Used with permission.

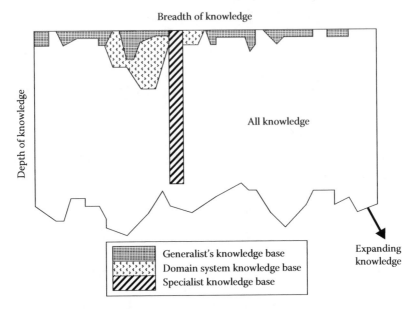

Figure 3.1 Knowledge basis for system development.

trend rapidly. But, the information explosion made possible by the computer and military competitions of the twentieth century have dramatically encouraged the evolution of more finely specialized engineers and ushered in the need for effective integration practices. This trend will surely continue since man will not suspend a quest for new knowledge and there is little reason to expect that man will change significantly in terms of mental capacity in the near term.

Integration is defined in the dictionary as the act of combining parts into an integral whole. It is only necessary if you are dealing with separate parts and wish to combine them in some organized way. Systems are, by their very nature, combinations of parts that together perform some function that no subset of the parts could achieve. But, it would be possible for a single human being to conceive, design, manufacture, and assemble a small system. Would integration take place in this case? Certainly, the individual would have to integrate the consequences of a broad range of ideas within his or her own mind, but would not have to integrate the work of two or more people or organizations focused on different knowledge domains.

We tend to think of the need for specialization only as a negative. Would we not be better off if one human mind could attack a problem and avoid the chaotic environment sometimes caused by people trying to work together or needing to work together but insisting on independence? Well, it simply is not so. However painful your memories of past programs may be, there is a positive benefit hidden in the specialization and integration scenario. Why is this? There really should be little doubt that if a single individual was in competition with a team of specialists who were well led, the system created by the team would be superior. Even if one person could master all knowledge, that one person's mind would function in a particular fashion constrained by his or her experiences in life. Given that no two people are the same, the team must collectively have command of a greater body of knowledge and experience. What is engineering but the translation of today's technical and scientific knowledge into designs for useful things. More knowledge and broader experience well used cannot help but fuel more possibilities from which to select an ideal solution to a particular engineering problem. Some of these ideas will be better than others.

We find it necessary to specialize in a competitive situation because our limitations prevent any one person from understanding and applying everything known by man or generalists from efficiently taking advantage of knowledge in a timely and competitive fashion. The added benefit is that we are forced to apply more than one person's mind to a problem and gain the advantage of expanded experience. To realize this benefit, however, we must be able to integrate the work of these teams of people and they must be well managed toward common goals.

3.2   The fundamental integration mechanism

The specialization answer to complexity in excess of an individual's capacity for knowledge requires many people working together to solve a common problem that none of them can solve independently. Obviously, human communication, therefore, plays a vital role in system integration. Integration takes place in the minds of the people working on a system development effort. It occurs, for example, when one specialist understands the need for design features required by another specialist that are in conflict with features he or she has identified, and together they reach an understanding that satisfies both of their requirements adequately.

Somehow, one of the specialists has to learn about the conflict and communicate it effectively to the other. That happens through visual or verbal communications in a neutral and common language, shared by all of the specialists, followed by the application of reason, analysis, and synthesis on the part of the other specialist. The opposite communication path is also exercised so that both understand the other's perspective. There follows a conversation where both try to understand the real constraints from the other's perspective in search of common ground and the true extent of the conflict. Integration, despite the availability of powerful computer tools and elaborate checks and balances, is largely an interpersonal relations problem and solution. The bad news is that people doing this work have to employ the worst interface on planet Earth to be successful—human communication.

Jointly, the two specialists are able to find common ground about two different requirements that neither could have mastered independently. This is the strength of the systems approach involving decomposition of big problems into smaller ones on which teams of specialized persons may work effectively together. This process of working together requires effective communication of ideas and information. The greatest opportunity for breakdown in integration occurs at the interfaces between team or individual responsibilities. These are the critical interfaces alluded to earlier. Unfortunately, many people tend to withdraw from these interfaces instead of plunging across to understand the other person's problems. A good rule of thumb in integration work is to encourage or require engineers to understand all of their product interfaces from the perspective of those responsible for the other interface terminal as well as their own. We take this matter up again in Chapter fourteen.

The whole teamwork notion that the integrated product development scenario promotes revolves around human communication because it is through communication that we understand together how to solve our common problems on teams. There is no grand mystery here and nothing very complicated. Specialized humans must communicate with each other to form the equivalent of one all-knowing person. The sports analogy is irresistible. In the years 1998 and 1999, the Denver Broncos captured the National Football League Superbowl because over their season, in the playoffs, and in the final game they were able to act as one toward a common goal better than any other team. In developing solutions to complex problems, we seek to work together with this same intensity over the season of the program. In our society, we must also do this while treating each other in accordance with our traditions of freedom and respect for law. One might think that things would go better if program managers were able to treat the workers as prisoners or privates, but this simply is not true, as the people of the USSR took 70 years and Nazi Germany far less than the intended 1000 years to find out.

Implementation of integrated product development is, therefore, a study in how to improve communication between people. The most important step, aside from a common language, is physical collocation of the people working on a team. Research has shown that the amount of communication between people is clearly inversely related to the distance between them in the work environment. Meetings, good telephone and electronic mail systems, and networked computer information systems offer other useful integration stimulants, but do not fully replace the need for physical collocation. In Chapter seven, we will discuss an alternative to physical collocation that is rapidly becoming possible, but for now, let us accept that physical collocation is the best way to encourage communication.

We also need a degree of order for integration to be successful so as to ensure that necessary conversations do take place. To facilitate this, we could introduce a very rigid structure with forms and checklists and rules, as some have done. It is not uncommon for system engineers to fall into the trap of over organization in this area to the extreme even of stifling creativity. Our goal must be to establish a degree of order that maximizes our ability to communicate while minimizing the need to communicate and to do so with the least negative impact on the creativity of our people.

3.3   Integration from a link perspective

The science and mathematics of edges, networks, or links offer another insight into the integration process. Albert-Laszlo Barabasi's book *Linked* (2002) offers a readable nonmathematical treatment that gives useful insights into how we can stimulate cross-functional teams to be more effective. Within the mind of any team member there is a network of neurons at work dealing with the team problem from a very specialized perspective. Teams are formed by individual human beings each with an individual and independent thought process. Each of these individuals forms a network node on a higher team plane following the pattern discussed earlier—they are all specialists who master some subset of man's knowledge base and communicate using some form of language like English. Figure 3.2 illustrates a program staffed by three small teams representing three layers of networks connecting specific nodes: (1) the neurons in the mind of the individual engineer (not illustrated), (2) the several team members linked into a communications pattern encouraging the paths that will advance the team toward its goals, and (3) the teams as nodes in the program network.

Each team should form a network cluster characterized by intense internal communications between nodes. If the team is well formed, then it contains people who collectively cover the knowledge base needed to solve a design problem defined in a specification. Some people in a team will act as hubs connected in a star configuration with the work of all or many of the team members. These people will include the team leader as well as one or more system engineers. But, it is insufficient to simply push this correct collection of people into a space. We must find ways to stimulate human communications. We want strong links between team members who are dealing with related problems. We want the intensity of the communications to be proportional to the degree of complexity of the interfaces between the entities for which the people are responsible. We will return to this problem in Chapter seven.

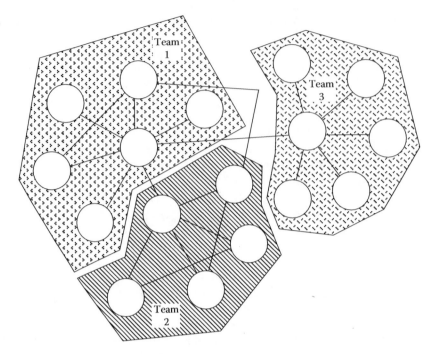

Figure 3.2 Intra- and inter-team links.

3.4   *The dark ages of system engineering*

The system engineering community must approach facilitation of both communication and team structuring with some care. If we gain benefit from the diversity of our specialized people, could we not defeat ourselves by being too successful in organizing and standardizing our process? That is a real fear. It is entirely possible that we could so rigidly standardize our process that there will be no room left for creative thinking. Some system engineers feel that during the 1970s and 1980s the Air Force Systems Command, responsible for the development of new Air Force systems, dangerously approached this condition with its Air Force Systems Command Manual (AFSCM) 375.

The procedures in 375 called for otherwise intelligent engineers to fill out what seemed like an endless array of forms to document the results of the engineering effort. Some companies actually had one team of engineers working on the design problem and another team working on the forms so as not to contaminate the creative minds of the "real engineers" with what some considered busy paper work. Many surviving system engineers from this period feel that the 375 series poisoned the well for years against the practice of effective system engineering. Engineers in this period who graduated to management came to feel that the system engineering process, they associated with those damnable dehumanizing forms, did not contribute value to the product design process commiserate with its cost and effects. In some organizations, the 375 series was the rationale for forming a system engineering department with the assignment of doing the paper work freeing the real engineers to do creative design work.

It is during this same period, and certainly not solely traceable to the 375 series, that the system engineering profession lost its way, overcome by the ease with which we can more finely specialize and the difficulty with which we improve our ability to integrate the work of many people. Specialized stovepipe disciplines developed with dedicated support within Government customer ranks as well as in industry. Standards were prepared that defined in great detail how each stovepipe specialty would be accomplished. Many engineering organizations evolved into transom engineering shops featuring serial work performance. Designs were prepared in one shop, thrown over the transom for independent assessment by a crowd of specialists, and this cycle repeated until the exhaustion and consumption of all of the available budget.

As an example of this process, while the Advanced Cruise Missile (ACM) was in development at the General Dynamics Convair Division in 1983, a design engineer responsible for the design of a portion of the airframe repeatedly offered his design solution to the nuclear survivability and vulnerability engineer for review. Each time, the design engineer was told that it would not satisfy his specialty requirements. After the third turndown, the design engineer asked the specialist, "How are we going to know when I get it right?" To which the specialist replied, "I will know it when I see it?"

This was a particularly good example of transom engineering because one solution to the nuclear survivability and vulnerability requirement—added mass—was in clear conflict with the weight requirements of the mass properties specialist. The chief engineer had earlier encouraged the mass properties lead engineer to organize an active program of weight control that even included prizes for the most weight removed. The structures engineer in this true story was trying hard to satisfy the weight requirements allotted to him, which he clearly understood, resulting in conflicts with the nuclear survivability and vulnerability requirements that he did not understand. This whole episode was a glaring example of a failure in system engineering of which the author was a part. Unfortunately, all too often we learn best from our most glaring errors. A good general solution in this case would have been for the nuclear survivability and vulnerability engineer to have offered a class or discussion for the design and mass properties engineers to explain his requirements including some ways that engineers had solved these design problems in the past. In this particular example, the nuclear survivability and vulnerability engineer did not speak the English language well, probably making the integration challenge more difficult. The fault mode here was actually among the system engineers on the program, including the author, who should have observed the symptom of the design engineer in difficulty with a specialty engineering discipline. Communications and teamwork can overcome transom engineering and we will see these same powerful and very inexpensive tools brought to bear throughout this book for the same purpose, but there must be a troubling spark of insight observed by a perceptive system engineer in order for this process to work its magic.

3.5   Order versus creativity

Now, we should not hang this whole guilt trip on the U.S. Air Force. Actually, the 375 series of procedures contained a lot of good ideas, many of which were perpetuated in a series of Intercontinental Ballistic Missile (ICBM) weapons system requirements analysis standards and later used by the author as a major component of the model for *System Requirements Analysis* (Grady, 2005). The 375 series attempted to capture information for which a computer was really needed but we humans had not yet figured out how to take full advantage of in association with this work. In the early 1960s, when the 375 series was published, the computer apparatus was not available. The principal mainframe computer input/output device was stacks of punched cards. There was no easy way for a team of people to interact with the contents of the mainframe through terminals. Early attempts to mate the process to distributed computing were also flawed requiring the people to deal with precisely the same forms on the computer screen that were used in the paper implementation.

Computer applications are beginning to evolve that apply the computer's strengths to eliminate human drudgery while helping to identify the work where human creativity and thought can most profitably be brought to bear. The product RDD-100 (requirements driven design-100) (no longer available) created by Ascent Logic, SLATE by SDRC, CORE by Vitech, DOORS by Telelogic, and other tools in the works by other companies offer graphical user interfaces (MAC and Windows desktop screens) that expose the human to useful fragments of the whole task and from which you can draw integrated views of the information contributed by a team of experts using workgroup computing. These tools are just scratching the surface compared to what we will be using in a few years to automate the system engineering process, but they are an absolutely necessary step in the evolutionary process. They help to expose new directions that can be exploited, and without them we would be unable to see new opportunities. In Chapter seven, we will reopen this discussion.

In our zeal to implement sound system engineering processes, we must guard against upsetting the balance between order and creativity in our organization. In hammering out an orderly process, we must not lose sight of the benefits to be derived from diversity, dispute, intuition, and creativity. We are often told in human resources courses on concurrent engineering that we should find ways to work together in a pleasant fashion on teams, but the author can recall times when doing system engineering at 100 dB was quite effective. When confronted with the thought that the fledging International Council on Systems Engineering (INCOSE) was unable to get anything done, the first President of INCOSE, Dr. Jerry Lake of Defense Systems Management College at the time, responded that there was value in diversity. As much as we may be infuriated by the workings of Congress, the balancing measures established by our founding fathers have protected us against a lot of bad ideas. And so it is in system development. Good ideas may sometimes have to prove themselves in honest debate about alternatives while surrounded by a lot of bad ideas that are well supported by people who seem to be very intelligent.

While this book encourages close teamwork and problem solving by consensus, the process of arriving at a conclusion may not, and need not, be completely blissful. This process may be characterized by heated debate, wild excursions of the human mind, and all of the potential interpersonal relations problems that define human affairs. People who feel passionately about their work should not be easily dissuaded from this attitude. Better to tolerate and take advantage of that passion. It may be contagious. After all, we must work with the normal slice of humanity our company is fortunate to have in its work force. To give this force the best possible chance at success, we must be conscious of a need for balance between order and creativity and a need for all of these people to feel fulfilled in their work. Either extreme is detrimental to the evolution of a sound system development capability.

Marty Wartenberg of the University of California at Irvine teaches courses in program management. He likes to begin with a question, "How many of you have worked on programs where you satisfied your cost and schedule goals?" A few people will raise their hands to which he will respond, "Now, of those who raised their hands, how many of you have in the process avoided damaging your people?"

We must establish sufficient order and discipline that everyone working on a program has a common framework for their work such that the information product of each individual fits into a pattern supporting one and only one configuration (parallel development of multiple alternatives in early program phases excepted). We must respect the schedule agreed upon with our customer and avoid analysis paralysis. But, we must also allow design engineers the freedom to conceive wild and crazy ideas consistent with the requirements, some of which might actually work.

Programs need to place the balance point far over toward creativity in early program phases and far over toward order and control subsequent to a design freeze during the detailed design phase. The author likes to picture this as a mixer facet with hot- and cold-water valves. Early in a development activity, you should have the cold water (order) on only slightly and the hot water (creativity) on in a mighty torrent. As the program progresses, these facets must be readjusted to protect the integrity of the evolving baseline through critical design review (CDR), or commercial equivalent, and thereafter. Late in the program, the creative juices for change must be tightly under control, and pure cold water flows. At the same time, the earlier appeal to creativity should have previously led to the best possible cross-functional solution that will not later require changes.

3.6   Mathematical chaos as an alternative

Rather than being two extremes on a linear scale, it appears to the author that order and creativity are arcs on a circle where, if you go far enough in the creativity direction, you descend into the true opposite of order—chaos. But we are finding now that there is a beautiful order in chaos.

The author has long speculated, from a perspective of imperfect knowledge, on the potential for derivation of a system development model from mathematical chaos that might be superior to structured, top-down development. Chaos helps to explain evolutionary processes that have done a pretty fair job of designing us and our natural world. The question is, "Can this same process be applied consciously by us to develop man made systems?"

Right now, the answer seems to be that we do not have the time and money to spare for mathematical chaos to work its magic. Chaos appears to require a lot of trial and error that would translate to time and money generally not available in sufficient supply to encourage anything but an organized direct approach to a rational solution. It can be argued that chaos is actually at work today in a macro sense shaping our systems of systems. Weak system solutions to problems are replaced eventually by new systems and subjected to the trial and error evaluation of the real world. This process keeps repeating, and when it works out well in a particular field, we call it progress.

But, could the principles of mathematical chaos be applied with more immediate effect at the micro level of system development in our day-to-day decision logic, trade study approaches, and the pattern of human interactions now referred to as integrated product development? The author does not know the answer to this question. What is apparent though is that for the time being, the structured development approach is the best machinery we have to evolve what is required to satisfy our needs. And system integration is an inseparable part of that process because of our need to decompose large problems into sets of small ones worked on by teams of specialists. This latter will be reopened in Part seven of this book as we look forward into the future of synthesis.

chapter four

Integration components, spaces, and cells*

4.1 Setting the stage for integration decomposition

It would be convenient if we could describe all of the facets of system integration in terms of a single entity. The author is convinced, after years of work, observation, and study in this field, that many of us accept incorrectly that integration is one activity. Few of us are able to describe how that single activity is performed, however. Many people assign the term "system integration" a mystical quality. Whatever it is, it is the answer to every system engineering problem, and seemingly just connecting the term in the same sentence with the problem is sufficient to define an approach to the concept in our proposals and conversations.

We will see in this chapter that the integration process can be decomposed into several parts, and these parts are explained in an uncomplicated way. System integration then becomes the sum of those parts hopefully with nothing lost between the cracks. We must, of course, be watchful that we do not lose contact with important information about the whole when thinking in terms of the parts as is the case in any decomposition application.

We begin with acceptance that an engineering organization that must deal with multiple system development programs should organize in a matrix structure, as discussed in Chapter six. The matrix has the advantage of focusing day-to-day work on specific program problems while providing a good environment within which to improve the organization's skills, methods, tools, and knowledge. We should also accept the good sense that we cannot beat the odds on specialization. Our engineering organization must select its personnel from the same pool as everyone else, the human race. We have seen that we are knowledge limited and we solve that problem through specialization.

Therefore, we will organize our personnel into functional specialties led by functional department managers. These managers will be responsible for providing all of our programs with qualified personnel, skilled in using a particular toolset and following the standard department procedures proven effective on past programs. We insist on internal practices,

---

* The material in this chapter is from Grady, J., *System Integration*, CRC Press, Boca Raton, FL, Chapter 4, 1994. Used with permission.

which are continuously improved, because we wish to take advantage of the practice-practice-practice template used by great athletes.

Our product will be organized into sub-elements each of which can be worked on by an integrated product and process team (IPPT) and we will assign our personnel to these teams that will form the principal program personnel supervisory structure. The work of all of the IPPTs on a program will be coordinated by a program integration team (PIT) that could be called the system team.

We will organize all program work into process steps linked to product entities under the responsibility of one or more IPPTs. Each process step will have a set of goals and a simple task description. We will map our IPPT responsibilities to the processes and identify leaders for each process. All of the processes will be laid into our integrated schedule with clear start/stop dates and budgets that reflect back into the IPPT structure. Each process will have clearly identified information and/or material work product outputs that are needed in other processes as inputs. We will apply that great system engineering saying "input-process-output" to describe how our programs will function. From the process steps all the way to the whole program, we will depend on this simple tactic to define and describe a program. The unique approach we will take is to link the program plan content into the generic common process employed by the enterprise to define itself.

4.2   Integration components

These assumptions and selections leave us with the problem of combining, or integrating, the work of many people in different functional disciplines, working on different product system components, in many different process steps over time. We define these three fundamental integration components as function, product, and process respectively. Be very careful that you understand that we have used the word function to mean functional organization and not product system function in this case.

In each of these components, we have to concern ourselves with two fundamental kinds of integration: co-component and cross-component integration. In addition, we have to account for the possibility that one or two of the components are not involved in a particular integration action. This means we have a three-valued situation: co, cross, and null values. With three variables, each with three values, it is obvious we are working with 27 (3 cubed) different integration possibilities. It is helpful to have a picture of this problem to better understand all of these possibilities.

Unfortunately, we cannot easily illustrate the three-valued relationship for each of our three components, so our visual model will disregard the null possibility for each component. Figure 4.1 illustrates the three components as a three-space coordinate system. We can imagine that we can

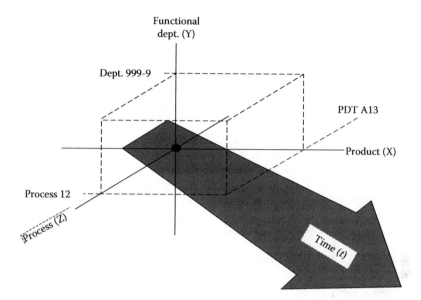

Figure 4.1 Integration components. (From Grady, J., *System Integration,* CRC Press, Boca Raton, FL, pp. 43–45, 1994. With permission.)

assign positions on the function axis to discrete organizations in our functional organization such as reliability, structural design, quality assurance, etc. Similarly, we can assign positions on the product axis to the elements of the system (avionics system, on-board computer, circuit card, transistor, etc.) in a hierarchical fashion, Finally, the third axis can have positions marked out corresponding to the processes on our program process diagram, such as identify item X requirements, design item X, and test item X.

For a given functional organization, all of the work that a specialty discipline does can be thought of as being within the plane passing perpendicular to the function axis at that function position. The task of integrating all of the work on that plane is called co-function integration, the integration of work accomplished by two or more specialists in that one functional discipline. Similar planes can be constructed perpendicular to the other two axes at different points on those axes to capture all of the co-process and co-product integration work. The attentive reader will observe that every point in the three-space coordinate system, then, corresponds to some combination of co and cross component integration for the three components. Let us now define the three integration components in terms of their possible values. Three combinations are nulls, meaning no integration for that component: null-function, null-process, and null-product integration. The other six possibilities are co and cross combinations with each of the three components. Let us take each of these six cases in turn assuming, in each case, that the other two components have a null value for the moment.

Co-function integration coordinates the work of two or more persons from the same functional department (specialty engineering discipline, for example) to ensure they are all using the same tools, techniques, and procedures in an appropriate fashion and that their results are consistent with other work on one or more programs or systems. This task is the responsibility of the senior program functional specialist for the project or the functional supervisor, in the case where all company projects are the target of the integration work. The author believes that functional managers should not be permitted to provide day-to-day supervision of their employees assigned to program teams, but auditing the work provided by employees in their department on programs to ensure that programs are applying the enterprise common processes is fair game.

Cross-function integration coordinates the work of persons from two or more functional disciplines in search of suboptimal design and specialty engineering solutions needing rebalancing, mutual conflicts between specialty requirements or the corresponding design solutions, available unused margin to be repossessed and applied more effectively, and wayward interpretations of the project or product requirements that may lead to conflict. This is a program responsibility that may fall upon an IPPT leader, task principal, or person from the PIT as a function of the relationship of the work to process or product and the integration level.

Co-process integration coordinates the work of two or more persons working in the same program process to ensure that they all are focusing on the same process needs within available budget and schedule constraints. This task is accomplished by the assigned task leader who is responsible to achieve the task goals on time and on budget.

Cross-process integration coordinates the work of two or more task teams working on different processes. It seeks to ensure that the work of the two or more teams is mutually consistent and driven by the same program goals. This integration task is the responsibility of a system engineer assigned to the PIT.

Co-product integration coordinates the work of two or more persons developing the design solution for a particular product item. This may be to develop an integrated set of product item requirements, evolve an optimum synthesis of those requirements in terms of a design concept, ensure that all of the cooperating specialty views are respected in the design solution, or integrate the test, analysis, and design work associated with that product item. Where an item requires a special test article, such as a flow bench for a fluid system, this work should also ensure that the test article properly reflects the same requirements and design embodied in the product item. This work will commonly be accomplished by team members on an IPPT.

Cross-product integration is the commonly conceived integration component that most people would first think of in response to the word integration. It focuses primarily on integration of interfaces between product items to ensure that all interface terminals and media are compatible both physically and functionally. This work could be as simple as ensuring that all items are painted the correct color, have satisfied a particular specialty requirement, as in being maintainable defined as an ability to remove and replace each part on a list within an average of 10 min. It is difficult to talk about this form of integration without linking other integration components.

4.3   Integration spaces

Integration work seldom falls into one of these pure cases. Commonly, we have to deal with more complicated situations involving combinations of two or three components with mixes of cross, co, and null values. Figure 4.2 offers four particular examples of these possible combinations, or spaces, for further consideration by the reader. Once again, we disregard the null-value case to enable simple graphical portrayal in three dimensions on two-dimensional paper.

In Table 4.1, we use a simple tertiary counting scheme to ensure we have not omitted any combination of the three variables. The reader

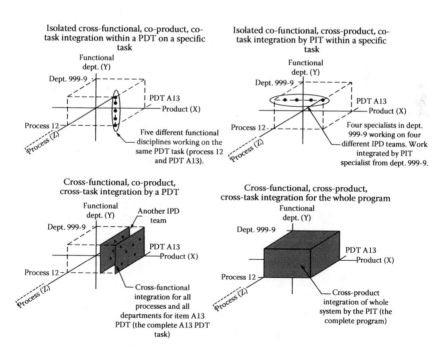

Figure 4.2 Integration space examples. (From Grady, J., *System Integration*, CRC Press, Boca Raton, FL, pp. 43–45, 1994. With permission.)

Table 4.1 Integration Space Identification

ID	Function	Process	Product	Integration type name
0	NULL	NULL	NULL	INDIVIDUAL EFFORT
1	NULL	NULL	CO	ISOLATED CO-PRODUCT
2	NULL	NULL	CROSS	ISOLATED CROSS-PRODUCT
3	NULL	CO	NULL	ISOLATED CO-PROCESS
4	NULL	CO	CO	CO-PROCESS & PRODUCT
5	NULL	CO	CROSS	CO-PROCESS/CROSS-PRODUCT
6	NULL	CROSS	NULL	ISOLATED CROSS-PROCESS
7	NULL	CROSS	CO	CROSS-PROCESS/CO-PRODUCT
8	NULL	CROSS	CROSS	CROSS-PROCESS & PRODUCT
9	CO	NULL	NULL	ISOLATED CO-FUNCTION
10	CO	NULL	CO	CO-FUNCTION & PRODUCT
11	CO	NULL	CROSS	CO-FUNCTION/CROSS-PRODUCT
12	CO	CO	NULL	CO-FUNCTION & PROCESS
13	CO	CO	CO	ALL CO
14	CO	CO	CROSS	FOURTEEN
15	CO	CROSS	NULL	CO-FUNCTION/CROSS-PROCESS
16	CO	CROSS	CO	SIXTEEN
17	CO	CROSS	CROSS	SEVENTEEN
18	CROSS	NULL	NULL	ISOLATED CROSS-FUNCTION
19	CROSS	NULL	CO	CROSS-FUNCTION/CO-PRODUCT
20	CROSS	NULL	CROSS	CROSS-FUNCTION & PRODUCT
21	CROSS	CO	NULL	CROSS-FUNCTION/CO-PROCESS
22	CROSS	CO	CO	TWENTY TWO
23	CROSS	CO	CROSS	TWENTY THREE
24	CROSS	CROSS	NULL	CROSS-FUNCTION & PROCESS
25	CROSS	CROSS	CO	TWENTY FIVE
26	CROSS	CROSS	CROSS	ALL CROSS

Source: Grady, J., *System Integration*, CRC Press, Boca Raton, FL, pp. 43–45, 1994. With permission.

should understand that there can be no other forms of integration work than those listed in Table 4.1 given that we have included every appropriate integration component. It would be possible to add one or more components to our list in addition to function, process, and product. The effect would be to multiply the number of different integration spaces. The total number of integration spaces (S) is, of course, predictable as follows: $S = C^n$, where C is the number of integration components and n is the number of values for each variable.

If, for example, we concluded that there should be five components, instead of the three covered in this book, each having three values (co, cross, and null), we would have $5^3 = 125$ integration spaces. From an enterprise perspective, we could recognize program as an additional component and in combination with the three covered here, would require 64 integration spaces (4^3). These 64 spaces would include co, cross, and null program integration as well as all of the new combinations with the other components. In the context of the new ideas you have just been exposed to, the author elected to set this component at a fixed value of co-program and focus on a single program. As you can imagine, the identification of additional components could get out of hand very rapidly making the description of integration more complex than the work itself. Three components and three values appeared to the author to be a good compromise between completeness and understandability for the purposes of this book.

Of the 27 integration spaces, we can dispense with INDIVIDUAL EFFORT integration immediately. We must assume that a single individual is fully capable in their field of specialized knowledge and able to effectively apply this knowledge to a specific product item and in a particular single process. This is why we specialized, after all, to create a human task that is within the power of a normal, single, specialized individual to master. We assume that each specialist can carry on an internalized conversation with themselves and use the power of their specialized discipline to solve the small problems we have tried to frame in our decomposition efforts. But, the acceptance of this case is important lest we conclude that these specialists need no quiet time during which they can accomplish the work and thinking focused on their specialty.

Similarly, the ALL CO integration space is usually not very interesting to a system engineer since it involves people from the same functional department performing work in the same process step, for the same system element. True, on a very large program, we could imagine how this could become quite a challenge. For example, we might assign seven technical writers to the job of writing a single technical manual for a single item of equipment. Integrating the work of these seven writers would be a case of ALL CO integration and not necessarily an easy activity.

One other fairly simple integration space to explain, though the most complex of them all in practice, is ALL CROSS integration involving the combination of cross product, cross function, and cross processes. When a member of the PIT integrates the work of several members of two IPPTs developing the design of two different product elements and the work of a facilities engineer responsible for the factory that will assemble these two elements and a tooling designer responsible for the manufacturing equipment that will hold the elements during mating, we have an example of this kind of integration. The reader will be able to imagine several other cases of this integration space.

We have given examples now for 9 of the 27 integration spaces: including the 6 isolated integration cases, INDIVIDUAL EFFORT (or ALL NULL), ALL CO, and ALL CROSS. This leaves 18 remaining to be explained with at least one example. Some of these 18, you can see from Table 4.1, are very difficult to name so we will simply use the ID number for a name. Examples for many of these other 18 spaces will appear in the remaining chapters of the book.

4.4 Integration cells

In almost any given system development work situation, we will find it necessary for some combination of the 27 integration spaces to be applied. Very little system development work can be accomplished in total autonomy today. In Chapter three, we have explained that this phenomenon occurs because individual engineers have had to specialize very finely to master enough of the available knowledge base to be competitive. The number of these combinations is finite but quite large for a large development program.

We have many options in grouping all program work into unique combinations of integration spaces, but the best way to do this is probably driven by the program tasks that would appear on a program task network. Each program task has associated with it some set of functional disciplines performing work on some particular combination of product elements. Many of these tasks, possibly most, will require some form of integration in the context of some combination of the 27 integration spaces defined above. Let us call any one of these combinations of task, product, and functional organization an integration cell, and thus there are 27 different classes of cells.

The more finely we divide the overall program into tasks, the more unique integration cells we will have. The more finely we assign IPPTs to develop the system, the larger the number of integration cells. The more functional disciplines that must be assigned to the program, the more integration cells that will be necessary. At the same time, the larger the number in any of these integration components, the more simply we can describe each integration cell because they will consist of a less complex combination of integration spaces.

4.5   Program world line

So, the integration process is more complex than we might have first imagined, composed of a finer structure than we might have thought. But the complexity does not stop here. We must not only apply each of these integration spaces well within the context of the integration cells defined for the program, but we must apply them in a pattern coordinated with the program schedule. Figure 4.3 illustrates this need by placing the integration spaces coordinate system on a program world line. Throughout program passage on its world line (network or schedule), appropriate specialists must apply the appropriate integration spaces to the planned integration cells in accordance with an integration plan. To the extent that we can reduce aggregate program work by early identification and resolution of inconsistencies between product and process elements, we encourage success in system integration. Ideally, we should be able to do all work error free on the first pass, but that is an idealistic expectation.

Now we can answer the question, "What is system integration?" It is the rich mixture of three integration components applied in combinations defined by the resultant integration spaces to the work confined to a finite

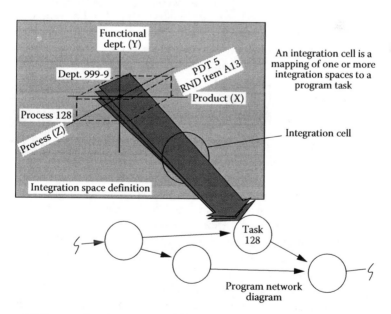

Figure 4.3 System development world line.

number of integration cells across the program world line that actually comprises system integration on any given program. Throughout much of the remainder of the book, we will consider some of these integration situations as they relate to process and product integration.

4.6   Integration principles

The principal work of the system engineer is integration and optimization at the higher system tiers. It is a reality that individual domain engineers and teams will tend to focus inwardly in solving their own development problems. They can generally be depended upon to coordinate the many elements relating to the solution of the design problem for which they are responsible. However, they will tend to back away from the boundaries of the solution space for which they are responsible rather than charge across and look at the problem from the perspective of other engineers and teams. Therefore, an individual engineer or team cannot be depended upon to accomplish integration work across these boundaries. If well led, they will commonly accomplish internal integration with skill, but their external integration work may not be well done unless aided from outside their team.

Integration is, as earlier discussed, the act or process of forming, coordinating, or blending two or more elements into a functioning or unified whole. When these elements are in the same domain for which there is an expert assigned to the responsible organizational entity, it is entirely possible that the integration work will be accomplished well and in a timely way. This integration work takes on two aspects: (1) where the elements are known, known to be related, and known to be in conflict, and (2) where the elements may or may not be known yet and are not consciously recognized as being related. So, there are exposed and unexposed integration cases. The simplest integration case entails two exposed elements in the same domain X that can be mastered by a single engineer in that domain. A very complex integration case might entail multiple (more than two) elements from multiple domains, at least some of which are yet to be exposed and exist in association with two or more product entities under the development responsibility of two or more different engineers or teams. This case will generally require a system engineer to identify and help resolve the related issues.

The reason that system engineers are helpful in identifying and resolving integration issues is that the domain people are limited in the breadth of knowledge that they can bring to bear on a problem. System engineers are every bit a knowledge space limited as domain engineers but they commonly have used that space they have mastered differently. An experienced system engineer tends to be broader in their knowledge than a domain engineer. This, of course, also means that they must be thinner in depth and not able to work very difficult domain problems best left to the domain experts. But this breadth of knowledge, an inquiring mind, and ideally some keen interpersonal skills will often permit a system engineer to see integration issues more easily than a domain engineer can. Experience with a product line also helps. These factors of experience and acquired knowledge breadth beyond one's original university education are reasons that engineers recently graduated may not have a great deal of success in doing integration work whereas more experienced engineers will more often have better success.

The integration work is relatively simple when the elements in question have been exposed. The solution then is simply to apply good management and sound engineering to encourage a compatible result across the two or more elements. An example of this kind of case is an interface issue. One side of an interface has been detected using a 12 pin male electrical connector while the other side is using a 16 pin male connector. Resolving this issue is an easy matter of conversation, reasoning, and cooperation. The real integration work that is difficult and very valuable is being able to first see that a conflict exists at the earliest possible time. Now, how do we train, educate, or otherwise bring a system engineer to be able to do this well? A vice president in Northrop Grumman wondered aloud astutely in a conversation with the author about the DNA of a system engineer that would permit them to do this kind of work and if known could perhaps be replicated through training.

Recognizing that there is much we do not understand about integration as it occurs inside the human mind, let us at least try to identify several unexposed integration principles that system engineers might profit from in their career path from young neophyte to older master.

Unexposed integration issue principle number 1 is to use more experienced system engineers in the most difficult integration situations. Do not throw an inexperienced system engineer alone into a difficult situation thinking that the experience will do him or her good. The experience will be better if that engineer can observe a professional at work in the same situation. Granted, that the new engineer eventually must lose the newness glow and become able to handle the difficult issues by themselves. As John Wayne's camp commander tells Wayne in one of his many western movies that he cannot return and save his troop, they will have to learn to ford a river in the face of hostile fire just as he had. But this is more easily done after having been a part of a prior successful experience.

Unexposed integration issue principle number 2 is to encourage the use of models in defining the system and its features. *System Requirements Analysis* (Grady, 1993, 2005) covers development models commonly calling for three facets: (1) product functionality that could be exposed in a flow diagram prepared by a system engineer, (2) product entities commonly exposed in a hierarchical breakdown diagram that tells what the system physically consists of, and (3) a behavioral diagram of some kind, that could be a schematic block diagram or n-square diagram and/or a state diagram, telling how the entity acts in accomplishing the functionality. These diagrams capture the essence of the system devoid of details that can distract.

The author, long before becoming identified as a system engineer came into the habit of using these kinds of diagrams while a field engineer working often alone on remotely piloted reconnaissance aircraft systems in the field far from the plant on incompletely documented systems. It was impossible to carry around a great deal of system data, even if it existed, on operational missions where it might be necessary to advise a launch control officer on prelaunch system performance or a remote pilot on post-launch flight performance or maintenance people in the hangar or on the flight line preparing a system for a mission struggling with a potential despised abort decision especially since much of the information needed was classified. A system engineer needs the facts stripped of any unnecessary details to support the reasoning process. These work products are models and have all of the characteristics of a model of reality. They separate the essential from the nonessential and present a useful abstraction of reality and provide a framework within which reasoning can be effective.

Unexposed integration principle number 3 is to recognize that there is a limit to the complexity that a system engineer can handle consciously. In reliability and safety engineering, in all but the most demanding applications, we attempt to prevent single-point failures where one part fails causing a catastrophic effect. It is possible to consciously think through how every part in a system can fail in isolation and to predict the effects of those failures. This is the result of doing a failure modes effects and criticality analysis (FMECA). It is much more difficult to discover all or even many of the two or three point failures where any of the single failures might not cause an observable problem but together two or more like that occurring at the same time may result in system failure. Yes, it is possible in these cases to consciously work through every failure that can occur assuming failure number 1 has occurred, and then move on to the assumption that failure number 2 has occurred. After perhaps years of analysis, one might finish the work, long after the system in question had entered service. As a result, one finds that most aircraft crashes are caused by two or more failures and one of them often involving human error.

The same problem arises in consciously trying to identify unexposed integration issues with more than two terminals. Given that a knowledge domain is a partition of man's knowledge base that is made the responsibility of a particular functional department that assigns its people to programs and who are members of teams on those programs and that other domains or spaces exist in association with the development of a new system, integration issues commonly will entail two or more terminals, each residing within one of those domains. As an example, a design engineer may think that his or her current design concept is beautiful and effective while a manufacturing engineer finds it unproducible.

Models can help us identify unexposed integration issues, and conscious thought about how the system will be used from both an operational and maintenance perspective within the context of these models will be helpful. We should encourage everyone on the program to come forward with any integration terminal issues they encounter in their work. Some of these will not connect to any other issues but each should be carefully considered. Avoid in all of these cases any public criticism of the messenger because it will reduce the chances of others coming forward with their issues.

While doing this work, the system engineer should think very intensely about potential problems encountered. This may not at the time expose any new integration issues, but it will prepare the subconscious for subsequent work. The subconscious is probably more powerful in exposing integration issues than the conscious mind where the complexity is high. The way to encourage subconscious success is to think intensely about the matter at hand and then give it a rest moving on to other issues. Principle number 3 rephrased is, therefore, to take full advantage of the subconscious.

Unexposed integration principle number 4 is to have some way of reporting and tracking integration issues. The safety engineer identifies hazards that properly receive attention on programs, and other domains have their own means of announcing issues from their domain perspective. The system engineer needs a similar mechanism to attract attention and focus work. The author tried at one company to float the notion of a system inconsistency notice, that could be known by the helpful acronym SIN. This is exactly what we are trying to identify. Some of these will fall away as we learn more about them and see that they are not really problems after all. Other issues will be easily dealt with by a pair of engineers having a useful conversation. A few may require tens of engineers working many hours to resolve. Each line item on our list of SINs should have a principal engineer assigned who is responsible for working and resolving the issue by some assigned date. These essentially form a subset of program risk.

Some unexposed integration issues will remain so because the connective tissue between the domain terminals is so obtuse that no one person can come to grips with it and express it and its consequences in a way that others can understand that it is a problem. In some cases, these issues will return when just the wrong set of conditions occur simultaneously in operation resulting in what we like to call an accident. By analyzing the accident and its causes, we may identify a serious problem that eluded us in development that can now be cleared so as not to cause another incident. Ideally, of course, we would find all of these serious problems during the development period, but there are limits to the amount of complexity with which our mind can deal.

It should be obvious that the critical problem is complexity. So, unexposed integration principle number 5 is to simplify wherever possible. We simplify the management of a program when we organize by physically collocated cross-functional teams oriented about the product entity structure and provide those teams with coherent planning data as explained in detail in *System Management: Planning, Enterprise Identity, and Deployment* (Grady, in press) in context with the integrated master plan (IMP)/integrated master schedule (IMS). Part two of his book offers an introduction to these ideas. In particular, this method of organizing encourages good accountability in the development of cross-organizational interfaces where most integration issues reside.

4.7   *Domain of the system engineer*

System engineers should live and thrive at the boundaries between knowledge domains and product entity domains. Given a collection of knowledge pertinent to a potential problem from domain X and Y perspectives as illustrated in Figure 4.4, the specialists in those domains may not be able to link together these two elements in conflict because they can only observe the world using their own knowledge, which simultaneously permits them to deal with that space in depth but prevents them from seeing it as part of a larger problem. System engineers live in a domain that seeks to understand the linkage of ideas across domain boundaries.

The integration work in a particular situation will be more difficult directly as a function of the number of product entity domains involved, the number of the functional domains in each product entity domain, the number of points of interest in a functional domain space, the number of

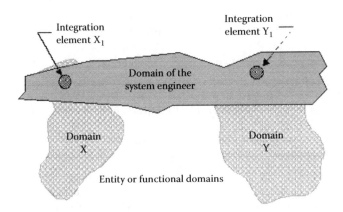

Figure 4.4 Domain intersections.

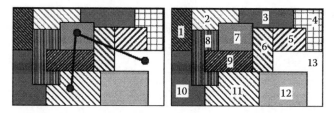

Figure 4.5 Degree of integration difficulty.

relationships between points of interest in a set of domains, and the difficulty of the relationships between those points. In Figure 4.5, we see a Venn diagram identifying 13 entity spaces one of which, entity domain 13, includes six functional domains (not illustrated). One of these functional domains has been identified as having one point of interest that is linked to a point of interest in entity domain 7 that is further linked to an interest point in entity domain 11. The first difficulty in this integration situation is that if these three interest points reflect a conflict, someone must observe the conflict potential as a prerequisite to resolving the conflict. It is doubtful that anyone in entity domains 7, 11, or 13 will identify the conflict. The system engineer will use the techniques discussed in this chapter to gain insight into the potential for a problem and through discussion with the domain experts come to a conclusion about the conflict.

4.8   This may be a little mad

Prior to leaving this subject, we should question one of our assumptions as a way of cementing our understanding of the integration spaces concept. Perhaps we were being premature in dispensing with the INDIVIDUAL EFFORT space so easily. Suppose that Company XYZ has acquired an engineering staff composed of 20 multiple personality individuals and that each of these people have 20 personalities.

Let us say that it will require knowledge of 20 specialized disciplines on an IPPT of Program ABC. Let us also accept that a person with multiple personalities enjoys inter-personality communication of knowledge and that all personalities can be simultaneously functioning. That is, anything known by personality 1 is available to personality 2 and time-sharing of the senses and mental processes is not required. We have to accept that speech is a single channel function for these people, however. Finally, let us assume that the members of the ABC program staff all graduated from MultiMind University where they have perfected the technique of educating each personality in a different engineering specialty.

Might we not have to rethink the nature of the INDIVIDUAL EFFORT integration space in Company XYZ? One person could completely populate an IPPT if the 20 specialties they had mastered happened to match the

20 disciplines needed for Program ABC IPPT. Let us assign 2 engineers to staff the PIT leaving the other 18 to work on 18 IPPTs. Perhaps this will allow Company XYZ to take advantage of the ultimate downsizing opportunity while also accomplishing the work necessary to develop the customer's product. The author was told by the training manager not to use this story in teaching a course at his facility in that a downsizing was in the works.

We have said that the need for integration is driven by the need for humans to specialize. In this case, the specialization process has been internalized and while external integration may not be necessary, there must be a means to integrate the information each personality is working with.

Exactly how does integration take place? It takes place through a human interface called communication, appealing to our senses of touch, sight, hearing, and, to a lesser extent, smell and taste depending on the product. We specialists must share a common information base permitting communication of our ideas and an ability to understand the consequences of information communicated to us in association with our own specialty.

If MultiMind University had fouled up the class schedule and educated each of the 20 personalities of Joan Doakes in a different language, Joan might be hard-pressed to take advantage of her formidable internal integration possibilities. A common language, logic, and mathematics form the principal integration instruments at the most primitive level. Whether we are dealing with graduates of MultiMind University or Georgia Tech, these people must enjoy a common basis of communication and reasoning through which they can exchange ideas that spring from applying their specialized knowledge base to a problem they share in solving. Our goal must be to create the equivalent of one great mind composed of the mental capabilities of all of the team/program members. And we must do it while causing our employees to be of greater value to the enterprise from each program experience doing no damage to them in the process. The author offers an apology to all multiple personality system engineers for any slight perceived because of the unlikely description of the multiple personality condition.

chapter five

The key roles

5.1 Specialization run amuck

At the time this book was being written, the American education system was generally not adequately preparing engineers and managers for service in enterprises doing system development work to solve complex problems posed by their customers. There were many excellent, but inwardly looking, programs preparing domain engineers and managers for service in industry. There were many excellent programs to train good engineers to be effective engineering managers. A few institutions offered systems engineering master degrees or certificate programs. Most universities offered excellent engineering programs for the many domains that were required in development programs with one exception. It struck the author as odd that so many universities had good programs for writing code (manufacturing of software) but so few had sound software engineering programs addressing the whole development process including analysis and modeling, design, test, and management.

The problem appears to be that universities are suboptimized in the various departments into which they have specialized. There is seldom a system influence in these institutions. Thus, they continue to develop good engineers capable of independently working within their specialized design domain but unsuited to interact effectively with other specialists to achieve objectives that none of them can achieve independently.

Universities are not, in general, preparing engineers and managers to work in the intense cross-functional environment that industry is finding necessary in order to competitively develop products within difficult cost, schedule, and performance constraints. What industry needs is groups of specialists who can individually see the need for a cooperative effort and enthusiastically join together for a season of intense work similar to the way that the team that wins the superbowl must perform for their whole season. The season-long focus and intensity that is necessary for NFL team members is exactly what participants in industry should be providing to their teams. Clearly, this is not easy to do. It would be easier to assemble these teams if engineers and managers were educated so as to encourage and support these ideas. It is generally left to the enterprise to provide an on-the-job education in this matter and every enterprise is not ready to do that. The author learned his craft largely through mistakes on programs. This is a very effective training method but enterprises deserve better.

There are several roles in industry that are critical for the successful implementation of system development. These people must come to each program fully committed to the cross-functional teaming approach and the need for strong communication links between the people and teams and take action to set up these teams in accordance with the demands of the program schedule. All too often, this collection of people do not approach programs from the collective "we" perspective leading to a lot of program risk so easily avoided. These critical roles are program managers, functional department managers leading some of the critical engineering domains, computer software engineers, system engineers, and program team leaders.

5.2   Functional managers

Part two of this book covers the development of a sound generic process depending on effective functional department managers focused on supplying all programs with the best possible resources in terms of well-qualified people, good practices coordinated with respected industry standards, and good tools.

It is too much to expect of individual functional department managers that they will place the enterprise at the forefront of their thinking and work. This is a fundamental reality that applies to all elements of the enterprise. We should not expect anyone to integrate and optimize at any level higher than their own level of authority, knowledge, and responsibility. Every organizational entity needs a system agent at every level of indenture. Thus, an enterprise needs an enterprise integration team (EIT) with overall responsibility for the enterprise process, the ultimate process owner.

The EIT must maintain a matrix showing how the process maps to the functional department structure and demand the functional department managers contribute to the aggregate process definition by providing best practice write-ups, tool development and selection, and workforce training.

The functional managers should be prohibited from becoming directly involved in any program leadership role and should be prevented from trying to manage their people in any way while they are employed on any program. Program managers should, however, take full advantage of the skill and knowledge available in the enterprise functional managers to serve as review team members, members of special investigation teams, and as document reviewers. The functional managers should also stay current with what their people are doing on programs and provide encouragement to them as well as technical support for specific tasks where the people are having trouble on a program.

Functional managers should be allowed to bring their people together periodically, once a month for an hour perhaps, to discuss how their work is being accomplished on the different programs, new ideas that appear to be working to advantage, and things that have not worked out well.

All programs should be instrumented with functional metrics leading to periodic numerical data on how the program is doing relative to each functional department at some level of indenture. Functional management should receive this data and use it as a basis for evaluating improvements that might be considered. Functional managers should coordinate their improvement efforts through the EIT, however, so that the improvement actions can be coordinated across the enterprise, and suboptimum situations avoided.

5.3   *Program managers and team leaders*

Program managers are in the best position to encourage effective system engineering in the enterprise. They are responsible for the overall program process, the budget, and the schedule. As a result, program managers are also in the best position to destroy the systems approach in an enterprise. So often the people attracted to program management roles are risk takers, and sound plans prerequisite to accomplishing work and good risk identification and mitigation work are not always fully supported by the program manager since these actions remove some of the thrill for people with these interests.

One of the greatest impediments to minimizing program risk is the program manager's fear that money spent early will not be available when needed later in the program. This result in a self-fulfilling prophecy, of course. Money spent well early reduces the overall program cost and risk. At the same time money spent cannot be unspent. This causes program managers to hoard funds early in the program, to the extent that they can, despite the evidence generally accepted that the later a problem is uncovered the more it will cost to correct. A sound system engineering job at the front of the program will generally be repaid by lower later program risk. However, this is true only if the programs are well served by system engineering professionals who understand how to do their job and are funded adequately to perform early on the program the known techniques that reduce program risk. The worst of all worlds is for the system engineering work to be funded but accomplished badly. Each time this happens, it pushes back the chance of improvements further in time.

5.4   System engineers

System engineers are expert in handling complexity whether it is of a product or process nature or both. System people have to be brought into the program early to define the problem that must be solved producing top-level specifications and supporting the identification and selection of a preferred concept. System engineers are generalists and should not be relied upon for specialist knowledge. They are skilled in piecing together the consequences of particular combinations of specialized knowledge that the specialists may not be able to see.

System engineers must reside at the product interfaces, or what some people call the edges, where different engineers or teams are responsible for opposing terminals. Historically, these are the locations of the most serious program failures. But the problem is greater than just the product in isolation. There are similar interfaces in the processes the system will employ in the field, the program and its teams, and the enterprise as a whole. This book is concerned with integration across this whole range of activities, all of the programs, and the enterprise generic process. Therefore, the EIT must include system engineers to accomplish this work at the grand system level.

5.5   Domain engineers

Domain engineers, of course, need to be skilled in their domain knowledge base and experienced with the enterprise product line and customer base. These engineers and managers must recognize that they are incapable of being successful alone without the cooperative work of people from other domains. The age of the independent, ad hoc, rugged individualist is long over in enterprises that develop systems to solve complex problems. This is not to say that all domain engineers should withdraw into a condition of meekness interacting with their peers only in the most polite and restrained fashion. No, good engineering can be a rough and tumble activity with many gifted people doing their best to influence the aggregate outcome in the best interest of the customer from their perspective. This conversation may become very intense stopping just short of physical confrontation without compromising the integrity of the cross-functional team process.

However, wherever possible, the many specialists must cooperate to the best of their ability to achieve group goals. When a decision has been reached even if it is contrary to an individual specialist's recommendations, everyone should join ranks in supporting the team in continuing to expand the team knowledge base within the context of the evolving decision stream.

It is especially important that software domain engineers be accepted into the mainstream of technical development rather than isolating them in separate teams. This works both ways, however, as software people have to cross over the boundary as well and interact with hardware people. Often, we have a choice on a program to populate one software team that may have an interest in the software components of several teams or distribute the software people into the several teams that have software entities included. Either arrangement can work provided it is well managed. Another alternative is to form a team responsible for the software running in the central computer but distribute software people into the teams to develop embedded software in the lower tier team areas of responsibility.

5.6   Top management

It is unlikely that any enterprise will ever improve its capability without the energetic demand of its top management. If the energy comes from the bottom up, it will eventually be exhausted by all of the negative forces at play against changes of any kind. The only way a bottom-up approach can work is through a conspiracy that shifts its leadership as a flock of geese does. The leader expends the greatest amount of energy so has to be rotated to one of the wings from time to time. Long distance bicycle racing teams do this as well with the team members drafting on one member alternately. In an enterprise improvement situation, the leader will be worn down because of the constant hostile stress. If top management does not see the good sense in continuing to improve the enterprise process and have a clear vision of how this will be accomplished, they should be the first to depart. This will likely not occur voluntarily, of course, so to the extent that top management does not take the lead, a conspiracy may be necessary to ignite the process.

The author has observed one other possibility for introducing and advancing the system engineering capability in a company. The division manager of an aerospace company that had no system engineering department or heritage formed a system engineering council with members from several departments. He directed that this council develop a system engineering manual telling what had to be done, and to the extent they could do it, tell how the work would be done. All of the members of the council were enrolled in a system engineering certificate program that the author happened to teach. After each course was complete, the council would meet and determine how they would deal with the subject matter they had just been exposed to in the system engineering manual they were building.

In another example of top-level leadership, the president of an aerospace company came to the conclusion that their system engineering capability was not adequate. He had the training department bring in a system engineering certificate program that happened to be the author's presented through the University of California, Irvine extension program and later through Indiana Purdue (for a total of six complete program presentations). Before the first class, the president put together a video that was to be played prior to the first class in the program. In the video, he laid out his concerns for the performance of system engineering work and how it related to the success or failure of their company. At the completion of the first program, the engineering vice president told the students that he wanted them to apply what they had learned on programs and that if they encountered anyone who prevented them from doing so, they were to bring that person to his office for a reeducation experience. Companies with these kinds of people in leadership roles are indeed fortunate.

chapter six

*Organizational structures**

6.1 *Updating matrix management*

In order to discuss synthesis, we must make some assumptions about the organization through which we will deal. There are three principal organizational structures common in industry: (1) projectized, (2) functional, and (3) matrix. In the projectized organization, the personnel are assigned to each program that has hiring and termination decision-making authority. In the functional organization, all personnel are assigned to departments that specialize in particular kinds of work, such as engineering, production, quality assurance, etc. The matrix organization approach imperfectly attempts to marry the positive aspects of each of these approaches and avoid the negative aspects.

The author supports the matrix management arrangement for large companies with multiple programs. The matrix is characterized by (1) functional departments led by a supervisory hierarchy, possibly including chiefs, managers, directors, and vice presidents, that provide qualified personnel, tools, and standard procedures, and (2) program or project organizations responsible for blending these resources into a set of effective product-oriented cross-functional teams called integrated product and process teams (IPPTs) in this book, and managing these teams to achieve program success measured in customer terms.

The functional departments provide a pool of qualified specialists trained to apply department-approved standard best practices and tools in the development and production of work products appropriate to the company's product line and customer base. The functional departments are charged with the responsibility to continuously improve the company's capability through small improvements in training, tools, and procedures based on lessons learned from program experiences and continuing study of available technology, tools, methods, company needs, and the capabilities of competitors. Company programs are internal customers of the functional departments.

Program IPPTs should be organized about the product entity structure derived through an application of functional analysis of the problem space reflected in the program work breakdown structure (WBS) overlay

---

* The material in this chapter is from Grady, J., *System Integration*, CRC Press, Boca Raton, FL, Chapter 3, 1994. Used with permission.

of that structure. The teams are identified and formed by program management from personnel assigned to the program from functional departments, and they are led by persons selected by program management. It was a popular topic in the management literature of the 1990s that the team leaders should be selected by their fellow team members from personnel assigned, but the author has not met a program manager who would surrender this responsibility. However appointed, once identified, the team leader is responsible for (1) molding assigned personnel into an effective team, (2) concurrently developing the assigned product requirements, followed by (3) concurrently developing a responsive product design, test, manufacturing (tooling, material, facilitization, and production), operations and logistics, and quality concepts.

IPPT personnel must first focus on team building matters in order to form an effective cross-functional group. Next, the team must focus on product and process requirements analysis. The program integration team (PIT) or other parent IPPT should have developed the specification for the top-level item for which the new IPPT is assigned responsibility, but the new team should be responsible for developing any lower-tier specifications some of which may be the top-level specification for still lower-tier teams. The product of this work should be documented by the team and approved by the program management. The team must, concurrently with requirements work, develop alternative concepts responsive to the requirements as a way of validating and demonstrating understanding of the maturing requirements. Where there is no clear single solution, the team should trade off the relative merits of alternative concepts. The team must be very careful not to influence the requirements identification too strongly by concept work but should take advantage of valid concept-driven requirements identified. This requires a delicate balance that should be assertively monitored by a system IPPT, called the PIT in this book, and by program management at internal reviews. The requirements must also drive manufacturing, quality, logistics, material and procurement, test, and operational process design work as well as product design.

Only after approval of the requirements and complying concepts at an internal design review, and at appropriate customer and/or independent verification and validation (IV&V) or oversight panel reviews, if required, teams should be authorized to proceed with concurrent product-detailed design work and related test, manufacturing, tooling, material, operations, logistics, and quality process design work. The team must implement all of the assigned tasks within budget and schedule constraints producing documentation that clearly defines what must be produced and how it shall be produced, tested, verified, and used to comply with the predefined requirements.

Functional department managers must staff all programs with qualified personnel appropriate to the tasks identified on the program period of performance and for which the program has funding covering those tasks. Personnel must be assigned to programs with a reasonable degree of longevity because the team efficiency will drop off to some extent each time there is a significant change of personnel. The reason for this is that the principal work of teams is communications and communications networks are fractured by personnel turnovers and these fractures require time to heal.

Day-to-day leadership of personnel assigned to programs should be through the IPPTs and their leaders rather than the functional departments. Personnel assigned to one team throughout a complete quarterly personnel evaluation cycle could receive an anonymous evaluation from their fellow team members coordinated by the team leader. Personnel assigned to more than one team in one evaluation period can receive evaluations from all teams within which they served. Each functional chief should review all of the program team quarterly evaluations for persons from his or her department and integrate this data into department ranking and rating lists used as a basis for all administrative actions (training needs, compensation adjustments, promotion/status quo/setback decisions, and program assignment considerations).

Functional department supervisory personnel, and senior working personnel under their guidance, should monitor the performance of the personnel assigned to programs from their department and provide coaching and on-the-job training where warranted. Functional management also should provide programs with a source of project red team personnel used by the program manager to review the quality of program performance in the development of key products. Functional department chiefs can then use this experience as a source of feedback on improvement needs appropriate to current standard procedures, tools, and training programs. Functional chiefs, managers, and directors should be encouraged to follow the performance of personnel from their department on programs, but should be forbidden to provide program work direction for those personnel. That is, functional management may provide help and advice for their program personnel in how to do their program tasks but should not direct them in terms of what tasks to do nor when to do them. Only the program-oriented IPPT leaders and lower-tier supervision, if any, should be allowed to direct the work of team members. Functional management personnel, in this environment, are rewarded based on the aggregate performance of their personnel on programs (in terms of CDRL submittals, major review results, budget and schedule performance, and noteworthy personal efforts recognized by project management) and the condition and status of their department metrics, which may include depth, breadth, and quality of standard procedure coverage for the department charter, toolbox excellence, and personnel training program effectiveness.

Program budgets are assigned to IPPTs by program finance based on the agreements between estimating, program, and functional management arrived at when the tasks identified in the IMP/IMS were estimated and subsequently influenced through negotiation with the customer on the way to contract award. The IPPT leaders are not only responsible for the technical product development tasks defined in an integrated master plan and integrated master schedule, but team budget and schedule as well. Top-level IPPT leaders report to the program manager. Teams are generally, but not necessarily, led by engineering personnel during the early project phases (when the principal problem is product concept development) and later by production personnel (when the principal problem is factory oriented). Some companies have solved this leadership problem with engineering led development teams through first article inspection and factory-oriented production teams thereafter with engineering in a liaison role. Once the functional configuration audit (FCA) has been completed, however, there may not be enough budget remaining to staff lower-tier teams, and the team structure should retract eventually perhaps only consisting of the PIT continuing to support manufacturing through special teams led by engineers might subsequently be formed on a temporary basis for large engineering change proposals.

6.2   A model program organization structure

In our model of the perfect world, each program includes two or more IPPTs and one PIT. Some people prefer the name system engineering and integration team (SEIT) in place of PIT, and these readers can insert that term or any other in place of PIT. The PIT should be led by a deputy program manager drawn from available senior personnel. During the early phases, this person should be someone with extensive engineering experience. In later program phases, this position should be reassigned to someone with extensive production experience. During the early program phases, the senior engineering person assigned to the PIT, or the PIT manager, could be called the program chief engineer with the responsibility for monitoring all engineering work on the program and coordinating changes in engineering personnel team assignments through the IPPT leaders. The chief engineer is the principal product technical decision-maker for a program.

The PIT is responsible for technical direction and integration of the work products of the IPPTs toward development of a complete product. This includes (1) the performance of initial system analysis and the development of program-level documentation (such as the system specification and other high-level specifications, program plans, system product entity structure, interface block diagrams and dictionaries, and system-level analyses); (2) the mapping of the evolving system product entities to IPPTs and the formation and staffing of the teams; (3) the review and approval of IPPT requirements documentation; (4) granting authorization for IPPTs to begin design work based on an approved set of requirements, concepts, schedules, and budget; (5) monitoring the development of interfaces between team items; and (6) the development of interfaces between the complete product and external elements (system environment and associate contractor items).

In the author's view, the PIT, like the IPPTs, should report to the program manager as illustrated in Figure 6.1a. A project business integration team (BIT) includes all of the program-level administration functions such as scheduling, finance, configuration management, data management, personnel (human resources), program procedures, project-level meeting management and facilitation, action item management, program reference document libraries, information systems services, and the program calendar of events.

IPPTs organized at the system level may have to further decompose items for which they are responsible in order to reduce the problem space to workable proportions. In this event, the system-level IPPTs (those illustrated on the first tier of Figure 6.1, who report to the program manager) will create their own sub-teams, like IPPT 2 in Figure 6.1, and acquire any additional personnel needed to accomplish sub-team tasks. Sub-teams may be fully staffed for independent work or rely heavily on the parent-level team for specialists. Sub-team leaders are referred to either as IPPT

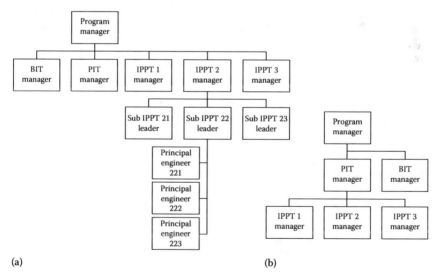

(a)   (b)

Figure 6.1 Program team structure: (a) recommended team structure and (b) alternative team structure. (From Grady, J., *System Integration*, CRC Press, Boca Raton, FL, p. 26, 1994. With permission.)

leaders or item principal engineers as a function of the magnitude of the task. At the lower system levels, such as an on-board computer or valve assembly, the person assigned development responsibility is referred to as a principal engineer (as under sub-team 22 in Figure 6.1). The principal engineers draw upon the specialty engineers assigned to the parent IPPT (or sub-team) for related specialty work. IPPTs are responsible for integration of the work of all sub-teams and principal engineers created under their authority. The PIT has this same responsibility relative to all IPPTs at the system level. However, in the author's view, while the lower-tier IPPTs should report through their parent team leader, the system-level IPPTs should report in parallel to the program manager with the PIT leader.

During the early development phases up through FCA, where proof is offered that the design solution has satisfied the predefined requirements, the chief engineer should normally lead the PIT and be the principal product system-level technical decision-maker subject to management by exception by the program manager. If the PIT manager has very demanding customer interface responsibilities and management problems with which to deal and has little engineering depth, the PIT manager would probably do well to appoint the senior engineer on the team the program chief engineer with the concurrence of the program manager. The PIT is the internal reviewing and approving authority for the system IPPTs. PIT must review each system team's requirements and concept sets before the team shall be allowed to proceed to concurrent design.

Each IPPT manager/leader is the principal technical decision-maker for the IPPT and should review and approve subordinate team requirements, concepts, and designs. Each IPPT is responsible for the development of all of its internal interfaces and jointly responsible, with the team responsible for the opposing interface terminal, for all of their external interfaces. This responsibility extends from PIT at the system level down through IPPTs, sub-teams, and principal engineers as a function of the scope of their responsibility.

Phantom IPPTs can be formed for items to be developed by associate contractors and work accomplished with this team through what is commonly called an interface control working group (ICWG). The senior member of this team will be selected for their interface development and interpersonal skills. On small programs, the ICWG may be a subordinate task of the PIT.

Teams or sub-teams may also be formed around major supplier items such as a rocket engine, guidance set, or other high-dollar or schedule-critical item. Leaders of these major supplier items may report to the program manager like other system-level teams or be organized as sub-teams to a propulsion or guidance system-level team. The team leader of a supplier team might be selected from procurement. These teams might be treated like principal engineer items as a function of the dollar amount, development complexity, schedule criticality, or program policy. No matter the organizational relationship, the program manager can manage all of these teams through the IMP/IMS with a subcontract laid on top of this arrangement in procurement cases.

6.3   An alternative team structure

The reader will note that the author chose to arrange the PIT at the same level as the top-level IPPTs in Figure 6.1a requiring the PIT to work with the teams through consensus rather than directing the IPPTs to conform. There is an alternative approach, often applied, inserting the PIT between the program manager and the IPPT leaders placing the PIT in a position to direct IPPT performance as shown in Figure 6.1b. The author maintains that all of these team leader positions (top-level IPPT, BIT, and PIT) are very responsible positions and the program manager needs input from all of them as equals. If the program is large, properly involves many top-level IPPTs, and the program manager will be very involved in customer interaction, the Figure 6.1b configuration may be the best arrangement because it will unload a large management focus from the program manager. A compromise would be to appoint a program technical teams leader to whom all top-level IPPT report and who reports in parallel with the PIT to the program manager.

The author is a system engineer who worked on several programs where system engineering was not as respected as he would have preferred but he still believes that the top-level design-dominated IPPTs should have a strong voice in the evolving development work. The PIT should integrate across the top-level teams from a position of equality in the author's opinion. In general, the IPPT members will have a more in-depth understanding of any problems that they are a part of that have been identified by PIT members, and in some cases the IPPT members will be right. The PIT should not be in a position to filter the input from the IPPTs rather should have to carry their position by force of reason where there is a disagreement. Also, placing the PIT in series makes the organization too vertical and reduces the opportunities for development of new program mangers among the IPPT managers.

The author noted above that he thought that lower-tier IPPT leaders should report directly to their parent IPPT leaders rather than through some lower-tier PIT, and this may appear arbitrary in view of the encouragement for parallel reporting at the system level. The logic behind this apparent conflict is that lower-tier PIT are not going to be financially supportable generally and the problems faced at this level can be handled by system engineers embedded within the parent IPPTs.

6.4   Resistance to IPPT

There are many reasons why the organizational structure described here is very difficult to put in place in a company currently organized to perform autonomously within their functional organizations on programs. These reasons include the following: (1) it disturbs existing organizational power relationships, (2) IPPT creates staffing problems for the functional organization, (3) IPPT interferes with an efficient personnel evaluation process, and (4) at one time it was felt by some people that there was a conflict between the cost/schedule control system criteria and the IPPT concept. Let us look at each of these reasons.

6.4.1   Human resistance

Many of those currently in functional management roles will perceive this prescription for distributed power as a personal, professional, and career threat. If they remain in functional management, they will see their relative power in the organization decline as it must to implement concurrent development. This will commonly result in resistance from those now in functional management roles. This resistance is inevitable. There is no best way to make the transition other than to announce it as far in advance as possible and openly discuss opportunities for those currently in power under the new system. It is, of course, entirely possible that at least some of those in functional management should not be allowed to survive the change.

The long range view of this shift, which many will not be open to see, is that a shift to the kind of organizational structure described above will attract a different collection of people to functional management positions. This position should focus on providing programs with qualified people trained in the standard techniques defined for a functional department and skilled with the standard tools supportive of those methods. Personnel selection and training, continuous process improvement, and tool building are the principal roles of this new kind of functional manager.

Many of those now in functional management positions will not wish to continue working in the new environment. Some will prefer to migrate to program centers of power (program office, IPPT leadership, or other program leadership roles). This is not a disaster, though some may perceive it to be, it is simply rebalancing of the power centers. Those for whom power is a motivator will migrate to it wherever it resides.

Those who have worked in a matrix management structure for many years have observed many shifts in the balance of power between the functional organization and program organizations. These are not uncommonly driven by the personalities of those in power rather than by some sound rationale for improvement in the company's capability. The change we are discussing here requires an enlightened management whose first priority is to satisfy the company's customers and needs for company profitability. They must have a willingness to make changes toward these ends despite a possible temporary setback in attainment of personal goals. This is a lot to ask of perfectly normal human beings populating management positions and it will not often be satisfied in the real world. On the other hand, the change may be a necessary prerequisite to the company's survival in a very competitive world, and this can be a very good motivator.

Obviously, movement into IPPT requires high-level support and leadership in an organization. Anything short of total commitment at the top will prolong the transition or destroy best efforts from any courageous and selfless people at the lower level. Some in the organization will attempt to defeat the change without appearing to do so.

6.4.2 IPPT-stimulated personnel staffing problems

Some people hold that IPPT creates staffing problems for the functional organization, and this is true if in moving to IPPT the functional organization is stripped of any knowledge of future budget availability. The functional organization needs this information in order to be able to ensure that the right number of people with the right kind of training and experience are available when the time arrives to use them on a program.

Some of the alternatives we considered above, frankly, can deprive the functional organization of the information it needs to satisfy future program needs. The last one considered and recommended, however, does preserve functional management access to this knowledge.

Cross-functional teaming can also result in IPPT leaders contracting with the wrong functional organization for a particular kind of work. Some small companies may be able to handle the resultant volatility, and it can even be a source of good restructuring ideas in a rough and tumble, but potentially effective fashion. Large companies will generally have difficulty with this and should have a clear definition of functional department charters with energetic enforcement of those charters by functional management when program management attempts to deviate.

The reason that these boundaries have to be jealously guarded is that you wish to hold the functional organization responsible for continuous process improvement concurrently with high standards of performance. A functional manager cannot be held accountable for maintaining a company's proficiency in a particular specialty if a different department is being contracted to do that work on programs. Suppression of deviations from the functional charter responsibilities is not necessarily, however, the best response in every case where this problem arises.

It is not uncommon that these occasions will be driven by a fundamental flaw in the way the company has assigned charters for the specialties. So, when an incident occurs, functional and program management should first consider if there is some value in considering an alternative organization charter responsibility map. If there is not, then the functional perspective should almost always win out. At the same time, during periods when one or more disciplines are understaffed with respect to the demand for work, they may find it useful to establish temporary agreements with other departments to provide personnel to do their work. This same arrangement can be useful at other times as a means to create effective system engineers or simply to increase sensitivity to the needs and concerns of other disciplines while creating personnel with broader qualifications to improve flexibility in satisfying changing program needs.

6.4.3   Personnel evaluation problems

We know that some people including Dr. Edward W. Deming, quality and production expert in the mid-1990s, was among those who believed that the devil himself designed the personnel evaluation system used in much of industry. This system results in functional management ranking all department personnel based on the functional manager's perspective of relative worth. This ranking is then used as a basis for salary increases, promotion lists, and, to some extent, work assignments that can influence future evaluations by virtue of experience gained through assignments. It is not uncommon that the evaluation criteria is flawed and applied in an irrational or uninformed fashion besides. Commonly, functional managers simply wish to get through this exercise as quickly and as easily as possible.

If you remove people from their functional organizations and physically collocate them with IPPTs, you make it difficult for functional managers to observe their performance. If we believe the IPPT/TQM literature, we should accept that performance evaluation should be done by team members and not by functional management anyway. So, how do we provide for personnel evaluation?

If we preserve the functional axis, we must accept that functional management be responsible for evaluation because they have the responsibility to provide programs with qualified people. However, functional management need not be the only source of input for evaluation data. If an employee were assigned to one and only one IPPT for a whole evaluation period (typically a year), then the evaluation could conceivably be done by the IPPT members in some fashion as Dr. Deming would suggest. Alternatively, the functional manager could request an input from the IPPT leader where the department member served.

What happens when a person works fractionally on two or more teams over the evaluation period? It will not be uncommon for some specialists in traditionally low budget specialties to work like this. We simply need some mechanism for the accumulation, merging, and distribution to functional department chiefs of team-derived evaluation data. Also, it would likely be useful to increase the frequency of evaluation events to perhaps quarterly, with this data merged in some way into annual figures for pay and promotion determination.

The author understands this problem from first-hand experience at General Dynamics Space Systems Division where he was the manager of the Systems Development department. Before engineering chiefs were installed in the department, the author had up to 95 people reporting directly and the annual review was a challenge in particular because his department members were spread out on the division's programs with only a secretary and a few people temporarily between jobs in the "home room."

This is a valid concern for entry into an IPPT work environment, but we should have little difficulty working up a computerized approach to acquiring the data from people assigned to teams, collecting and assessing that data, and providing it to functional managers in a fashion that encourages rational decision making in rewarding personnel for performance. It will require a change from present methods that generally are arbitrary, counterproductive, and otherwise just terrible. Good riddance!

6.4.4 C/SCS criteria conflict

At the time this book was originally published under the title *System Integration* (Grady, 1994), there was reason to believe that there were conflicts between what was called at the time the cost/schedule control system criteria and work team concepts that are referred to in this book as IPPTs built around the product entity structure and thus aligned with the WBS of the product. The author frankly never understood these concerns because the teams should be aligned with the product WBS such that the other axis of alignment with the functional department structure would simply coordinate cost and schedule between the organizational axes of the enterprise. Perhaps in enterprises where the IPPT were not aligned with the WBS, it may have been difficult to deal with three-axis mapping (functional departments, WBS, and team structures).

In many organizations, the use of the C/SCS criteria called for program persons called WBS managers who were responsible for managing cost accounts relative to the participation of a functional department on a program. This was a flawed management concept in the author's opinion because it inhibited the program from placing all of the management responsibility in the hands of the IPPT managers who should be held accountable for cost and schedule performance of their team relative to the WBS element for which the team was responsible in combination with responsibility for the performance of the product design relative to the content of the specification for that item. The author does not believe that there is any conflict between what is now called the earned value management system (EVMS) and the team concept nor any rules that would prohibit the IPPT managers from being given the complete responsibility for the development work related to their product entity.

6.5 Model matrix for this book

Figure 6.2 illustrates the overall organizational structure we will assume in later chapters dealing with synthesis specifics. One of several program planes is expanded with programs reporting in the vertical dimension and people from the functional departments reporting in the horizontal dimension. The program managers report directly to the company or division executive (or through a programs manager). The top people in the functional organizations on the left margin report through a functions manager. The top-level managers taken together are called an enterprise executive to emphasize their principal function. An enterprise integration team (EIT) performs day-to-day enterprise resource integration through the PITs under the oversight of the executive composed of senior management, and coordinates functional department participation in enterprise common process development and maintenance.

Each program has two or more integrated product development teams aligned with the product structure that is also coordinated with the WBS. Each of these teams is staffed by personnel derived by the program from the functional departments as a function of team budgets coordinated with IMP/IMS content. In many companies, the functional chiefs would control the tasking of the people staffing the intersections of the matrix. In our model, we will insist that the functional department chiefs specifically not provide program task assignments or work direction. That must come through the program channel traceable to an IPPT leader who also has WBS responsibilities.

Each program plane also includes a PIT responsible for system-level integration between the IPPTs and a business team responsible for overall program management, administration, finance, contracts, and program office support.

In Chapter seven, we will define one of the many integration activities as cross functional. This activity takes place in the vertical axis of Figure 6.2 performed by system engineers and leaders in the teams. The PIT members from each functional discipline perform cross-product

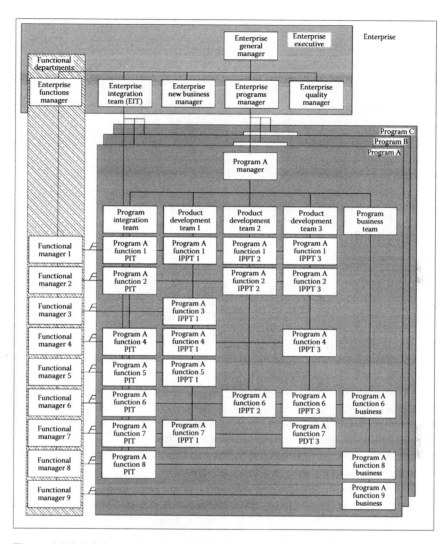

Figure 6.2 Model matrix organization. (From Grady, J., *System Integration*, CRC Press, Boca Raton, FL, p. 35, 1994. With permission.)

integration within their discipline in the horizontal dimension of Figure 6.2. Functional management, along the left boundary of Figure 6.2, should not interfere in the leadership of each program as we have said. But they should perform cross-program integration work in the depth dimension of Figure 6.2 for their function to stay abreast of their internal customer needs for personnel, tools, and practices, and to ensure all programs are applying their best practices well. This book focuses on program integration and does not expand on the need for cross-program integration work by the functional management. It could be argued in Chapter four that we should have identified it as one of four components instead of the three included (functional, product, and process). In the context of Chapter four, cross-program integration is nulled. If it had been included, it would have resulted in 64 rather than 27 integration spaces to discuss and this would have distracted from the intended focus. We will return to this matter in Section 6.7.

The EIT shown in Figure 6.2 is intended to monitor the resource needs of the programs and to proactively seek out future potential conflicts and risks and then mitigate those risks through cross-program integration work seeking a condition of the greatest good for the aggregate customer base. One principal function of this team will commonly be the coordination of shared manufacturing and test facilities.

The matrix structure shown in Figure 6.2 is not intended to encourage matrix management where it makes no sense. The structure illustrated is a limiting case on the complexity extreme where one company has many programs each involving systems with sufficient complexity to warrant multiple development teams. Let us consider some other limiting cases creating an envelope within which every company will find their particular situation.

In the event the company in question only has one program in the house during a particular period, it only needs one program plane and the PIT can be merged with the EIT. If a company has multiple programs but each program is for a relatively simple element, they may need only one team on each program. In this case, the PIT is absorbed as a part of each team in the form, perhaps, of a single system engineer responsible for requirements definition and integration. The EIT is needed because we have multiple programs, but the EIT will coordinate through the program system engineers performing the PIT function.

In the case of the ultimate simplicity of one simple program not warranting multiple teams, only one plane shows, the general manager and program manager become one, and there is one team composed of the functional managers and their personnel. The EIT is composed of the management staff who may have one person to implement their decisions. So, the author is not advocating the maximum matrix for all situations. Rather, a flexible application of this structure is encouraged that matches the particular business situation.

A corporation may also find that it needs to add one more layer to this structure to integrate the work of several divisions. This same concept of structural flexibility can be extended to this higher level as well as we will do shortly. As particular divisions become very successful in expanding their product and customer base, they may have to be partitioned into more efficient business units. As the business level retracts, it may be necessary to recombine some separate business units in search of a critical mass.

6.6 The virtual team concept

We have just painted a picture of necessity for the members of product development teams to be physically collocated in order for them to be able to take advantage of the resultant synergism. The reader should note that even if a team is physically collocated, they can be badly led such that this desired synergism of group effort does not take place. The team could conceivably interact in a series of one-on-one discussions in which the other team members did not participate. So, physical collocation by itself does not a concurrent development success make. It also requires good leadership that encourages effective group communications and use of effective communications tools, about which we will say more in Chapter seven.

But, let us question the absolute necessity of physical collocation. Is there a way that a team can function comprised of people at least some of whom are physically or geographically separated by some distance to the extent that direct interaction is not possible without special communications aids? Yes, it is possible and some companies are already working at the fringes of what will become a very normal condition once the necessary interfacing wiring and fiber-optics networks are extended to homes and enterprises through the information superhighway.

On the day that this passage was written, it was possible for the author to interact with others over vast distances via telecomputing in one-on-one situations. But, we know we must empower the group dynamic, with groups larger than two, because of our narrow specialization forced by our limitations and expanding knowledge. In Chapter seven, we will return to this problem looking for solutions. For now, let us assume that an effective team telecomputing capability is a near-term reality. While working on this chapter, the author was working on the Army Future Combat System, often engaged in teleconference meeting using meeting software that permitted very effective meetings to be held where the meeting population included someone who had mastered this medium.

Some companies that have manufacturing and design facilities in two or more countries are working effectively with integrated product development today empowered by good teleconferencing, videoconferencing, and telecomputing sharing of electronic data. But in these cases, while the full team is composed of nodes of people at physically separated places, they are collected into nodes at companies. The leap that will soon be possible is to form teams of people working out of their homes. The difference between then and now is that few homes are wired into the kind the information grid with the kinds of circuits that can handle the bandwidth necessary for simultaneous near full motion video, voice, and computer data. These capabilities are actually available today using cable modems.

Picture for a moment our IPPT of the future. Our team members work at their home computers under flextime rules with an agreement that they must all be on their machines during the hours of 10:00 to 11:30 AM and 12:30 to 2:00 PM each work day based on the time zone of the principal work site. Throughout their workday, all of these people will be tied into their program central database (see Chapter seven) and perform their assigned tasks adding to the central data store from their area of specialty.

At 10:00 AM (in the principal site time zone) each meeting day, all team members connect to the conference bus and join a televideo conference discussing team status and progress. As each person speaks, his or her picture comes into view in a corner of the screens of all team members. The screens all reflect the view that the speaker is addressing. All members are simultaneously hearing and seeing the same information stimulating new thoughts in the team members. These ideas are put into words and pictures and communicated to other team members. What results is exactly the same effect that occurs if these people had all been in one room at the principal site. The team members are virtually in that single meeting room and have become a virtual team.

The author first became aware of this possibility through discussions with a colleague Gene Northup, executive vice president of Leading Edge Engineering, Inc. based in San Diego, California, which was founded on the principles discussed in this section to provide companies with engineering services through remote computer communications. The concept was also covered in the February 8, 1993 issue of *Business Week* in an article titled "The Virtual Corporation" (Byrne, 1993) and in a book by the same title (Davidow and Malone, 1993). These are not new ideas but they have been relatively slow in coming into popular practice. A higher speed Internet will greatly facilitate commerce in the country as discussed above.

As this book was being written in 2007, the next great leap into the future can be described as model-driven development. Prior approaches could be described as document-driven where we prepare paper documents that describe the results of our work and pass these documents around between team members. Most engineering disciplines today actually employ some form of modeling within which they accomplish their work, but these computer models are, in general, disconnected and connected only through conversations between people from the different disciplines within the context of cross-functional teams. The model-driven approach in its most out-in-front manifestations could actually feed information automatically between models or through some form of human approval process.

For example, let us assume that the mass properties engineer has concluded that the mass properties will have to be set higher than currently calculated for a particular end item. This will have many effects that should be considered by people responsible for those other areas of interest some of whom could actually fail to respond to this mass properties change within the document-driven paradigm. In a fully automatic implementation of the model-driven approach, the computer system would make the changes in the aggregate model that can be followed by the specialists responsible for the several models that are interacting. Clearly, a semi-automatic application is also possible where the changes that are needed based on the mass properties change are presented for team and management approval. Given approval, the changes would be implemented in the model set. The computer advances to become an important member of the program teams.

Many companies are working now to introduce and perfect their application of the integrated product development approach with physical collocation thinking that it is some kind of a terminal condition of excellence. The reality is that change will continue to flow over us in powerful surges with decreasing periodicity. The virtual team and the model-driven development are good examples of that reality. While some are working toward IPPT as the ultimate structure, others are working toward the next step of the virtual team and the flexibility it provides. In creating and maintaining our process, we must not fall into the trap of thinking in terms of discrete terminal events. We should think of our process improvement activities as a continuum upon which we have constructed a planned growth path that takes advantage in the near term of available communication, computer, management, and process technology while planning future possible paths based on research that we consciously seek out and experiment with.

An excellent way to stay abreast of new paradigms is membership and active participation in a national society that shows an interest in this area. Some that come to mind are the International Council on Systems Engineering (INCOSE), American Institute of Astronautics and Aeronautics (AIAA), Institute of Electrical and Electronics Engineers (IEEE), Electronics Industry Association (EIA), Society of Automotive Engineers (SAE), and National Defense Industry Association (NDIA).

6.7 Extension to large organizations

The organizational structure suggested in this chapter can be extended to larger structures including business areas and sectors within a corporation. One should be very careful, however, how broadly one extends the common process idea. The author recalls one car company that included styling as a very early step in the development process. In another division not involved in automobiles, an engineering manager was heard to say, "What are we going to do with this, we haven't cared about styling since the 50s." The structure we are about to set up will include integration entities at every level of indenture, but a sensible rule would be to graduate downward the permissible integration rigor or zeal as we move higher in the organizational structure. That is, the integration goals should become fewer in number and focus on the absolutely necessary as determined by the higher-tier organization. This higher-tier organization must keep in mind that it is the viability of the lower-tier organizations that encouraged higher-tier success.

Figure 6.3 extends the concepts contained in Figure 6.2 to the corporation composed of sectors and business areas by whatever name, that is, two layers of organization between corporate management and programs. In this model, the word project is reserved for a major part of a program. The diagram generalizes three organizational entities at each level, but the actual number could vary over fairly wide limits. One sector could consist of two business areas and another of seven, for example. One new program may stand alone while another more mature one could consist of three projects dealing with the legacy system, an advanced development project for the next generation product, and a new business project to sell the next generation product.

One of the key decisions in this structure is at what level to focus the functional department structure. In this case, the program level was chosen but it could be at almost any level. The decision should be driven by the breadth of application of the functional disciplines. The ideal case would be if all of the people in a particular functional department could be applied to any project/program within that enterprise level. Inefficiency starts to creep in when a functional department manager must maintain the resource base to serve two or more different kinds of programs that require different resource sets. At the same time, we are limited at the lower end by a need to minimize the overhead burden. Concentrating the functional department structure at too low a level adds functional staff personnel, possibly unnecessarily. Each functional department should be a bare bones operation with a manger, secretary/helper, and possibly one or two department members between assignments who might be temporarily working on process improvements. The purpose of the functional organizations is to provide projects/programs with resources (an integrated set of people, practices, and tools). One might add a department member to audit projects/programs for compliance with the department's common process component. Alternatively, this could be a function of the next higher-tier integration team.

The reader will note that in Figure 6.3, all of the IPPT are staffed from the lowest-tier functional organization in that organization's hierarchy, which has a functional department structure. These connections are not illustrated on the figure in the interest of clarity of the remaining drawing elements. In the example used in this book, the SEIT for Project 1133 and Sub-team 11331 as well as Program 113 are staffed from the same functional organizations staffing the program and project structures. It is entirely possible that corporate, sector, and business area levels may

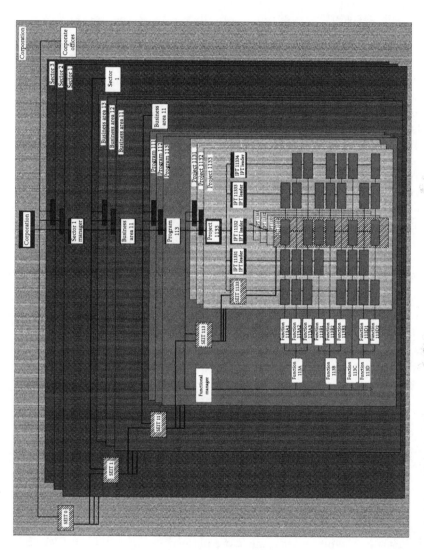

Figure 6.3 Corporate integration structures.

require some form of functional department structure though these will likely not be so heavily staffed with engineers and technical people as the program-level functional structures noted in Figure 6.3. Perhaps, the functional organization could be thought of as being within the corporation, sector, and program manager offices box in each of these cases. These functional structures would be heavy in contracts, finance, administration, and legal.

It may occur in some enterprises that there are specialized disciplines that are required across several organizational structures above the level of the lowest functional organization structure, but that the numbers of people required at any one time at any one level is small. In this case, there may be functional departments at higher-tier organization levels that provide lower-tier program/project personnel from such departments in the interest of efficiency and providing a critical mass of personnel for which a functional manager may exercise resource maintenance and improvement. The higher the degree of commonality in organizational structures across all of the corporate elements the greater the flexibility in personnel movement between those elements and the better the support for common process.

The integration teams identified on Figure 6.3 are called SEIT rather than PIT and EIT as used elsewhere in this book in the interest of simplicity in this discussion. The program and project SEIT correspond to the PIT and the corporation, sector, and business unit levels correspond to what is called an EIT in other chapters.

Each SEIT except the corporate SEIT has horizontal and vertical integration responsibilities. Its primary role at the lower tiers where programs exist is to integrate horizontally across the IPPT in its own organization (not shown in Figure 6.4) and the sister SEITs within its parent organization (as illustrated in Figure 6.4), but it must also integrate vertically across the SEIT within its parent's immediately lower tier and the SEIT in its parent organization following the patterns noted on Figure 6.3. The corporate SEIT has only lower-tier vertical integration tasks unless the enterprise applies it to external factors as well as internal.

It is possible that a project may require sub-teams and one of the three IPPT illustrated in Figure 6.3, Team 11312, has a subordinate SEIT (which may consist of a single system engineer), and three IPPTs, one of which is a supplier team. Each team, project, and program manager can manage his/her teams using exactly the same pattern. Assuming an IMP/IMS is employed with summary-level SOW and WBS work definitions, each team is organized relative to some architectural entity that has a corresponding WBS identification that sits within the SOW and IMP/IMS structures as a coordinated entity. This entity and team has associated with it a specification defining the problems space design must respond to: WBS funding

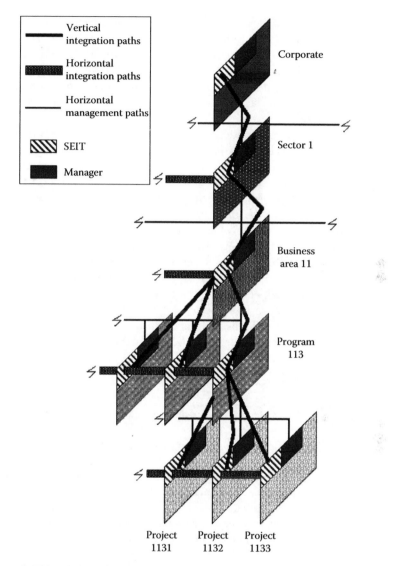

Figure 6.4 SEIT relationships.

and schedule responsibilities, general work instructions in the SOW, and IMP/IMS detailed work requirements.

Some very large and important entities procured may be assigned an IPPT, and the work accomplished by that team controlled by the parent team they report to, supported by procurement people in the lowest-tier SEIT that the parent team receives horizontal integration from. In that case, the IMP/IMS may include detailed instructions for the supplier team, but in any case the supplier IPPT must have a contract executed that implements their IPPT activities. The contract that an in-house IPPT has with the program/project manager consists of the WBS/SOW/IMP/IMS and the top-level specification for the team item. Lower-tier supplier items can be worked by the responsible IPPT with the suppliers essentially providing team members as necessary, coordinated through a contract or purchase order.

Figure 6.4 focuses our attention on the SEIT within the corporate structure encouraging a discussion of the SEIT responsibilities at the various levels. Table 6.1 summarizes these responsibilities, but the implementing organization should build the real table for their use based on their own appreciation for their business situation. This example places the functional departments at the program level so the enterprise common process is applicable for this level as well.

It is important to differentiate between integration and management responsibilities. It is not intended that the SEIT replace the level manager, rather to support the level manager with integration of level interests beyond the level manager's authority and to handle the day-to-day cross-team integration under the level manager's supervision. As suggested in Figure 6.4, the managers at all levels are responsible for all of the activities of the organization that they lead. These managers

Table 6.1 SEIT Responsibilities

SEIT level	Responsibilities
Corporate	Adjudicate inter-sector conflicts relative to corporate resources. Ownership of corporate practices.
Sector	Adjudicate inter-business area conflicts relative to sector resources. Ownership of sector practices.
Business area	Adjudicate inter-program conflicts relative to business area resources. Ownership of business area practices.
Program	Adjudicate inter-project conflicts. Ownership of program practices and coordinates with functional department managers to cause needed changes. Program functional quality assurance audits program and project IPPT and SEIT work relative to common process and reports results to program IPPT for corrective action if needed. Integrates and optimizes across project SEITs.
Project	Adjudicate inter-IPPT conflicts. Owns any project practices that should be minimized in the interest of supporting the program common process.
Sub-team	Adjudicate IPPT inter-sub-team conflicts. Generally, sub-teams will not possess their own practices because they will be implementing the program common process in accordance with the program/project contract and planning data.

are guided by corporate, sector, or business area procedures as appropriate and higher-tier manager direction or program/project contracts, WBS/SOW/IMP/IMS, and top-level specifications and interface control documents.

The SEIT (blue cross-hatched boxes in Figure 6.3) are responsible to their level manager for all inter-entity and intra-entity integration relative to the product and the process. All of the SEIT are charged by their level manager with cooperative work with other SEIT to reach mutually beneficial decisions that encourage mutual success. The SEIT are collectively trying to bring about a condition of optimum solution of the problem their level is responsible for. The IPPT at that level may independently arrive at a set of solutions that in the aggregate is not optimum. To the extent that this condition can be moved toward a better condition of optimization, the SEIT must interact with their program/project/team leaders to reach agreements supporting that new condition.

The heavy line in Figure 6.4 links the SEIT for integration purposes while the vertical hierarchy lines join the teams/projects/programs to their parent organizations signifying management authority. The heavy line does not signify a stronger influence than the management connections, rather this chapter deals with integration so that aspect is emphasized. To illustrate the relationship in another way, if Program 113 SEIT through its integrating work determines that Project 1132 has identified a design solution that is damaging to the Project 1133 IPPT 11332, it does not have the authority to direct Project 1132 SEIT to change its design for optimization at the Program 113 level. Rather, the SEITs for Program 113 and Projects 1132 and 1133 must negotiate a mutually acceptable solution entering into a trade study if necessary. The recommended action that arises from this work must be approved by the program manager of Program 113 and implemented by the project managers for Projects 1132 and 1133 as a function of the direction determined appropriate.

The SEITs should conduct themselves like ICWG recognizing that they do not have the authority to direct work on the part of their own or subordinate IPPTs. That direction must come from the appropriate program/project manager or IPPT leader. The SEITs may encourage IPPT changes that are within the cost, schedule, and performance goals affecting product form, fit, or function of those team's products that are contractually binding and must be addressed through a configuration control board (CCB) and program manager's decision and direction with possible customer coordination. To the extent that the SEITs can influence change for the better with the force of SEIT leader personality so much the better. Otherwise, an appeal to the several managers to fix observed problems may have to be resorted to with attendant hysteresis and cost impacts that could be avoided by the good works of talented SEIT leaders.

The discussion on organizational structures covered in Figure 6.3 has actually been based on a development program. In actual practice, the organization will have to flex as a function of where the program is in its life cycle. During the earliest program/project activity the emphasis may be on producing a winning proposal and the IPPTs may be focused on proposal volumes and the SEIT on integrating across the volume teams. During the proposal work, a clear definition of top-level product system architecture will evolve down to the level appropriate for development IPPT assignment with leadership by people skilled in marketing and proposal work. When the win is announced, the program/project is ready to staff up the IPPT as planned, and those IPPTs function for the duration of the development period. Throughout this period, IPPT leadership by people with engineering skills is generally appropriate. Eventually, the program/project focus will transition to production when the teams and their leadership will focus on manufacturing facilities, and leadership should move to people with manufacturing knowledge and skill. SEIT should transition from marketing to engineering and remain engineering oriented so long as engineering changes are being processed. This may be quite a long time as engineering change proposals and variants evolve. In fact, this can be the most complex period for engineering because it is necessary to maintain currency on several variations while evaluating new changes, the cut-in point, and the article span of those changes. From critical design review on, the SEIT will have to focus on both engineering integration work in support of the manufacturing-led IPPTs and engineering integration for evolving changes driven by the customer and marketing.

There is a period between the development and shipment of product systems to the field for user implementation that will be very intense in test and evaluation, and SEIT leadership might do well to switch to someone with strength in that area. This period will last throughout item qualification and system test and evaluation (both development and operational test and evaluation). It is possible that the author has emphasized replacement of the SEIT and IPPT leaders too much. But he has observed it happen so often that team leaders tend to remain in their role longer than is good for their career and the program/project health. Many readers will have experienced the case where the person who brought in the program to the company stayed on too long and was found too late to be the wrong person to manage the program in production. This has happened to some pretty great engineers and managers with bad effects on their career opportunities. There is another effect at work as well. Leading programs, projects, and teams is very hard work and even the strongest among us can be worn down by the necessary stress. The parallel with leadership in a flight of geese is appropriate where the goose in the lead position changes from time to time because it requires more energy than the other positions in the "V" as any NASCAR driver in the draft can tell you.

As product starts flowing from the production line and moving into field use by the user community, program emphasis will switch to sustainment of the system in the field. SEIT staffing should evolve toward logistics strength with some residual engineering support for changes driven by field experience.

As discussed in other chapters in this book, the author maintains that the personnel supplied to the program/project SEIT and IPPT as a rule should be physically collocated in those program/project teams rather than with their functional department. The day-to-day management should be from the program/project team leaders and program/project managers. Ideally, the programs/projects will provide opportunities for the functional department managers to staff their special evaluation and red teams. Also, the programs/projects should be faithfully following the common process practices. To the extent that they may not be doing that, this fact should be brought to the program/project manager's attention either through conversation by the SEIT leader or as a result of a quality audit. Functional department managers also should follow the work occurring on program/projects by participation on design reviews, mentoring their personnel on programs/projects, and discussions with department members assigned to programs/projects. Program/project SEIT and IPPT should implement the common process, but there may be cases where a functional manager wishes to pilot a new practice on a program/project leading to a possible process improvement.

All program/project SEIT and IPPT should be watchful for problems, especially recurring ones, that result in program/project inefficiency or unnecessary cost and schedule hits. For example, when every program/project finds that they cannot prevent weight growth, or a particular requirement is always carried as a TPM parameter, someone should come forward and open a discussion with the responsible functional department manager. Lessons learned should have an effect in stimulating process improvements. Functional departments should also identify implementation metrics that signal program/project health or the lack of it. The lowest-tier SEIT involved with the source of the observed problem should be given first right of refusal in dealing with problems.

The real job of the SEITs is to knit together the connections between the human intelligence nodes that appear in the IPPTs. Yes, they must network the people in the IPPTs. As discussed elsewhere in these books, man long ago pushed the aggregate knowledge base to a size larger than his own maximum capacity. The solution to this problem, applied for thousands of years on planet Earth, is specialization. This solution solves the knowledge coverage problem for a team of people, but it is not the whole solution. In order to be effective, the knowledge one person provides to the team must flow through the team network in some fashion so that the aggregate effect is the equivalent to one great mind. Whenever we differentiate or partition a whole into parts (knowledge into specialties and departments in this case), we must integrate and optimize at the level of the whole, and this is the purpose of the SEITs.

The principal network mode is human conversation, but it also includes documents whether in paper or computer media form. The currency of the SEIT is information exchange between members of the IPPTs that results in action that improves the whole. One of the most serious faults in us humans is a common lack of skill in effective human communication. Add to that a general tendency of us to withdraw from instances of boundary conditions experienced by teams and we have a potentially very serious problem. SEIT members must find ways for teams to exchange information and to motivate their members to cross over those boundary conditions to experience them from the other side. The work of the SEITs is at the very heart of the systems approach. As industry moves toward model-driven development, some of these interorganizational and interpersonal relationships will be improved after we suffer through a round of model relationship faults, but others will likely come into being that are very resistant to identification and correction.

6.8   Lowest common team concept

It is important to clearly organize the teams relative to the product entities, but we must remember that systems consist of entities and relationships between those entities that are satisfied through interfaces. In that some of the most difficult development work will occur in association with a particular kind of interface, we should ensure that our teaming arrangement has the necessary machinery to deal effectively with these interfaces.

The author uses a trio of words to refer to special subsets of system interface that collects all interfaces as a function of the viewer's perspective. Given a Team 1 on a program that is responsible for the development of particular items, how might the interfaces be partitioned relative to the Team 1 perspective? The word innerface is applied to an interface that has both of its terminals resting on items under Team 1 responsibility. Outerface is applied to an interface where neither terminal is connected to an item under the Team 1 responsibility. The third kind of interface has only one terminal under the control of Team 1 and is called crossface. Some other team has responsibility for the other terminal in this third case. These are what have been named cross-organizational interface in the prior discussion. These are the interfaces in an system development effort where system engineering has to focus its budget and work because it is at these interfaces where development risks will commonly be concentrated.

Product entity responsibility is simple in that there is a one-to-one correspondence between things in the system and responsible teams where a program employs cross-functional teams oriented toward the architecture. Innerface responsibility is also clear in that there is only one party involved who is responsible for both terminals. Outerface will show up somewhere as crossface or innerface in all cases except in those cases where an external interface touches nothing in the system but actually completes a path needed by the system. The problem in interface development is, of course, crossface. The author has concluded that the receiver should generally be identified as the principal party for any interface, meaning that they must trigger the conversation and insist that it continue until the interface is fully characterized. But there is one more concern and that is who should act as the integrator in all cases.

Product entities and interface relationships both require an integrator at all levels, the latter especially, because there are two parties to the work and people tend to turn their backs on their external interfaces and focus inwardly on their architecture responsibilities. The architecture integrator is very simple to identify: it should generally be the common parent team responsible for the item of which both team elements are part. Interface integration is a somewhat more difficult issue in that items in far-flung parts of the system architecture can have interfaces joining them. Therefore, the rule must be altered slightly to read lowest common parent. The author first encountered this term with Raytheon system engineers working on the DD(X) Program, later referred to as the DDG 1000 Program. Figure 6.5 illustrates an organizational structure recognizing teams organized relative to the product entity structure. It also identifies a series of ICWG through which interface is defined, integrated, and managed. If, for example, item A11 interfaces with item AX3 then ICWG AX is the lowest common team (LCT) and would have the integration responsibility.

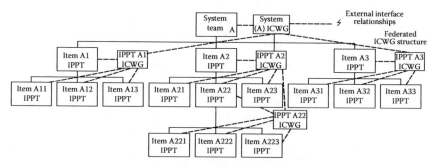

Figure 6.5 Federated interface control working group structure.

If item AX1 interfaces with item AY3, then the integrating team would be ICWG A. The parent teams, except for the lowest-tier teams, all operate an ICWG that is active whenever active problems involving it as the LCT exist. The ICWG should also audit all interfaces for which it is the LCT. This same concept can extend to the external interfaces where one or more ICWG (group) may exist reflecting a super system IPPT beyond the control of either terminal contractor independently perhaps operated by a SEIT contractor.

Figure 6.6 and Table 6.2 offer a view of interfaces in a complex system involving five product entity levels: (T) total system, (S) segment, (E) element, (C) component, and (I) item. Figure 6.6 identifies every kind of interface relationship between the various product entities in the system no matter how many of each kind of entity actually exists in the system. Table 6.2 identifies the LCT in each case.

It is an inherent responsibility of an IPPT as well as a PIT to engage in cross-organizational interface identification and development. The system agent (or LCT in the context of Figure 6.6) should audit both terminals of all cross-organizational interfaces, interfaces with different organizational responsibilities on the two terminals. This can be accomplished by simply reading the corresponding requirements for the two terminals and reaching a conclusion about whether or not they are describing the common interface in a compatible fashion. Generally, mating interfaces will have some form of plug and socket relationship but otherwise they should be the same or at least compatible.

Each non-lowest-tier team should operate its own ICWT as illustrated in Figure 6.6 and report up through the ICWT hierarchy any interface

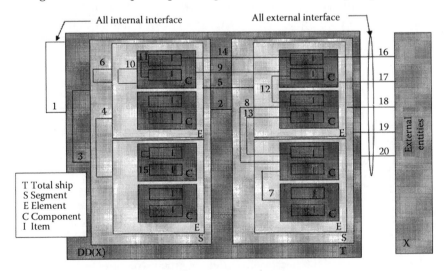

Figure 6.6 Interface integration categories.

Table 6.2 LCT Identification

NBR	Name	LCT
1	All-system innerface	T
2	Segment-to-segment crossface	T
3	Segment innerface	S
4	Element innerface, same segment	S
5	Element-to-element crossface, different segment	T
6	Element innerface	E
7	Component-to-component crossface, same element	E
8	Component-to-component crossface, different element, same segment	S
9	Component-to-component crossface, different element, different segment	T
10	Component innerface	C
11	Item-to-item crossface, same component	C
12	Item-to-item crossface, different component, same element	E
13	Item-to-item crossface, different component, different element, same segment	S
14	Item-to-item crossface, different component, different element, different segment	T
15	Item innerface	I
16	Item external	T
17	Component external	T
18	Element external	T
19	Segment external	T
20	All-system external interface (crossface relative to environment)	T

inconsistencies. The owning IPPTs will have to address these interface inconsistencies discovered and the responsible ICWT should follow up to ensure that the fix is implemented and that it results in a sound interface relationship.

Below the system level, interfaces should also be audited. Within the architectural responsibility of any one cross-functional team, the team should audit lower-tier interfaces where both terminals are items under the team's responsibility. Ideally, each non-lowest-tier team would act as the system integrator for all of its lower-tier teams and principal engineers by operating an ICWT. This approach will work for the cases where the team and its sub-teams are on both terminals of an interface, but it will not work for the case where one or more interfaces have terminals outside the team responsibility. Here is where the LCT concept must be applied. Thus, a particular ICWT has both internal and external responsibilities.

At the system level, the reader will note that the interface team is referred to as an ICWG that will have the same external interface responsibilities as the members of any ICWG but also the internal interface responsibility where the system team is the LCT. Many programs only have a two-layer team architecture, system and bottom tier. But, the tremendous complexity of large system development programs today will result in increasingly deep product entity structures encouraging a need for the LCT concept.

6.9   A process-based organization

Some enterprises have attempted to build their department and program structures based on the process they apply on programs. One enterprise at which the author presented a course had renamed its functional departments based on their processes. The research and engineering department was renamed the Define department. Looking at this technique from the author's perspective, the Define department would be the primary department responsible for the requirements function (F41 of Figure 1.1) as well as the design portion of function F42. We could make an Acquire department responsible to the procurement portion of function F42 and a Produce department responsible for the manufacturing component of function F42. A Verify department could be made responsible for function F44 of Figure 1.1. One would suppose we would have to have Manage and Quality departments as well.

It appears to the author that it is better to focus the functional department structure on an interest in knowledge rather than process and to allocate process-related functions to those knowledge-based departments. In any case, we have to establish a relationship between processes, knowledge needed to accomplish those processes, and organizational structures.

6.10   The organizational structure for
the thoroughly modern person

There is another alternative all together for determining how to organize in the enterprise as well as on its programs. That other alternative is to reject all forms of organization and permit employees to form self-directed nodes that appear to get things done. This may actually work in a limited sense where a problem is extremely difficult involving what today might be called complex systems. More on this later.

chapter seven

Information systems and communications*

7.1 The critical nature of communications

Successful system synthesis can only occur when the humans performing the work communicate effectively. This simple idea is the single-most
important notion in this whole book. It is so because the work must be
accomplished by many people, not any of whom knows everything. The
knowledge and conclusions held by each member of a development team
must be shared with other team members. This process of sharing information between humans is called communication. Communication connects the many specialists into a network of ideas that all can tap, and
it transforms the individual skills and knowledge of each team member
into one powerful force. The teams form the entities of our process system
called a program, and information flowing between the members of the
teams within and across team boundaries forms the interfaces of the process system called a program.

We seek to build, in our IPPTs, a series of subsystems, not unlike those
of the product system we intend to develop. Our product system will
include subsystems that must interface (communicate) with other subsystems to exchange information, and the components of each subsystem
must interact in the same fashion. We seek to weld the humans working
on the IPPTs into a temporary system as well. The second most important
point in the book is that there should be a strong correlation between the
product system composed of entities and the program system composed
of IPPTs.

The principal mode of communications between humans is language
composed of words and symbols, about which we have reached agreement, arranged to express ideas. As we know, these ideas can be expressed
as marks on paper or their spoken equivalent. So, communication takes
place through our power of speech and sign combined with senses
of hearing and eyesight. It is hard to imagine effective communication
appealing to other senses except in the isolated but important case of the
blind using tactile sensing. The media that we have found most effective

---

* The material in this chapter is from Grady, J., *System Integration*, CRC Press, Boca Raton,
  FL, Chapter 5, 1994. Used with permission.

in communicating include language on paper, speech (direct, recorded, transmitted), and, more recently, with computers using keystroked and video displayed language, and even the synthesized spoken word.

The single-most important communication channel on a program is probably direct, face-to-face human interchange with verbal language. In order to get full benefit from this mode of communications, with the technology available at the time this book was written, the team must be physically collocated such that members are close together facilitating easy verbal communication. It is possible to get some benefit from excellent telephonic communication between physically dispersed team members, but there is an added dimension to the actual direct conversation where eyesight as well as hearing can come into play.

Unfortunately, it is not possible to develop complex systems using only verbal communications, however. We need records and we need instruments through which to communicate complex ideas. It would not be possible to communicate the design of a rocket engine propellant turbopump from the design community to the manufacturing community using only verbal conversation. The richness of the idea defies simple treatment.

Engineering drawings and reports are needed to communicate the designer's conclusions for translation into manufacturing instructions appropriate to the operator of a machine tool whether it be a human being or a computer program. The requirements even for a valve are too many and too complex to trust to conversation between a buyer and a supplier. Without belaboring the point with a blizzard of unnecessary examples, there is a valid need for the capture of a tremendous amount of information in the development of a system. This information must be organized and stored in a fashion that it can be retrieved easily and quickly. It must also be protected against unauthorized change, loss, or damage.

Those familiar with the great fictional detective Sherlock Holmes, will recall that he felt that a detective should not burden his mind with useless information unrelated to crime. Nor should system engineers or the information systems that serve them suffer useless information. We must be careful to select the essential information for retention and allow the useless to be lost in order to ensure that our means of retaining information is not overcome in the storage of data and that the process of retrieving needed information is not made unduly complex.

The big question is, of course, "What is important and who decides?" Clearly, the customer has a big stake in what is retained. A DoD customer will define their formal information needs in a contract data requirements list (CDRL). For other data that exists, the customer may require the contractor to list it in a data accession list (DAL) and share the list with them periodically in the form of one of the CDRL items. For an agreed-upon fee, the customer is allowed to obtain any item on the list.

The things on the DAL should be every significant data item that is not a CDRL item. The criteria for significance could be a matter of judgment on the part of the program data manager or defined by a written DAL criteria. Some programs require a copy of every data item produced by anyone be sent to the data manager that results in that item being listed in the DAL. Here is a partial list for a criterion (all of these items are listed conditionally on not being CDRL items). The reader will be able to add many other items to the list and should do so as an exercise.

1. Procurement specifications and in-house requirements documents none of which are configuration item specifications so are not CDRL items
2. Internal design review meeting minutes
3. Supplier correspondence and supplier data requirements list (SDRL) items
4. Engineering drawings
5. Analysis reports
6. Verification requirements documents including plans, procedures, and reports
7. Program memos
8. Manufacturing planning and stamped planning cards
9. Quality inspection data including deficiency reports and responses
10. Lower-tier program plans (assuming the top tier are CDRL items) and procedures
11. Generic company procedures that apply to the program with a customer understanding of their proprietary nature
12. Program plans, controls, and schedules
13. Design decision traceability database
14. Requirements traceability data

Many program managers have agonized over whether particular items should be placed on the DAL due to their sensitive nature. An internal memo might indicate a serious flaw in the design that would be damaging to the company's position with the customer and even end up costing money when there is already a thin profit margin. It is very hard, under these circumstances, to do the right thing. It is very hard to accept that it is in your long-term best interest as a company and a person of integrity to provide this kind of information to the customer. However, it is true. A habit of openness with a customer will generally be repaid, if not in this life (program) then in the next. Any deviation from this position requires you to keep two sets of books with the ever-present possibility that the second set will come out in a very damaging fashion during the subsequent investigation or litigation over the consequences of the design flaw. There are, of course, sensible as well as self-destructive ways to initially inform a customer about the content of a damaging memo or report.

This kind of memo contains valuable design rationale information that completes the design decision trail for future research on the same product or another. One of the side benefits from being open with the customer in defining DAL content in the broadest possible way is that the combination of CDRL data, DAL data (including items on the SDRL), and contract correspondence defines the materials that should be available in the program library for internal access by everyone within the bounds of security prescribed by the customer for the program. It is possible that another program library category could be identified for proper company private information dealing with competition sensitive information of value in the next phase of competition, such as marketing reports. But this category should not be extended to accommodate a place to hide cover-ups of our mistakes.

Okay, so we have decided that we will retain everything of lasting value and we have a criterion for deciding what is important. Now we need to decide in what media we shall retain the data. Those familiar with the data management field understand that the volume of information just from suppliers can become overwhelming in a paper media. The Data Manager on the General Dynamics Space Systems Division Titan Centaur Program had to move into the electronic age when the paper-filled file cabinets were about to literally push the desks of the data management people out of the available space. The obvious alternative today is to place all data in electronic media and make it available to all via networked computer workstations.

Once you are in electronic media, whether you arrived there driven by physical space limitations or a conscious plan, you will find that many new possibilities present themselves that might not have occurred instead of routing paper supplier data in serial fashion through your departments for review: you will find that it is possible to send it in parallel to all reviewers via the network with no reproduction and distribution delays. Engineers and managers can also call up this data and project it onto a screen in a meeting powerfully making a case for some recommended course of action.

Communication is the exchange of information and, while the most important communication that takes place on a program is direct verbal conversation, the prepared material that will become part of the program library is an essential part of the complete program communication picture. Information can become lost in our paper filing system and the same can happen in our electronic equivalent. So, not only do we need to capture information for future use, we need to be able to access it on demand when that future time arrives.

7.2   A common database

Some companies have made serious and costly attempts to establish a database that contains every piece of information related to their business and product line. One term to describe this entity is a common database. Few, if any, of these efforts have been successful. It is simply very hard to know at the macro level what information should be retained and how all of that information should be interrelated. Most companies have settled on a less ambitious approach involving growth to a common database condition through evolution of databases useful for subsets of the total set of information organized about their process or functional departments. As time progresses, interfaces between these subset databases, initially accomplished by human communication and sneaker networking, become automated leading to an increasingly grand and complete database. This can also support a move to what is now called model-driven development.

Clearly, the medium to use for information systems involves computers and computer networks. The original computer systems featured very large computers operated by specialists in a central facility in batch mode. People who wished to use these machines had to bring their work to the central point, in some cases in the form of a box of punched cards. The development of microcomputers, powerful engineering workstations, and effective networking hardware and software have radically changed the way information systems are organized. Today it is not uncommon to find quite powerful Macintosh and Windows-compatible machines distributed within an organization where the working people are. Many companies have interconnected these machines with powerful mainframes and with each other via computer networks enabling a powerful degree of information sharing.

Is there some useful prescription for employing these resources already available to solve today's information retention and communications problems that also encourages growth toward a final solution? One approach the author has found effective is called an interim common database (ICDB) suggesting that it is on the technology growth path toward a true common database and eventual entry into a true model-driven development capability. This approach is also tuned to the dynamic period encountered during the early program phases, where rigid configuration management of data normally is not applied, with a built-in transition to production information retention.

7.3   Program interim common database

Several logistics engineers at General Dynamics Space Systems Division developed a shared information scheme that motivated an approach the

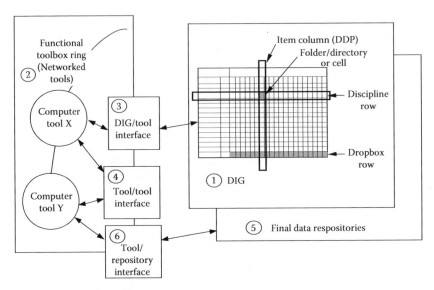

Figure 7.1 ICDB components. (From Grady, J., *System Integration*, CRC Press, Boca Raton, FL, pp. 55–66, 1994. With permission.)

author has named an ICDB in this book. This construct has six fundamental components illustrated in Figure 7.1. A grid of organized storage capability, referred to as a development information grid (DIG), provides a baseline set of information derived from all of the program specialists organized about the product structure. It is a matrix of information with a human interface. This information is accessible by everyone on the program in read-only mode from networked microcomputer/terminal machines. It is protected on a network server storage device from being changed except by the assigned expert. The information is entered into and maintained from workstations forming a toolbox ring surrounding the DIG and connected to it via the network. The five components of the ICDB are (1) the DIG, (2) the toolbox ring workstations, (3) the network resources that interconnect or interface the DIG and toolbox ring, (4) any connections between the tools that have been formed to date, (5) the final data repository, and (6) a data view connection between the toolbox ring and the final repositories.

7.3.1   The development information grid

The ICDB centers on a core DIG organized into columns by product entity structure (and, therefore, IPPTs) and into rows by specialty disciplines as illustrated in Figure 7.2. The vertical columns are called development data packages (DDP) under the responsibility of the IPPT leaders. The horizontal rows correspond to the several specialty disciplines assigned

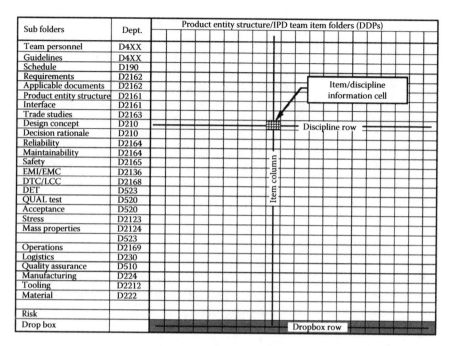

Sub folders	Dept.	Product entity structure/IPD team item folders (DDPs)
Team personnel	D4XX	
Guidelines	D4XX	
Schedule	D190	
Requirements	D2162	
Applicable documents	D2162	Item/discipline information cell
Product entity structure	D2161	
Interface	D2161	
Trade studies	D2163	
Design concept	D210	Discipline row
Decision rationale	D210	
Reliability	D2164	
Maintainability	D2164	
Safety	D2165	
EMI/EMC	D2136	
DTC/LCC	D2168	
DET	D523	
QUAL test	D520	
Acceptance	D520	
Stress	D2123	
Mass properties	D2124	
	D523	
Operations	D2169	
Logistics	D230	
Quality assurance	D510	
Manufacturing	D224	
Tooling	D2212	
Material	D222	
Risk		
Drop box		Dropbox row

Figure 7.2 DIG organization. (From Grady, J., *System Integration*, CRC Press, Boca Raton, FL, pp. 55–66, 1994. With permission.)

to the teams, such as, reliability, mass properties, manufacturing, quality, and so forth. Compare the structure of Figure 7.2 with the structure in Figure 6.2, and you will see a strong correlation between the cells of the DIG and the cells of the organization forming IPPT on the program. The DIG may have voids at its intersections as the matrix organization diagram does correspond to IPPTs that do not require particular specialties. The program panes in Figure 6.2 suggest that a company actually needs not one DIG but many DIG panes to serve its several programs.

The ICDBs are sets of computer folders/directories installed on a computer server not necessarily located within the physical spaces used by the team/project. These shared areas are specifically set up for whatever kinds of machines the team has available: MAC only, Windows only, integrated MAC and Windows access, or combinations also involving engineering workstations. Each DDP folder/directory is composed of a column of sub-folders/sub-directories such that everyone on the team has a place to put their team information work product. The DDP, therefore, becomes a kind of joint, public engineering notebook for the product item for which the team is responsible.

This engineer's notebook analogy forces recognition of a real problem with the ICDB that must be disposed of as a prerequisite to success. The information in the DIG is accessible to everyone all the time, and that is its strength from a communications perspective. However, many of us humans find it very difficult to share our work product until it is complete and perfect. This attitude is encouraged by management persons who are quicker to criticize than to encourage. Before we inaugurate our ICDB, we must deal with the human issue of access to imperfect and incomplete work. The DIG content should reflect, at any moment, the current information baseline, the best that we know at the time. People will be reluctant to share their evolving work in this open fashion unless they can be assured that it will be a positive experience. It may be helpful to apply a credibility marker to the information early in a program. An example of such a marker is (1) postulated, (2) analytically supported, (3) validated, and (4) baseline. The owner of the information would include the marker with the data and advance or retard it as appropriate. This would give others an indication of how hard they should run with this information in their own work. By accepting less credible information into one's current work, the risk is included of having to update it more frequently. By delaying until the story is complete, there is some risk of being overtaken by events and not contributing to the baseline conclusion.

The information contained in the DIG cells may have been placed there through the use of any number of different applications: spreadsheets, wordprocessors, graphics packages, or analytical tools. Naturally, the program, and perhaps the company/division, will be well served to encourage the smallest number of different software products for the same function consistent with permitting the most freedom possible in choosing applications that people can be efficient with. As in so many other things there is a broad continuum here with a large overlapping mutual conflict space. Frankly, some companies have found that it has been advantageous in the long run for them to have allowed employees and departments a degree of freedom in the past in selecting applications with which they could be efficient. It forced them to become more expert in data connectivity skills than they might otherwise have become. But, ideally, it is probably advantageous to work toward a condition of minimized applications with maximum interoperability. All Windows and MAC applications can share information via a clipboard, for example. There are an increasing number of applications that can also share data across the Windows/MAC valley. Microsoft Office applications are a prime example as are pdf and rdf file formats.

The immediate past suggests that the future of computer software will continue to be characterized by dynamic change. Some companies have attempted to adopt rigid software application standards for wordprocessing, spreadsheet, and graphics applications only to find that other applications arriving on the market were more effective. The U.S. Air Force and NASA managers of the National Launch System program in the late 1980s and early 1990s encouraged an excellent standard for software that we could all benefit from. They encouraged several contractors, competing in early stages of the program for later hardware contracts, to evolve program and product information system components that consisted to the maximum possible degree of off-the-shelf or over-the-counter software modules that could be replaced easily as newer, higher performance software became available.

Clearly, there is no magic solution in the near term to the question of the right hardware and software upon which to base an ICDB tool ring. You simply have to begin where you are this moment and adopt a continuous process improvement approach to making progress toward a set of goals for your information system needs. It is possible to enter into a condition of paralysis on this matter that is driven by a fear that anything you purchase now will be dated by new improvements very soon. Therefore, you should wait for another few weeks that turn into months with no change in the situation. We have all agonized over this effect when deciding on the right time to buy a personal computer.

The author finally understood that the reason he was reluctant to start using a personal computer at work in the early 1980s was that he was afraid he would look as stupid as he felt he was about these machines. He realized that he would have to buy one and experiment with it at home to overcome this inhibition. It was months before that happened, however, because of concern that whatever he bought would be immediately outdated. He finally bought an Apple IIe. Very soon there were a host of better machines on the market. The author finally graduated to an IBM-PS2 with a 286 chip upon which this book was originally written in its original form (*System Integration;* Grady, 1994). The 286 chip was way behind the times immediately, of course. But, in this dynamic evolution of computer resources that we will be trapped within for a long, long time, you have to put your foot in the water at some point and grow as fast as you can thereafter. It is not possible to intellectualize this process; you must be a participant.

7.3.2   The toolbox ring

The computer applications that your enterprise very likely already owns, like wordprocessors, spreadsheets, graphics packages, databases, and specialized analysis tools, reside on the toolbox ring machines. The toolbox ring is formed simply by having them on a network and saying, "Hey machines, you are a toolbox ring." These machines are used to create and maintain program information initially populating the DIG. The engineer uses a particular tool to create a data file and copies the data file into an assigned cell of the DIG via the network interface. This information is immediately available to everyone on the program in read-only mode from all other workstations on the toolbox ring.

The machines need not all be the same type nor must the software be standardized initially. You can simply start with what you have. The machines must be networked together, however. You should work toward an early point where the software does follow some minimal standards to assure maximum access and use of the information. It is possible to configure servers so that both Windows and MAC machines can access the same data. Someone can write a report on a MAC in Microsoft Word and someone else on a Windows machine can call it up for review in Microsoft Word, even cut a portion of it and paste it into a different document they are preparing.

There is no science to creating the toolbox ring. You can start, if your machines are already networked, by simply inventorying the machines you have and the software they are licensed to operate. Otherwise, you need to first introduce networking. Then you must decide what software you feel you should be using for each task that commonly has to be accomplished on programs and comparing that with what you have in each application type (such as word processing or spreadsheet). The result is an error signal that can drive your efforts in continuous process improvement.

The functional organization should be held accountable for developing or selecting appropriate computer tools for use by their department members on programs and ensuring that their department members are skilled in the use of these tools in accordance with a standard procedure. Over time, the functional organizations should be improving their tool set in concert with other departments with which their tools have information interfaces. This is a valid responsibility for the EIT suggested earlier. Each program will provide opportunities to experiment with slight changes in the department toolbox, and these opportunities should be seized upon as part of the department continuous process improvement process. As new tools are proven on programs, they should become new department toolbox standards for future programs.

7.3.3   *Toolbox ring-to-DIG interface*

Early in a particular program, the DIG is used to capture requirements and design (product, manufacturing, logistics, and operations) concepts for joint use throughout the program or during some fixed time span. As the program progresses to a more stable condition of the design and plans, information gradually drains from the DIG via the toolbox ring into the final repository databases where rigid configuration management policies restrict the ease of making changes, but the data remains accessible from all workstations. Team members reach agreements on who is responsible for creating and maintaining each piece of information needed by the team. Further, team members agree on which team members must be current on the content of the different team DDP folders. The DDP, therefore, becomes a means to communicate information about the product baseline and encourage consensus. Team members approach their concurrent engineering activities from a position of common knowledge of where they have been, where they are at any moment, and where they are going.

For example, early in a program, the designer for the vehicle hydraulic system uses the folder in the designer row and the hydraulic system column of the DIG to retain his or her concept sketches (using a simple tool like Mac Draw, Excel, or Autocad). This information is available to all of the specialty engineering disciplines concurrently as a basis for their analytical work. The specialty engineers prepare their data and load it into their corresponding folders/directories within the hydraulic system column. The manufacturing people working concurrently with the designer define a manufacturing process for the hydraulic components that will be manufactured and assembled by the company, and this is placed in the manufacturing row folder of the hydraulic system column. Procurement does the same for things that will be purchased.

IPPT leaders reach agreements with team members on who is responsible for each folder/directory for that team. Note Figure 7.2 includes a functional department (DEPT) column indicating these agreements. It is not suggested that the functional department managers would become involved in this process on any program, rather that someone from that department on the program would be responsible for the folder. At any moment in time, the DIG contains within its directories/folders the joint conclusions of the IPPT responsible for the vehicle hydraulic system. Everyone on the team and on other teams can gain access to any of this information from their workstation. Two human interface modes are suggested for the DIG: (1) open mode and (2) control mode.

The open mode is useful on early program phases when there is a great deal of volatility and a real need to initially load and change the data often to reflect changing concepts. The IPPT leader, in the open mode, will allow people free access to the DIG folders corresponding to their area of responsibility in order to quickly load team data into the folders/ directories. Once a baseline is established, reviewed, and approved by the team leader, chief engineer, or PIT, DIG access should be controlled. You reach this point when you become more concerned about unauthorized changes to the data than you are about stifling creativity. This change is accomplished by providing read-only rights for everyone and restricting writing rights only to the IPPT leader or a person of his or her choosing for their corresponding column of the DIG, that is, their DDP.

Any team member who wishes to change the content of his or her folder after a baseline has been established (control mode) based on the results of new work must copy the file in his or her folder/directory to their local workstation over the network, make needed changes using

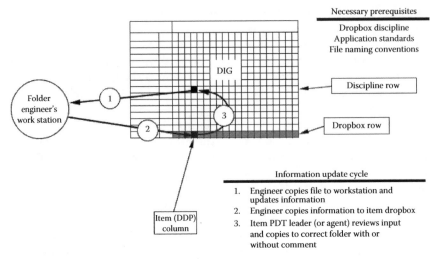

Necessary prerequisites

Dropbox discipline
Application standards
File naming conventions

DIG

Discipline row

Folder engineer's work station

Dropbox row

Information update cycle

1. Engineer copies file to workstation and updates information
2. Engineer copies information to item dropbox
3. Item PDT leader (or agent) reviews input and copies to correct folder with or without comment

Item (DDP) column

Figure 7.3 Data update in control mode. (From Grady, J., *System Integration*, CRC Press, Boca Raton, FL, pp. 55–66, 1994. With permission.)

their toolbox, and then copy the folder or file into the appropriate DDP Dropbox as illustrated in Figure 7.3. The Dropbox is a folder/directory the contents of which can only be viewed by the team leader or authorized agent but one into which anyone may deposit (write) a file. It is set up so that anyone may write but only the authorized person can see that the file exists and see into its contents. This is essentially the reverse of the controls on the domain cells in the DIG.

The authorized person opens the team Dropbox periodically and opens new files deposited in the Dropbox. The agent reviews them for acceptance. Notes may be written upon this material as to its quality, appropriateness, and timeliness and then the file is copied back into the appropriate DDP sub-folder from where it came before modification. Alternatively, the DDP agent could send the file back to the person who filed it with a note about the poor quality of the information or specific errors that must be corrected.

Clearly, the DIG depends on a prescribed file naming convention that is strictly enforced. Some operating systems that permit only very short file names provide a real challenge to make file content and proper DIG cell location obvious from the file names. Current operating systems are much less restricting in this regard.

Now, how might the IPPT members take advantage of the DIG to support concurrent engineering? The team members can periodically gather around one workstation and discuss DIG content making changes in real time based on the synergism of their conversation. If the team is too large for this and they are physically collocated, they can get together about several workstations in their common work area to view the data in common while carrying on a somewhat louder but no less effective conversation. If the group is too large for this, they can assemble in a meeting room and project the DIG images on the screen making real-time changes to the content. The projection scheme could also be used in their work area using a clear wall as a projection screen. If some or all of the team members do not share a common facility (some degree of non-collocation), they may gain access to the same imagery for this meeting via the network and converse via a conference call using a speaker phone as the data is changed in accordance with the team agreements or the team leader decisions. The telephone arrangement will work for hooking up supplier, or even customer, representatives into these concurrent team meetings in any of the alternative meeting styles discussed above provided they are networked with you.

Meeting software has become very sophisticated permitting very effective virtual team meetings with participants spread across intercontinental distances. It does require very good meeting leadership to make these meetings effective. Everyone who has something to add should be respected though it may be discovered in the process that the input was not very useful or it was even incorrect. It is human nature no matter whether a meeting takes place in a common room or with participants spread across the globe that some people will tend to dominate the meeting if permitted to do so and others will fail to come forward with even critically needed information. A good meeting coordinator will recognize these traits in people attending regularly and act to control inputs without suppressing needed information.

The author does not claim to be one of these gifted people but has observed some masters including Warren Smith, president of Execuspec based in Scottsdale, Arizona. In these meetings, control of the desktop can be transferred to any party in the meeting. Anyone can "raise their hand" asking for temporary control of the desktop to make a point. Ideally, there would only be a single person talking at a time but that is not always the case even in a single meeting room. Commonly, everyone can talk at once when using these applications, and this can become a problem if self-discipline is not exercised by the participants. If there were five people at one site all logged on independently by phone and several of them were talking simultaneously, the meeting coordinator could require that they all use only one phone (or the mike of a single computer if using a built-in televideo capability) with the rest listening on a speakerphone on mute. If one person has been silent for some time, the coordinator can ask, "John how does this sit with your appreciation for the requirements?" The author is reminded of a class he taught for a company that had two other facilities linked by this kind of capability with views of the other rooms available to him. He noticed that there was not a lot of movement at one of the facilities and asked someone to call them. It happened that they had placed the system on freeze frame and all gone on a break. A lot of the same techniques that will work in a meeting in one room will work in this environment but there are some new challenges that we should all work to perfect our skills in handling because this is a rapidly approaching reality.

Regular internal team meetings should be held often in a very informal atmosphere with no requirements for preparation of formal presentation materials. All of the needed materials should be available in the DIG. Without a DIG, the only means you will have for the team to interact concurrently is to hold frequent meetings. Without a DIG, the team data must be changed based on meeting results in a two-step process. With a DIG, the data can be refreshed with new ideas during the meeting with everyone in attendance. This is truly a concurrent engineering environment.

With a DIG the team information is available to the team at all times. Without a DIG, the team can only gain access to team information when it calls together the team in a meeting because the team information is largely in the minds of its members. A DIG multiplies the team's opportunities for effective concurrent work because everyone's baseline is available at all times.

These techniques can also obviously be extended to formal customer reviews requiring some degree of preparation, as suggested in Figure 7.4, whether all of the participants are congregated in the contractor's facility

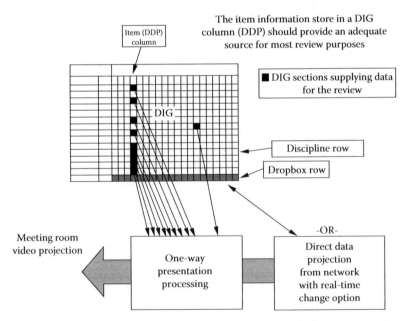

Figure 7.4 The DIG as a meeting presentation source. (From Grady, J., *System Integration*, CRC Press, Boca Raton, FL, pp. 55–66, 1994. With permission.)

or separated by some distance joined through their information and communication systems. The process of creating review data in a computer, only to print paper copies to feed the copy machine producing plastic presentation foils and paper copies is very inefficient. The data can be projected directly from your network with real-time changes during the review for relatively simple matters. Other expressed concerns may require assignment of an action item (entered into the network action item database by someone in the meeting room at a meeting support terminal before the action item discussion is complete) and further study.

Alternatively, the planned presentation data could be combined into an integrated data set using a presentation application like Microsoft PowerPoint. This approach permits a carefully planned presentation of the elements at issue at the time. The network connection in the meeting room also permits the presenter to call up detailed backup data to answer questions about concepts not clearly communicated in the selected presentation materials.

In order to take advantage of these powerful data projection features, meeting rooms must be wired into the network and the presenter's position in meeting rooms must provide access to computer workstation controls or the presenter can be supported by a computer operator. Over time, you should evolve to a condition where you have presenter stations with built-in computer terminals, possibly with the screen flush with the desktop like those used in TV news anchor stations. In addition, provisions should be made for a computer support station, perhaps in the back of the room. This would be used for action items, minutes taking, and other record purposes during the meeting. A program or company might have one or more workstations installed on carts that can be wheeled into meeting rooms or used in work areas to satisfy this need. The author has presented courses at many companies and universities using elaborate meeting room systems allowing the presenter to control the whole room from the speaker's position with recording of the presentation and any questions asked or statements made by those attending.

7.3.4   The tool-to-tool interface

Initially, your ICDB may not permit interaction directly between the toolbox ring tools. As we said at the outset of this chapter, however, our intent is to use the ICDB concept as a stepping stone toward a true common database. Requirements tools offer an example of how this can be implemented. Figure 7.5 illustrates a requirements tool on the ring that can be updated from a reliability database. The requirements database would have to be structured to assemble the requirements statement from fragments one of which is a numerical quantity field. In the case of reliability, this field would be updated for all items periodically from the reliability

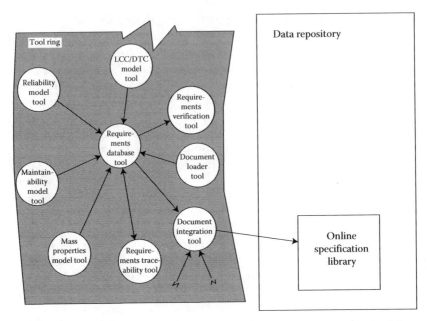

Figure 7.5 Tool-to-tool interfaces. (From Grady, J., *System Integration*, CRC Press, Boca Raton, FL, pp. 55–66, 1994. With permission.)

model database maintained by the reliability engineer. Reliability requirements data would always reflect the single authoritative current reliability baseline. This arrangement can be extended to weight, quantitative maintainability figures, and other quantitative parameters. In these cases, the specialty database quantities may be passed on to the requirements database at particular times when baselines are refreshed or the requirements database can be made to go to the specialty databases whenever it must produce a complete requirements statement using those numbers.

Other examples of possible tool-to-tool interfaces are

1. Traceability relationships between databases for statements of work, product requirements, and WBS Dictionary.
2. An interface between the company personnel database and all the personnel fields in product development databases identifying people as principal engineers, responders on action and issue closure items, and team responsibilities.
3. A system product entity database that all tools relate their data to. Reliability, mass properties, life cycle cost, and maintainability math models all hook to the same definition of the system such that it is possible to gather the complete story for a given item from databases maintained by different disciplines.

Where a tool-to-tool interface exists, it makes the use of a DIG for the data treated this way unnecessary. The utility of a DIG is in providing an organized repository for data that you have not yet figured out how to relate directly in a common database structure. As a company progresses in its climb toward a true common database, its need for the DIG should decline during the development period, being replaced gradually by the databases joining the tools on the toll ring.

7.3.5   DIG content evolution to the final data repository

As a program matures, the design sketches in the DIG will be converted into engineering drawings located in a protected computer-aided design (CAD) database and the sketches are no longer needed except as a possible aid in presenting design information at reviews. The manufacturing concepts flow into formal planning data. Procurement data flows into supplier contracts, statements of work, and procurement specifications. Specialty data is transformed into reports conforming to a customer data item description. Requirements database content becomes a pile of specifications generated by a requirements database. Most of the data in the DIG can be aligned with some terminal form of formal documentation.

Immediately after preliminary design review (PDR), much of the content of the DIG should begin flowing out to permanent information residencies. The design concept data in the design sub-folder becomes the source of engineering drawings; the requirements sub-folder content, possibly phrased in primitive terms, evolves out of a database into specifications stored in an online specification library; and some analysis data matures into CDRL reports stored in an online CDRL library.

By CDR and onward, the toolbox ring is connected to these final information repositories, rather than the disappearing DIG, with changes regulated by the strict information configuration management rules applied to those systems.

Clearly, companies dealing with DoD, DoE, and NASA will have to make this conversion of informal data retained in our ICDB into formal data. But, if we are a commercial firm, the ICDB content may serve our formal needs perfectly and provide us with time-to-market advantages over formal conversion of this data.

7.3.6   Common database approach

As experience is gained in the use of the ICDB, it will be improved over time through continuous process improvement to become a fully networked set of interoperable databases from which selected data views may be called by program and functional personnel and, with configuration controls and security provisions, accessible selectively by customer and supplier personnel as well. The tool-to-DIG (item 3 in Figure 7.1) interface will be replaced by the tool-to-tool (item 4 in Figure 7.1) interface and the DIG will wither away being replaced by the final information repositories that the toolboxes provide selected data views of.

The author, while employed by General Dynamics Space Systems Division in the 1980s and 1990s, worked for years building an integrated toolset he informally called Rosetta Stone. The real Rosetta Stone included the same message in several languages that allowed us to finally understand one of these languages that had defied deciphering for years. The author's system was intended to capture the information needed by many specialized groups during the system development process and make it available and understandable to all, hence the name. The author uses this environment to experiment with system engineering information concepts to determine what information might be useful in a final common database solution and how that information might have to be interrelated. The requirements components of this toolset are sometimes used in classes taught by the author to demonstrate generic requirements database concepts without endorsing a particular tool available on the market. The organization of this experimental system may be of interest as one small example of the potential scope of a common database.

Figure 7.6 illustrates the principal interfaces between some of the tools that comprised Rosetta Stone. It shows the top-level relationships between the information residing in each system. Each can represent an information system that has one or more companion databases that share certain key, or index, fields with other databases in patterns driven by needed relationships exposed through years of experience in industry.

This system actually included 62 component subsystems. Only those related to product and process requirements work are shown in Figure 7.6. Some of these are also supportive of product and process integration work. Other integration and management systems not illustrated included risk management, project definition (defines all projects for which the system provides information in one or more subsystems), configuration tracking, major review, action items, trade study management, personnel, meeting management, system analysis tracking, test and evaluation tracking, product identification, process audit, program events, product representation, and design rationale.

The user first selected a program defined in the project system and then selected an application. This combination defines the computer network server where the corresponding database files were located and the system connected these files to the calling workstation. If the person logging on had authorized access to the server directory within which the files are located, they could change the data, otherwise only read it.

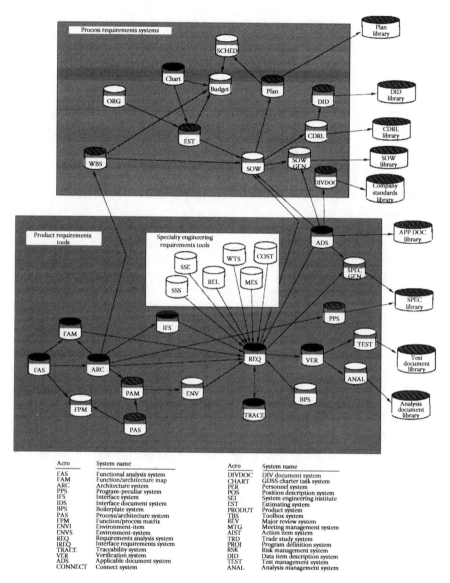

Figure 7.6 Rosetta Stone information relationships. (From Grady, J., *System Integration*, CRC Press, Boca Raton, FL, pp. 55–66, 1994. With permission.)

Some systems provided online access to document libraries, as illustrated in Figure 7.6. The applicable document system (ADS), for example, provided legalistic tailoring for several hundred documents, but the same database could be used to call the actual text of several documents that had been scanned. The operator could edit and highlight the text and store it as an in-context tailored document for a specific program. As new products came online permitting Microsoft Windows access to dBase data (such as Microsoft Access and Foxbase), the author intended to extend this experiment into the graphical user interfaces (GUI) world but retirement and a new career in consulting and training moved this work to a lower priority.

These systems were all character based and none provided a graphical front end. Some available requirements tools today provide very effective front-end functional analysis capabilities and even simulation features, which are very hard to master but powerful in the hands of a skilled user. Some companies find it very hard to maintain staff proficiency with this full capability because of the low capture rate on new unprecedented programs. They really need only a small subset of the complex tool as a repository for requirements within which traceability and verification data may be retained and revisions accomplished, and from which specifications may be printed.

Some of the tools that provide extensive front-end capability do not provide effective specification handling capabilities. Some other tools that provide effective specification data retention do not provide any front-end help. The author, after years of exposure to this problem has concluded that a company needs a relatively simple database for retention of specification data, however acquired, that provides good traceability and verification capabilities. If it also provides features for technical performance measurement (TPM), margin management, specification change notices, and source references, so much the better.

The user access screens for this database should be very simple such that anyone, managers, design engineers, and analysts, can operate it with very little training. This simple system should have an interface with one or more powerful front-end analytical tools on which only a relatively few engineers need to maintain proficiency for performing concept development, functional decomposition, and system simulation work. The results of this work should be ported into the simple database, and further requirements work completed in the simple system environment.

The author's system did include some tentative and unfinished steps in the direction of interoperability. The requirements database saved requirements in a primitive format with fields for attribute controlled, quantity (a numerical field), units, and relation (less than or equal to, for example). The requirements data was formed into sentences by code for printing a specification to screen or in paper format using a Windows version of Microsoft Word. The requirements system (REQ) permitted specialty engineering disciplines (mass properties, design to cost, and reliability, for example) to update their data independently and a refresh capability such that there was only a single source for the requirement numerical values. Links existed between the requirements system and the applicable documents system for printing Section 2 of a specification with tailoring included. A development data system pulled together the complete story expressed in all of the data systems for any given item in the system architecture (ARC) that the author now prefers to refer to as a product entity structure.

7.3.7 Web-based common data

Some companies have satisfied the goals noted above by establishing a Web browser for each program. The software can be acquired very inexpensively but does require Web site building skills that appear to be in abundance in companies. The beauty of this arrangement is that it does not require the program population to maintain application skills in each application used by the many different specialists. The specialists can bring their tools to bare on their piece of the problem whatever tools they choose no matter how difficult they may be to master and maintain skill with, and anyone will be able to open up and study their results in a pre-determined format requiring no special skill.

For example, the FA18E/F program at Boeing (then McDonnell Douglas) in Saint Louis selected Ascent Logic's RDD-100 as the requirements database application. This was the top-level tool for this purpose but it required an effort to acquire and maintain skill in its use. This tool was the top of the line for requirements analysis and management but there was a big problem in acquiring and maintaining skill unless using it everyday as the requirements engineers would be doing in maintaining the requirements database, of course. This problem may have had something to do with the company going bankrupt. Vitech's CORE would be a better example today. A manufacturing engineer, program manager, or reliability engineer might have difficulty maintaining skill in the use of one of these tools especially if they also had to maintain skill in their own computer applications and all of the others used by the different specialists.

In the case of the FA18E/F Program, any specification could be called up by clicking first on a specification icon, then highlighting the specification of interest. These actions brought up the specification from the database in the form of an Interleaf document that the reader could page through with no special skills.

This same approach can be extended to all program, data. The data can be maintained by many different persons even in different applications and all of them could be opened and used by anyone in a common format. The power is obvious here. Specialists may independently and deftly move from one application version to another with agility in a dynamic environment of software application evolution. Independently, the population at large can easily gain access to the data no matter what the application of choice is by the data owner.

7.4   Model-based development

The current title for a broad program database is model-based development. Under this concept, a program uses a set of tools to accomplish all of its specialized work, but instead of connecting the results from the application of these many models through human conversation and shoe leather, the models are interconnected and updates can ripple through the system between the models in an automated or semiautomated fashion. Imagine the mass properties engineer concluding that the mass assigned to a particular entity must be increased and there is an effect on control system gains somewhere in the system. Clearly, we were not ready to permit this kind of action to occur on a program in an automated fashion at the time this book was published. In the beginning, the control system engineer will be notified of a need to evaluate system gains based on the mass change. Today, we are dependent on the mass properties engineer making the case for change known to all and the independent analysis of this condition for appropriate responses by all of the other team members. In a fully implemented model-based development capability, the computer network could conceivably act as a responsible team member adding value beyond the mere connection of all of the machine assets into a network. The gain change could be automatically made with an automatic notice to that effect sent to a list of interested parties.

Clearly, this kind of capability will require a substantial modeling capability behind the wall, but within a particular product line it is not unreasonable to expect that this capability could be assembled generically based on the way that data has to be dealt with on every program by the program team. Enterprises will likely have to pass through an evolution where the machines notify the right people of proposed changes that can be evaluated by the responsible team(s) before pushing an execute button. That is, we will likely employ a semiautomatic mode for some time before the management will permit any kind of automatic operation. Once an enterprise enters this automatic mode, it is entirely possible that a program may find that this back plane was constructed based on a past use of a particular technology making it difficult to move forward beyond some boundary condition no longer necessary due to the availability of new technology. What we are suggesting is that in the future, a good deal of quality system engineering work may have to be accomplished on and even by the computer network connecting all of the applications.

7.5   War room or wall

Each IPPT, and the PIT for the whole program, should be provided with a wall or room of walls (ideally with stickboard material applied to part of this space) upon which they can hang large drawings and materials that help team members visualize team problems and item configuration features that are in work. This can be as simple as a wall in the workspace assigned to the team. Even though a team has access to a DIG as described above, there are some things that just cannot easily be grasped from the small computer screen no matter how ingenious the computer application. A wall full of the right materials can be very effective in triggering new ideas and insights about team problems. There are stories, it is said in Hollywood, that cannot be properly told on the small screen of television, stories that can only be told on the big screen of a movie theater. The same is true in system development where the big screen is a wall full of information.

A powerful combination is formed by integrating DIG projection with the war room by placing a networked computer with projection capability in the team's war room. This provides a combination of the grand system view on the walls and any selected detailed view of system information via computer projection. The war room should also be equipped with a speakerphone for remote voice access to meetings held in the room. When the DIG data is fully networked, the speakerphone can expand the size of the war room to include your whole facility overflowing to the whole world where remote engineers on a conference call are looking at the same image on their workstation that is being projected in the war room. This makes for a nice compromise to cover those persons who cannot be physically collocated with the IPPT. The IPPT and PIT should use a war room for regular standup meetings and maintain the displayed data to support their changing needs.

7.6   Virtual teams in your future

If your enterprise intends to attain or retain a position of leadership in your market, you will have to migrate to a condition of information sharing along the lines discussed in this chapter. The ICDB, or an evolving common database moving in the direction of model-based development, will also provide exactly what you need to enable a more flexible teaming arrangement in the form of physically collocated teams extended to include distance attendance. Clearly, future teams will consist of people not all physically collocated. As noted at the end of Chapter three, we are actually able today to effectively interact in a synergistic way even if physically separated through the use of computer resources designed for that purpose.

Virtual teaming will require a common information deposit (our evolving ICDB can serve this need initially) around which your virtual team members may collect for their discussions connected via computers capable of multimedia performance. This will include full motion video of the participants as well as voice and data. These teams will have virtually the same experience as if they were all together in a single war room.

7.7   *Integration excellence = communications*

Integration of product system elements occurs through the human communications patterns within the teams responsible for product development. Everything we do to enhance the communications abilities of the people within their teams as well as across team boundaries will appear as benefits in the product features. These benefits will be nowhere more obvious than in the absence of conflict at the product cross-organizational interfaces coinciding with different team responsibilities. A sure route to success in system synthesis and in the development of successful systems is to maximize the ability of the people to communicate and minimize their need to do so.

The author experienced an interesting means of encouraging communications recently when his daughter gave him a subscription to Netflix through which one can obtain movies on DVDs fairly quickly through the mail and one of his sons introduced him to a Netflix service called "Friends Invitation." This service allows subscribers to gain insight into the movies their friends are interested in and may find interesting themselves. Now, the author first saw this as an idiotic practice paralleling the way some younger people in a restaurant sample each other's meals in an excessive form of togetherness. But, he overcame this prejudice when he saw that there was a potentially effective communication method buried in it.

Picture, for a moment, a program where someone has built a "friends matrix" that defines what the members of each discipline assigned to the program need in the way of communication patterns. This could be implemented with an n-square diagram useful showing relationships between n objects of interest (knowledge domains or source functional departments in this case) named on the diagonal of a matrix of a size corresponding to n objects on a side. The other squares are marked to indicate to what disciplines people must send their ideas. Figure 7.7a is marked up to suggest the clockwise intra-team required communication pattern. When someone in discipline 1 communicates his or her ideas to others on the team, people in disciplines 3, 4, and 5 will receive that information no matter to whom the information is specifically addressed. Similarly, someone in discipline 5 must send to people in disciplines 1, 2, and 3. If we combine this matrix with the matrix offered in Figure 7.2 showing the participation of disciplines on teams, we could have a pretty comprehensive view of required program communication shown in Figure 7.7b.

One solution to the interdiscipline communication problem is, of course, everyone sends everything to everyone. But, these attempts at

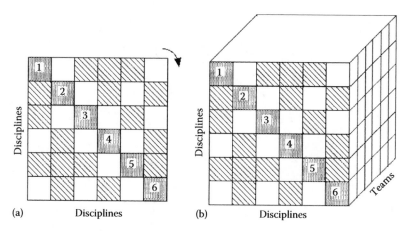

***Figure* 7.7** Program friends: (a) intra-team communications matrix and (b) program communications demand matrix.

communication approach the definition of noise as far as those on the receiving end are concerned. Communications must be targeted to those who absolutely must understand the ideas expressed. The communication rules would also have to include the rule that people must actually read and understand these messages addressed to them. To some fellow system engineers this may appear to be an idea that could easily have been hatched in a highly organized fascist or communist state, but it could encourage effective communication. What we are trying to do on a program is knit together the combined mental power available on a program into the equivalent of one great mind without damaging any of the pieces of that great mind. We have to depend on a very inefficient aggregate nervous system containing a spoken and written language.

All of us working on programs have an obligation to apply our skill and knowledge to the benefit of the program to which we are assigned and part of that job involves listening as well as speaking in spoken and written forms.

7.8 Program interfaces

The author is evolving an idea for inclusion in *System Management: Planning, Enterprise Identity, and Deployment* (Grady, in press) where the enterprise develops a clear understanding and definition of all of its development processes as discussed in Chapter eight and joins these processes together in a combination of sequence and network of work product flow. The interfaces between process steps or tasks are work products that flow from the completion of one task to another. Each task has certain required inputs and it accomplishes some process on those inputs producing one or

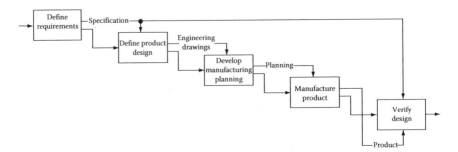

Figure 7.8 IDEF-0 depiction of work product flow.

more work product outputs. These work products can in the manufacturing case produce tangible metallic objects that eventually are shipped to a customer. But, they may also be information needed by people on another team in order to accomplish their work.

We have kept announcing in this book that we are depending on each other to communicate information of value from one to another, that we are all parts of one great mind working on a problem with a rather inefficient communication means between the mental nodes populating teams. On a program, the process steps on our flow diagrams are the entities of the process system that a program is. The interfaces in that system are the work products that flow from individuals and teams to others. An IDEF-0 diagram or enhanced functional flow block diagram (EFFBD) may be a very effective way of illustrating those interfaces. Figure 7.8 offers a partial IDEF-0 diagram showing the flow of steps illustrated in a kind of echelon fashion to better facilitate showing the controlling influences flowing from prior steps to later ones hitting the top edge of blocks or the flow of resources needed in later blocks with arrows striking the lower edge of blocks. The entities flowing to strike the upper or lower edges of the blocks are work products.

These work products can be very dynamic, of course, rather than the fairly stable entities illustrated. During the team work of an IPPT, the reliability engineer may communicate his or her conclusion that the current failure rate of the design is 0.00008 but that the requirement is 0.00009 meaning that the team design is not satisfactory. Some additional work by the principal designer over the next 2 days might change the failure rate assessment to 0.000093 and a fairly stable reliability work product output relative to that team for some time. Thirteen other instabilities may be in work within the team at the same time, of course.

The author is convinced that Vitech's CORE could be used to design an enterprise process flow using its EFFBD capability and depict the flow of work products using the secondary flow capability that can be used for data or commodities among other things, for example.

part two

---

*Enterprise common
process integration*

chapter eight

Enterprise process requirements definition

8.1 Process requirements sources

Enterprises are process systems that employ programs (that are also process systems) to transform knowledge and materials into useful product and process systems that are capable of achieving predefined goals defined by a customer in a need statement and system specification. Developing these systems under contract is the whole purpose or function of the enterprise for which the enterprise is paid in accordance with the contract. In that these programs and enterprises are systems, why should we not develop them using the same process we would apply within them to develop the product systems themselves? This process starts with requirements from which a design is synthesized.

System Requirements Analysis (Grady, 2005) describes how an organization can effectively define requirements for a system in a set of specifications and this book summarizes that process in Part three. This book covers the product design element in Parts four and five and *System Verification* (Grady, 2007) describes how product verification can be employed to determine the extent to which the product design satisfies the driving requirements captured in program specifications. Part five of the latter book also tells how we might apply the product system verification process to enterprises and programs to reveal the extent to which programs and enterprises comply with their requirements. This chapter deals with the application of the structured analysis process to enterprise requirements definition and Chapter nine deals with the design of the development process for an enterprise, hopefully in compliance with the requirements discussed here.

8.2 Enterprise structured analysis

The structured analysis approach can be applied to the development or reengineering of systems known as enterprises. In this case, the ultimate functionality is their vision or mission statement. It very seldom occurs that an entrepreneur starts an enterprise thinking in such organized terms, but every enterprise that survives will eventually come to a point where management is overtaken by complexity and the loose controls that worked previously no longer seem to encourage good results. The vision statement can be formed as of that moment when it becomes clear that they have to improve their organizational structure and rules of behavior. The process covered in this part can be applied to reengineer such an enterprise as well as craft a new enterprise to satisfy a predetermined purpose. In the reengineering case, one begins with whatever the current process is and whatever the current organizational structure is and seeks to improve it over time. The principal problem when this process is begun will likely be that the enterprise has not evolved a common process across its several programs and building a common process is a good place to begin.

The structured analysis process is accomplished on development programs to understand the problem corresponding to a need. The result is a list of requirements for the object of that analysis, a product and/or process system to be. When developing an enterprise, we are interested in a process specification as a prerequisite to designing that process just as in the case of developing product systems.

Several sources are available for a set of requirements appropriate for the development of a process design for the enterprise. We will explore an application of the structured analysis process beginning with the ultimate requirement, the enterprise vision statement. Next, we will apply structured analysis to identify needed enterprise functionality that will be allocated to enterprise functional organizational entities. Then, we will consider standards as a source of process requirements for the work that the functions of the common process diagram implies must be accomplished on programs. Finally, we will address the needed knowledge base to understand and respond to the requirements common to our chosen product line and customer base defined in the vision statement and expanded into our common process diagram.

8.2.1   The enterprise vision statement

In the development of product systems, we commonly think of beginning the development process with a customer need statement when developing an unprecedented system. In the structured, top-down development model encouraged in *System Requirements Analysis* (Grady, 2005) for the definition of product systems to be developed, this ultimate function statement is expanded into a multitiered expression of functionality using a functional model where lower tier functionality is identified using a functional flow diagram. Each block on the diagram identifies some necessary function the system will have to perform. Performance requirements are derived from this function and allocated to an entity in the product entity structure. This is the traditional structured analysis process for defining a problem of a complex nature following Sullivan's idea of "form follows function." This same process can be applied to precedented systems that have to be reengineered looking for functionality differential and the effects of that differential on the existing product entity structure, items that may have to be modified or deleted, and new entities that have to be developed.

The enterprise should have a mission or vision statement, analogous to the customer need statement for a product system that provides the ultimate statement of purpose or function of the enterprise. This statement will commonly have two components to it when the enterprise is formed within a country employing capitalist economy, one associated with providing customers some service and the other related to profit making. An example: "The SolderRite company will deliver the best soldering irons available to the construction industry while achieving a progressively increasing return on investment and treating its employees well."

8.2.2   Structured analysis of enterprise vision

The intent is to apply structured analysis to define the enterprise process, a common process, useful on all programs. The first step, addressed in the prior paragraph, is for the enterprise to define its ultimate function named the enterprise vision. This can be decomposed into a life cycle model that defines at a high level all work the enterprise must perform in the development and support of systems for its customers.

The author's company, JOG System Engineering (JOGSE), Inc., has developed a generic enterprise work definition regimen in support of its Grand Systems Development Training program that expands a vision statement, identified as function F, into a life cycle model containing functions F1 through F5 and further decomposes function F4 to lower-tier functions covering work that is common during the life cycle of all programs and systems developed by those programs. Figure 8.1 illustrates this common process at the life cycle level. This figure expands into several dozen process diagrams that collectively offer one way to define the common process of an enterprise. Each task illustrated can be further expanded as necessary to cover the enterprise work anticipated on all programs.

Each task in an enterprise life cycle model should be allocated (assigned) to one or more functional departments in an enterprise managed using a matrix structure, and those functional departments will be responsible for providing the necessary resources to enable that task on programs that function in accordance with the common process diagram. These resources include a coordinated set of people who know what they are doing, good tools, and good practices. Common process combined with lessons learned, phrased within that common process context, and conscious management action to respond to lessons learned so as to make corresponding improvements in the education of the people, the

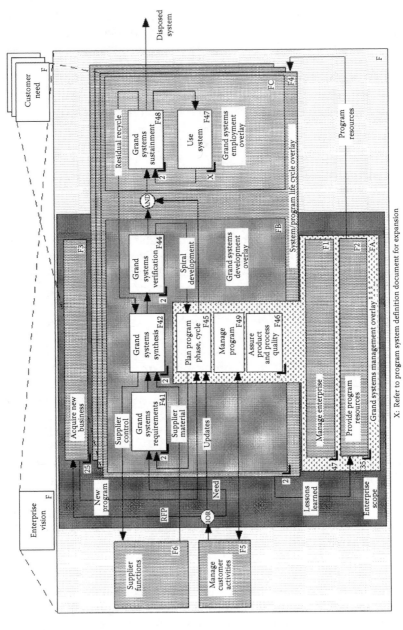

Figure 8.1 JOGSE common process diagram.

effectiveness of the tools, and the correctness of the process description, will result in continuous improvement of the enterprise resource base and improving program results over time.

An enterprise requirements analysis sheet (RAS) may be used to allocate common process diagram functions to enterprise functional departments. This is simply a list of all of the functions in one column and the functional department(s) responsible for each function in another column. This is obviously the same traditional structured analysis process that one would apply in developing a new system to satisfy an unprecedented need. The functional department structure is the hierarchical department entity structure of the enterprise. In addition, one can identify performance requirements corresponding to these allocations that when coupled together for one functional department form the department charter that can be translated into the content of a department manual as an expression of its design. Common process tasks in the JOGSE generic enterprise Grand Systems Development model are defined in the form of functional planning strings (or F-Strings) such as FY-DZ, where FY is the particular flow diagram task or function, the top level of which is shown in Figure 8.1 as F, and DZ is the functional department responsible for delivering a specific service related to that task to any program that pulls the task into its integrated master plan (IMP) during the program planning phase by associating a budget with the planning string. Programs buy the planning strings from the functional departments.

Some tasks are allocated to more than one functional department and this is, of course, more common at the higher levels of enterprise functionality. For example, many departments contribute to requirements analysis on a program. Thus, where a function is allocated to more than one functional department, one of those departments must be named the principal department with responsibility for initially preparing the process description, reviewing other department task descriptions, selecting tools to be used in performing this work, and providing for some way to qualify personnel to use the tools and techniques.

Figure 8.2 illustrates the formation of F-strings where a task requires multiple departments to contribute to work products created on programs. The string for department 230 in this case is F411-D230. The enterprise can capture clear definitions of each F-string in an enterprise work catalog. Each work catalog sheet, completed by the responsible department, identifies the related resources including: a generic cost and schedule estimate, in-house practices reference, reference to how-to information, input work products, output work products, and exit criteria. The input and output work products represent the common process interfaces between the responsible teams that are assigned task responsibilities on programs. Figure 8.2 also illustrates the assignment of principal responsibility to department D230 for task F411.

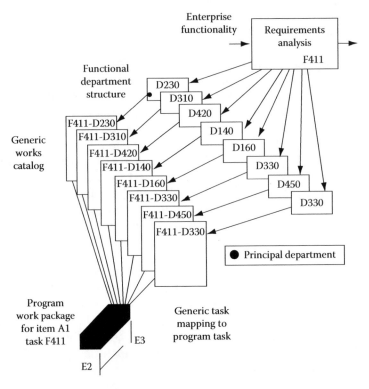

Figure 8.2 Task principal department concept.

In the ideal situation, our company will have been involved in a continuous process improvement program for some time and we will have developed a generic set of good practices coordinated with our tools, personnel knowledge, and skills base. This data should include (1) a generic process flow diagram that hooks common program tasks into relative time, (2) a functional department organization diagram, and (3) an organizational RAS that links the two. For each functional department named, the enterprise should have (1) a department charter listing all of the tasks for which the functional department is responsible for maintaining and improving company technology and capability (all the tasks allocated to it), (2) a task planning sheet for each charter task, and (3) a functional department manual at some level of granularity. For a system engineering department, the department manual will be a generic system engineering manual, of course.

Figure 8.3 offers an example of a generic planning sheet for one of the tasks illustrated on the flow diagram and listed in the task list. It happens to be a reliability task. Some of this information may have to be

Modeling ID	F151822312
Department	D834, RAM Engineering
Task name	Reliability Allocations (MIL-STD-756 Task 202)
Procedure ref	Company Procedure 42-8; MIL-STD-756B (tailored); Reliability Engineering, ARINC, Prentice Hall, Chapter 6.
Objective	Define a quantitative reliability figure for each system item
Description	The analyst studies the planned design concepts provided by the IPPT designer(s) and allocates the system or parent failure rate to sub-elements from top to bottom in each product entity layer and branch based on available historical data or predictions determined by the item composition.
Significant accomplishment	Reliability figures assigned to all system items.
Completion criteria	The task is complete when every configuration item and procured component has been assigned a reliability figure and the complete set of data has been checked for internal consistency, customer requirements compliance, and been approved by the PIT.
Events	Model structure complete at SDR, allocations loaded by PDR, and predictions loaded by CDR.
Resources	
Tools	Five IBM compatible computers each running RAMEASY, Microsoft Word, and Microsoft Excel for first two quarters. Thereafter, two machines adequate.
Personnel	At peak work load, six experienced reliability analysts skilled in failure rate prediction and reliability mathematics and familiar with equipment used in missile systems.
Facilities	Each analyst requires a standard company engineer's workstation. Personnel will be assigned to IPPT with two assigned to PIT and four to product teams one per team during the first two quarters of the program. Thereafter, two analysts on PIT only.
Inputs	1. Design concept information from item designer/team
	2. System product entity structure definition
	3. Maintainability allocations
Outputs	1. Reliability Model complete with a quantitative reliability number for each configuration item and procured item. Data users include: IPPT as a source of reliability requirements, Availability analyst as a source of reliability data for computing availability

Budget Est per quarter	1	2	3	4	5	6	7	8	9
Heads	2	5	5	5	2	2	2	2	1
Man hours	1008	2520	2520	2520	1008	1008	1008	1008	504

Figure 8.3 Sample functional task definition form. (From Grady, J., *System Integration*, CRC Press, Boca Raton, FL, 1994. With permission.)

adjusted or tailored as it is applied to a specific program, in particular the intensity of task application, but it is helpful to have this information as a beginning point improved over time based on experience on programs.

If you have these sheets available as a generic input to the program planning process, you can save time and introduce your company's standards for process quality into the program planning process commonly accomplished during the proposal development period. If you have no enduring generic planning data now, these forms could be filled in initially as part of your next proposal task estimate and later iteratively improved upon in each subsequent proposal process using the incremental continuous process improvement notion.

Some of the charter tasks expressed on the generic process diagram will entail personnel from only one department working alone to complete them. Many tasks will, however, require cooperative effort on the part of personnel assigned to the program from two or more departments as a function of how you are functionally organized and the business in which you are involved. Therefore, the atomic structure of generic task definition includes a task number and a department number. The generic task described in Figure 8.3 would be referred to as F151822312-D834. This identification number appears quite long due to the number of process layers included, but it is consistent with the generic task definition data in the enterprise process documentation suite and we will see that it can be easily coordinated with the program planning documentation as well.

With generic planning data in hand, we can apply this data in a tinker toy or erector set fashion to particular customer needs expressed in their request for proposal. To do so, we need an efficient transform process between generic planning data and specific program planning data. If you look at the planning data prepared in many companies, it includes a mix of functional department and product-oriented data. Our transform process must map the generic planning data into a program and product context understandable to our customers and valuable to us as a basis for managing the program through product-oriented integrated product and process development teams.

We have assumed throughout this discussion that the company is organized into a matrix structure. You may not be organized in a matrix, but you will still need to focus on the many specialties the technology base associated with your product line demands. The advantage of the matrix is that the functional organization can provide process continuity and a continuous improvement focus while the program organizations run the programs. The danger occurs when the functional organization is allowed to enter the program work supervision path.

8.2.3   Enterprise entity structure

Given a common process diagram for an enterprise, the process blocks so depicted represent work that must be accomplished. The work associated with these blocks can be assigned or allocated to functional department organizational entities that will be responsible for providing the resources programs needed to implement related tasks on programs. The author chooses to allocate the ultimate enterprise function, the vision statement, to an organizational entity structure called a matrix as described in Chapter six. The element of this organization to which this functionality is allocated is the functional organizational structure that happens to be structured around the knowledge base that the enterprise will have to maintain based on its vision statement and the technologies to which it will have to appeal to be successful in its chosen product line. So the first design step, at the system level, in this process is the selection of a particular organizational structure for the enterprise entity structure. As we continue to decompose the vision statement, we will continue to allocate increasingly detailed process functionality to lower-tier functional organization departments.

During this process, we will have to continue to integrate and optimize at the system (enterprise) level to ensure that the resultant organization is capable of operating multiple programs in accordance with a common process. Here is the heart of the whole grand systems development idea expressed in the first two pages of this book. An enterprise capable of implementing a generic enterprise process jointly optimum for all of its programs will realize a series of programs stretching off into the future, each of which performs in accordance with customer cost, schedule, and performance constraints while encouraging the continuous improvement of the resources available for use by all of its programs. With a common process, all of the programs can contribute metrics that have some common meaning toward process improvement. The enterprise will not be measuring apples and oranges. A common set of tools can be employed on all programs. A common set of how-to practices may be used across all of the programs. Training can be focused on this common process. The enterprise can use the several on-going programs as cauldrons for cooking up improvements for continuous incremental improvement of its process.

The application of structured analysis produces a graphically portrayed image of the process requirements in the form of a flow diagram and encourages the assignment of these process requirements to an organizational structure that will be responsible for implementing that common process. The flow diagram corresponds to process requirements and the functional department structure to the system design for the enterprise.

Next, we should coordinate our process descriptions with available standards of excellence found in industry standards.

8.3   Industry process standards

8.3.1   Standards as process requirements

Several good process standards have matured as a basis for the design of a process as illustrated in Figure 8.4. These standards offer requirements for a sound implementation (design) of a particular process. The U.S. Commerce Department has developed a way of measuring the quality of enterprise performance at the top level including its business practices and technical capabilities. This is the Malcolm Baldridge Award. This measurement approach focuses on the enterprise's capabilities to develop product and process systems for delivery to customers. The International Standards Organization offers ISO-9001 as a good standard for overall process quality. This is actually a quality standard, but it covers the whole development process generally offering characteristics of sound business processes appropriate to any organization.

A few years ago, system engineering was defined only in a military standard and poor software process standards existed in both military and commercial form. Now, the Electronic Industries Association (EIA) has developed the excellent standard ANSI/EIA-632 that simply lists 33 process requirements for a sound system engineering capability without offering a design solution or describing how to accomplish the

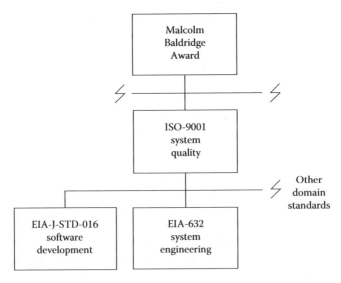

Figure 8.4 Process standards hierarchy.

tasks identified. IEEE-1220 from the Institute of Electrical and Electronic Engineers is an alternative system engineering standard in narrative form. IEEE/EIA-12207 provides a good software development standard. EIA has also developed an excellent software development standard feeding off from MIL-STD-498 (no longer maintained by DoD) in the form of EIA J-STD-016. An enterprise should select a set of standards across the range of functional departments it finds it necessary to identify, and work to design a common process for use across all of its programs that can be mapped to this set of standards. The standards selected should be ones that the common customer base prefers.

8.3.2   System engineering standard

EIA-632 began life as a slightly modified version of MIL-STD-499B when DoD decided to discontinue management and manufacturing standards as well as many others opting for the use of commercial standards developed by respected societies. The standard evolved over a period of a couple of years under the leadership of James Martin into a sound set of requirements listed in Figure 8.5 organized into requirements groups and subsets referred to as processes. These are further partitioned into tasks described in terms of expected outcomes.

The standard does not prescribe a particular process design, which is a departure from past attempts in standard development. This standard is strictly a process specification. The enterprise choosing to apply it to their process has the obligation to design a process that they can show is compliant with the standard. The reader familiar with the product system development process should see a parallel encouraged in this book of defining the problem in a set of requirements followed by synthesizing a solution through good design and verifying that the design satisfies the requirements. We should be able to apply this same model to any system including the one that is our enterprise as well as those that correspond to each of our enterprise's programs.

IEEE-1220 offers an alternative system engineering standard. ANSI/ EIA-632 is simply a list of process requirements without much of a story included. IEEE-1220, on the other hand, is a narrative. Figure 8.6 offers the table of contents for 1220.

8.3.3   Software engineering standard

IEEE/EIA 12207 partitions all software development work into three primary processes as shown in Figure 8.7 and further subdivides those into the secondary processes noted. Each of these is the subject of coverage in the paragraph numbers of the document phrased in the form of process requirements.

Requirement group	Processes	RQMT Id	Requirement title
Acquisition and supply	Supply process	1	Agreement satisfaction
	Acquisition processes	2	Product acquisition
		3	Supplier performance
Technical management	Planning process	4	Implement strategy
		5	Technical effort definition
		6	Schedule and organization
		7	Technical plans
		8	Work directives
	Assessment process	9	Progress against plans and schedules
		10	Progress against requirements
		11	Technical reviews
		12	Outcomes management
		13	Information dissemination
System design group	Requirements definition process	14	Acquirer requirements
		15	Other stakeholder requirements
		16	System technical requirements
	Definition process	17	Logical solution representations
		18	Physical solution representations
		19	Specified requirements
Product realization	Implementation	20	Implementation
	Transition to use	21	Transition to use
Technical support group	Systems analysis process	22	Effectiveness analysis
		23	Trade-off analysis
		24	Risk analysis
	Requirements validation process	25	Requirements statements validation
		26	Acquire requirements validation
		27	Other stakeholder requirements validation
		28	System technical requirements validation logical
		29	Solution representations validation
	Product verification process	30	Design solution verification
		31	End-product verification
		32	Enabling product readiness
	Product validation	33	Product validation

Figure 8.5 ANSI/EIA 632 process requirements.

8.3.4   Coordination between internal and external standards

You may not yet see how one could apply the structured analysis process to craft a new process where none exists today since we will not get to the details of this work until the next chapter. Your enterprise may already possess some form of internal process definition that just does not happen to be clearly documented, rather developed anew for each program from

1   Overview
2   References
3   Definitions and acronyms
4   General requirements
    4.1   Systems engineering process
    4.2   Policies and procedures for systems engineering
    4.3   Planning the technical effort
    4.4   Evolutionary development strategies
    4.5   Modeling and prototyping
    4.6   Integrated database
    4.7   Product and process data package
    4.8   Specification tree
    4.9   Drawing tree
    4.10 System breakdown structure (SBS)
    4.11 Integration of the systems engineering effort
    4.12 Technical reviews
    4.13 Quality management
    4.14 Continuing product and process improvement
5   Application of systems engineering throughout the system life cycle
    5.1   System definition stage
    5.2   Preliminary design stage
    5.3   Detailed design phase
    5.4   Fabrication, assembly, integration, and test (FAIT) stage
    5.5   Production and customer support stages
    5.6   Simultaneous engineering of product and services of life cycle processes
6   The systems engineering process
    6.1   Requirements analysis
    6.2   Requirements validation
    6.3   Functional analysis
    6.4   Functional verification
    6.5   Synthesis
    6.6   Physical verification
    6.7   Systems analysis
    6.8   Control

Figure 8.6 IEEE-1220 outline. (From Institute for Electrical and Electronics Engineers (IEEE)/Electronic Industries Association (EIA), 1998. IEEE/EIA 12207 *Software Life Cycle Processes*, New York. With permission.)

the corporate memory of your staff, and you may not yet see how that process could be coordinated with industry standards as a first step in improving your process design.

The first step is to make a list of your process steps at some level of indenture and then select a standard you wish to benchmark against. Establish a list of process requirements from that selected standard and form a matrix with the external standard content in an order dictated by the paragraph numbers in the standard. In the right-hand side column corresponding to each line item on the list, place a reference to a line item on your existing process list and mark that item on your list. Continue this process until arriving at the bottom of the external process list. In

Primary processes		Secondary processes
Primary life cycle processes	5.1	Acquisition
	5.2	Supply
	5.3	Development
	5.4	Operation
	5.5	Maintenance
Supporting life cycle processes	6.1	Documentation
	6.2	Configuration management
	6.3	Quality assurance
	6.4	Verification
	6.5	Validation
	6.6	Joint review
	6.7	Audit
	6.8	Problem resolution
Organizational life cycle processes	7.1	Management
	7.2	Improvement
	7.3	Infrastructure
	7.4	Training

Figure 8.7 IEEE/EIA 12207 structure. (From Institute for Electrical and Electronics Engineers (IEEE)/Electronic Industries Association (EIA). IEEE/EIA 12207 *Software Life Cycle Processes*, New York, 1998. With permission.)

some cases, you will be unable to find anything on your internal process list that corresponds to the external list and you have to conclude that your internal process does not fully comply with the external standard relative to that item. You may also find in the end that there are some elements in your process description that do not correspond to the external standard content (not marked on the list). The union of the external items and internal items that do not map constitute the extent of the compliance conflict.

It is not necessarily bad that elements of your internal process do not correspond to an external process standard, but you should reevaluate them to determine if they are really necessary. In the case of those external standard elements that do not map to your process, you have to decide whether or not you should be supporting those requirements. If your conclusion is that you should, then you have to change your internal processes to comply. Otherwise, you should generically tailor the external standard and develop the logic telling why it is not and should not be part of your process. This logic and tailoring can be used in your proposals to avoid a customer conclusion that you are not responsive to their preferred standard.

This process can become quite complex if your enterprise deals with several customers each of whom prefers a different external standard and they are not all in agreement. In this case, the enterprise should map each one to the internal processes and tailor each one for agreement with the internal process after having considered the noncompliant parts of each external standard for inclusion in the common process definition. The story that can be woven into a proposal about your performance relative to the customer's preferred standard is that

> ...we are interested in applying the best possible system engineering (or other function) capability to all of our programs using a common process that encourages staff repetition of work on all programs and progressive improvements in staff performance on all contracts. The map between your preferred standard content and our internal process standard reflects the degree of compliance and the logic provided relative to the tailoring we find necessary explains our rationale for our current business case.

8.4   Enterprise process documentation

Each functional department at some level of indenture must build a department manual describing what work it must do on programs when specific tasks mapped to that department are called in the program planning process. This manual is an expansion of the department charter but need not include the how-to information. The how-to information may be referenced in the work catalog and/or the department manual to textbooks and training courses.

The content of the department manual should be mapped to the related customer-preferred standard(s) and tailoring developed, if necessary, to cause the customer-preferred standard(s) to be equivalent permitting enterprise personnel to always follow the enterprise process description. It is essential that the enterprise have a common process, and it cannot do so if it must respond to three different customers, each of which prefers a different development process description. The way this can be achieved is for the enterprise to tailor each customer-preferred standard for equality with their own process. In so doing, they may find that there are one or more activities that are not now in their process and should be, resulting in an enterprise manual change. The result will be a single common process that may be followed on every program.

The results of this common process definition activity can be verified by mapping the process steps, at some level of indenture, to an industry standard of excellence. ANSI/EIA 632 offers a clear definition of 33 requirements defining an effective system development process. Figure 8.8 provides a map between the tasks in the life cycle common process diagram shown in Figure 8.1 and the content of two current system engineering standards, ANSI/EIA 632 and IEEE 1220. This same model can be

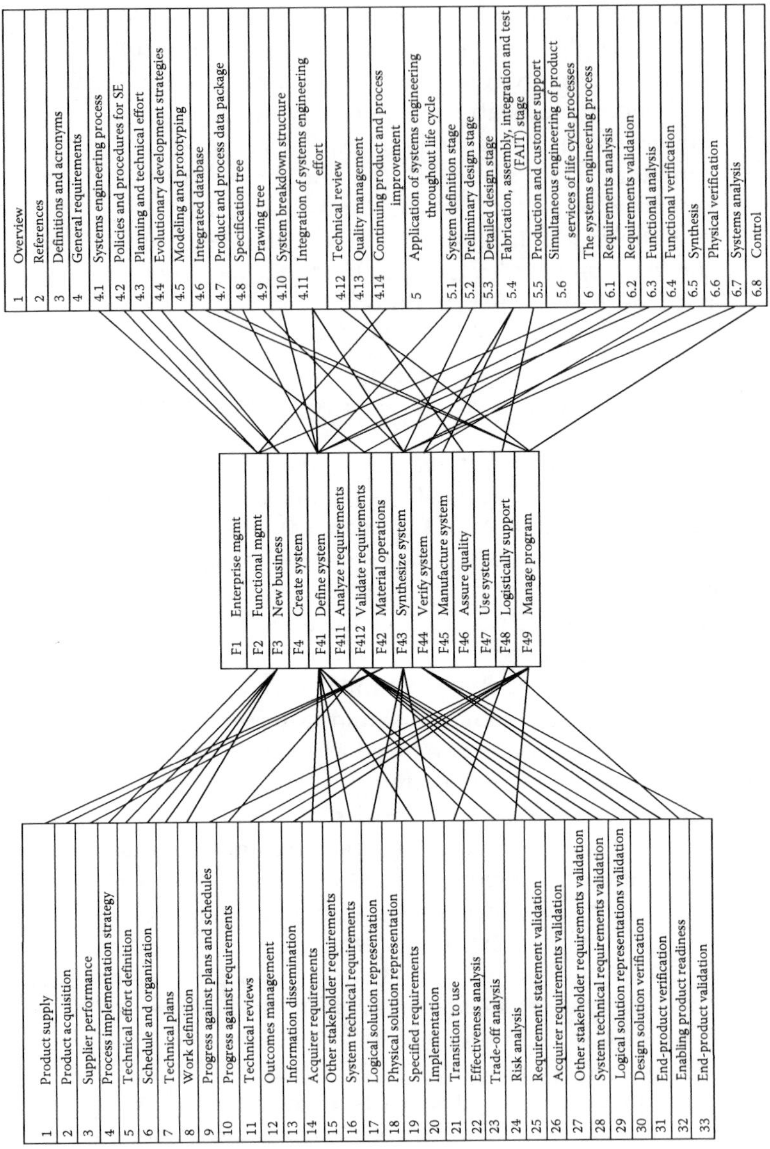

Right column list:

1 Overview
2 References
3 Definitions and acronyms
4 General requirements
4.1 Systems engineering process
4.2 Policies and procedures for SE
4.3 Planning and technical effort
4.4 Evolutionary development strategies
4.5 Modeling and prototyping
4.6 Integrated database
4.7 Product and process data package
4.8 Specification tree
4.9 Drawing tree
4.10 System breakdown structure
4.11 Integration of systems engineering effort
4.12 Technical review
4.13 Quality management
4.14 Continuing product and process improvement
5 Application of systems engineering throughout life cycle
5.1 System definition stage
5.2 Preliminary design stage
5.3 Detailed design stage
5.4 Fabrication, assembly, integration and test (FAIT) stage
5.5 Production and customer support
5.6 Simultaneous engineering of product services of life cycle processes
6 The systems engineering process
6.1 Requirements analysis
6.2 Requirements validation
6.3 Functional analysis
6.4 Functional verification
6.5 Synthesis
6.6 Physical verification
6.7 Systems analysis
6.8 Control

Center column list:

F1 Enterprise mgmt
F2 Functional mgmt
F3 New business
F4 Create system
F41 Define system
F411 Analyze requirements
F412 Validate requirements
F42 Material operations
F43 Synthesize system
F44 Verify system
F45 Manufacture system
F46 Assure quality
F47 Use system
F48 Logistically support
F49 Manage program

Left column list:

1 Product supply
2 Product acquisition
3 Supplier performance
4 Process implementation strategy
5 Technical effort definition
6 Schedule and organization
7 Technical plans
8 Work definition
9 Progress against plans and schedules
10 Progress against requirements
11 Technical reviews
12 Outcomes management
13 Information dissemination
14 Acquirer requirements
15 Other stakeholder requirements
16 System technical requirements
17 Logical solution representation
18 Physical solution representation
19 Specified requirements
20 Implementation
21 Transition to use
22 Effectiveness analysis
23 Trade-off analysis
24 Risk analysis
25 Requirement statement validation
26 Acquirer requirements validation
27 Other stakeholder requirements validation
28 System technical requirements validation
29 Logical solution representations validation
30 Design solution verification
31 End-product verification
32 Enabling product readiness
33 End-product validation

Figure 8.8 Common process ANSI/EIA 632 and IEEE 1220 map. (From Institute for Electrical and Electronics Engineers (IEEE), 1998. IEEE 1220 *Standard for Application and Management of the Systems Engineering Process*, New York. With permission.)

applied to all of the other functional areas of the common process using standards associated with the other organizational charters.

Given common process coverage of the requirements contained in one or more well-accepted external standards, the functional department manuals created by these functional departments should also be mapped to the content of those standards. To the extent that there is conflict between these sets of documents or omissions in the enterprise documentation, the enterprise documentation must be changed to provide coverage or the external standards tailored for equivalence with enterprise documentation. If the enterprise chooses to serve multiple customers each with a different preferred standard in some field, they must map to all of those standards and tailor all of those standards, if necessary, so that enterprise personnel may follow a single common process for all customers.

It is true that some customers may prove to be uncomfortable with your generic planning in that you can change it without their approval after a contract has been awarded. Some customers will want review authority on your internal plans and that can be a nightmare when you have multiple customers with very different interests. In such cases, you may have to run a copy of a generic plan and reidentify it for specific use on a program. This program-specific copy could thereafter be changed with customer review without causing chaos in your internal documentation. This will force you to deviate to some extent from the practice-practice-practice notion, but it is not a bad compromise when forced to respect your customer's requirements for review authority on all program procedures. The program-adjusted documents should contain the maximum generic content.

The party line in your proposals and in conversations with your customers should be that you are motivated to continuously improve generic procedures that will provide all of your customers with the best possible value. In cases where the needs to satisfy two or more customers are in conflict, you need to consult with those customers as part of the process improvement activity. You should also offer your customers access to your generic internal planning data on the basis that it is proprietary. Let them act as one of the pathways through which you can detect incompatibility using management by exception. One excellent way to do this is to provide internal read-only access to all planning data via computer network throughout your facility and extend this access to your customer base either at their facilities or only at your own.

In any case, the specific work required for each customer's program would be clearly defined in their program-unique and customer-approved program plan, work breakdown structure (WBS), statement of work (SOW), IMP, and integrated master schedule (IMS). These documents would reference the generic planning data that only tells how to do these tasks. The principal obstacle in implementing this approach is, of course, that every company has not done a fine job of documenting their current practices. In the author's opinion, this is not a valid basis for rejection of the offered planning method, nor does it represent an impossible barrier. It means that you must begin developing these generic practices now and keep improving them over time. Your competition may yet give you time to improve your performance for they are likely every bit as deficient as you are (if that be the case) in this respect.

In the case where a company is organized in a projectized fashion, or only deals with a single product line, you still have to acquire your personnel from the same source as everyone else, from the human race. We are all knowledge limited and are forced to specialize. So, even if you have no matrix and only a project structure in a company with one or more projects, you will have to accomplish program work using specialists whose work patterns should be standardized, one program to another. In this case, however, it will require a degree of cooperation between program managers that they consider heroic requiring the insistence of top management.

8.5   Knowledge base coordination with process responsibilities

The knowledge that the enterprise needs based on its selected product line and customer base could be used as a source of process requirements. Knowledge is acquired by the enterprise through hiring the right individual human beings who collectively have command of that knowledge. The principal question is how to organize this knowledge in the functional organization. One model is the university model that is probably a good place to start as the default condition. The benefit in roughly applying the university model is that most of your skilled position candidates will have been educated in that context. Those disciplines that are difficult to match with that model will have to be handled on a case-by-case basis.

The reader may revolt at this point believing that we should organize cross functionally for concurrent engineering. Well, that is a good idea on programs where the people should be collected into integrated product and process teams (IPPT) oriented around the physical product entity structure of the system. But, here we are talking about the functional organization that should, in the author's view, be knowledge based. The functional organization in a matrix-managed enterprise should provide programs with resources needed by the programs and base this structure on knowledge with the managers of these departments charged with maintaining and improving the knowledge base. Programs focus on satisfying customer needs organizing the resources provided by the functional departments into cross-functional teams oriented around the program product entity structure.

8.6   Requirements union

We should use all three of these process requirements sources (structured analysis, process standards, and knowledge structures) in the development of our internal process. The author recommends application of the structured analysis process expanding on the vision statement and allocation of expressed functionality to functional departments as covered in *System Engineering Planning and Enterprise Identity* (Grady, 1995) and *System Management: Planning, Enterprise Identity, and Deployment* (Grady, in press). One should then establish a charter for each functional department by aggregating all of the functionality allocated to each department in the context of the product line and customer base. This will determine the knowledge needs for each functional department. The result should then be mapped to the selected external standard and noncompliance noted and dealt with as discussed.

chapter nine

Enterprise process design

9.1 Preamble

In Chapter ten, we will identify program work in a very organized fashion permitting us to catalog all work required on any program in house. This can be done in any number of ways for any one program. What we desire is one way to do this for all programs allowing for the differences in products from program to program. We will discuss one cataloging method in this chapter that the U.S. Air Force developed called the integrated management system, but we will find that it does not easily allow for enterprise identification of work relative to its own process. Then we will discuss a variation providing a cataloging method that embraces the goals of the integrated management system but does allow linkage into the enterprise's generic work cataloging system keyed on their generic process description.

So, before we attack program planning, we must have our home court covered in the form of an enterprise design that is responsive to our enterprise requirements. Before we design a program, we should have a clear knowledge of the resources available in the functional organization structure. Once again, we intend to apply a generic or common process to every program oriented toward the program product entity structure.

9.2 Process component

In Chapter eight, we discussed a structured approach for identifying required activity. This should begin with the vision statement leading to a life cycle flow diagram, as illustrated in Figure 8.1. The vision statement is the ultimate function identified as function F in this figure. Each process step is named and given a modeling ID (MID) in the form FN where N is an alphanumeric string of numerals and letters using a base 60 system (all of the one-digit Arabic numerals, uppercase English letters less O, and lowercase English letters less l). Given a process step F14, it is expanded by identifying next lower-tier process steps, linking them together with directed line segments and logic symbols, and assigning F14N strings to each new step where N is one unique base 60 character. We could alternatively have chosen to use a base 10 decimal delimited MID system and some readers will prefer to apply that approach. The author does not like it because the periods do not sort or index well relative to the numerals in affordable databases.

Each block in the process diagram represents a work task that must be accomplished by someone or some group of persons. Each of these work elements requires certain input conditions, prior work products as inputs, and resources. It must operate on inputs and resources in some fashion and produce output work products. Each of these tasks can be associated with program resources of cost (material resources, numbers of people, and intensity of their effort) and schedule (start time, end time, total time, etc.).

Each of the major life cycle functions can be further expanded. At the lower tiers, these cannot be profitably decomposed from a generic perspective, but the grand systems development overlay functions can be. The reason the author has identified the life cycle tasks using a MID referring to functions rather than processes is that he wishes to employ a common process diagram that embraces all of the grand systems employment overlay functions that should be decomposed using some form of functional analysis applied on the related program doing the work corresponding to task F41, grand systems requirements. Therefore, this diagram represents every program implementation in the form of function F4 of which there may be several programs in different phases in the enterprise at any one time.

9.3   Responsibility component

In Part one, we established a functional organization structure based on the knowledge our enterprise must master, retain, and continuously improve our command of. These are the functional departments that we will hold accountable for providing programs with needed knowledge areas based on our chosen product line and customer base defined in our enterprise vision statement as well as good practices and good tools.

We named each department and assigned it a department number such as D1212. We arranged these in a hierarchical structure with all of the technical functional departments reporting to a technical functions manager and all of the administrative functional departments reporting to a business functions manager, the two of which reported to a functional manager or directly to the enterprise executive.

Every piece of work identified in our generic process must have a home in one or more of our functional departments. It is possible that some tasks will require more than one functional department and its related knowledge space in order to complete the task on a program. The higher in the common process diagram levels one looks, the more likely this is the case. Obviously, at the enterprise vision statement level, the ultimate enterprise process block, all functional departments at all levels map to it. Therefore, in Chapter eight, we established a way to assign primary responsibility for each process block to one functional department.

The responsibilities that a principal functional department has for each of its tasks are fivefold:

1. Acquire, train, and husband the necessary number of qualified personnel to do that department's work on all programs requiring those services. Where persons are found to be unfit to perform the work of a department on a program, that home functional department must withdraw and replace that person and find a suitable assignment for the person withdrawn through reassignment, training, or dismissal from the staff. In some cases, this may require the person be fired.
2. Provide programs with written best practices telling what must be done to accomplish generic process goals and provide reference to sources of information that tell how to do this work. The latter may be in the form of reference to textbooks, videos, courses, or be contained in the department's best practices directly.
3. Provide programs with good tools coordinated with best practices and personnel skills and knowledge.
4. Coordinate with all other departments that have some responsibility under the task in question to ensure that their personnel understand their part in the process, are aware of the aggregate process description involving the principal department's role and that of their own, and understand how to use the supplied tools. Any training that the principal department provides must accommodate the collateral responsibilities of other departments that also have a role in the process step.
5. Coordinate and cooperate with functional department managers whose departments have assigned a supporting role in the task to encourage that personnel from the aggregate of departments involved will work synergistically and cooperatively on programs to complete the task with an adequate work product within the available budget and schedule constraints.

Tasks 1 through 3 have to be supported by all functional departments while 4 and 5 are required only of those principally responsible for tasks requiring multiple departments to contribute.

9.4   Work responsibility fusion through allocation

We have applied functional flow diagramming (one could argue that it was actually process flow diagramming without author complaint) to identify the generic steps or tasks that would have to be accomplished on any program in the form of a life cycle flow diagram for any system that we might develop, produce, and deploy. Just as in the case of the use of this modeling environment on a program to develop the product entity structure by decomposing functions F47 and F48 of Figure 8.1 in function F41, we applied function allocation to associate process steps with the functional departments responsible.

As noted above, we also have to identify the role that a particular functional department would have in implementing the generic task description and resource base. Either the department would be called upon to lead the generic process development and maintenance or cooperate/contribute to its development under the lead of the principal responsible department. Knowing that in every case where there is more than one person or organization responsible for any part of a whole we must integrate and optimize, we should establish an enterprise integration team (EIT) at the enterprise level that will act as the enterprise process owner with the responsibility to integrate and optimize the generic process at the enterprise level leading the functional departments to contribute appropriate content.

9.5   Pull the strings

The intended generic process structure must connect all of the related resources to the process step MIDs such that when a program calls for the application of that process step to some part of the program work, all of the related resources flow into the program with minimum thinking required. As shown in Figure 9.1, when we pull the FN string into the program process flow for some item development to form program

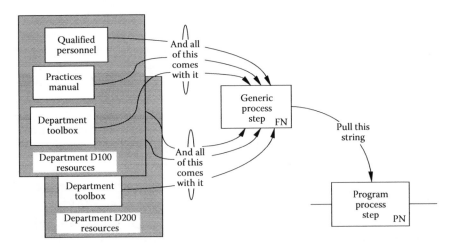

Figure 9.1 Process strings.

process task PN, all of the related resources are also pulled into the program from all of the departments that have a role in that process. For a particular generic task there could be a dozen contributing departments and there may be a good case on a particular program and program phase that only some subset of those would be called by the program. The way a call is made by a program is by listing a particular F-string in the program integrated master plan (IMP) and the establishment of a budget estimate associated with that call. Ideally, the enterprise work catalog would include a generic estimate that would be adjusted, if necessary, by a person from that department working on the proposal.

9.6   EIT efforts

The functional departments to which enterprise functionality is allocated are responsible for developing the corresponding resources. But, they cannot be depended upon to cooperate in optimizing these resources at the enterprise level without some leadership. Just as in the development of product systems, there must be an integration and optimization agent at the system level. The author refers to this entity as an EIT that is the ultimate process owner for the enterprise. This integration work should include assertively searching for incompleteness in the process and functional department resources, ambiguity or redundancy in resources, and lack of clarity in the details.

9.7   Functional planning strings

9.7.1   All of the functions

The generic process diagram identifies all of the work that must be accomplished on any program. If we form a process dictionary with a line item for each box on the diagram set, we will have a comprehensive listing of all of that work the enterprise is capable of accomplishing. We can then map these line items to the functional organization diagram resulting in a list of all of the work as well as knowledge of who it is going to be accomplished by on programs. For every function block on the process diagram, there will be at least one functional department identified. In some cases, many departments will be related to a single function step as noted earlier and we have a way to deal with this.

9.7.2   The organizational RAS

The result is an organizational requirements analysis sheet (RAS) that shows every combination of functionality and department responsibility. This is an expanded list of all of the work that must be accomplished in context with the departments responsible for that work. These line items are the work catalog numbers for that work and they can be linked to the corresponding resources. These are referred to in this book as F-Strings and they take the form of F141-D220 where F141 is a block on the generic process diagram and D220 is a functional department responsible for performing at least some part of that work F141 on any program. Figure 9.2 illustrates a fragment of the organizational RAS suggested based on an expansion of the common process diagram illustrated in Figure 8.1.

Each common process task is listed in alphanumeric order in the left-hand side column and named in the next one. The next column identifies the figure in an enterprise common process diagram. The task is then allocated to a particular functional department by number and name. The role column tells whether the department must take a leading (L) or contributing (C) role. In all cases where the task is allocated to only a single department, the role must be leading, of course. The text reference should tell where the reader would look in enterprise documentation to find out how to do this task.

For each line item in the organizational RAS, the responsible functional department must create a functional task description form such as that shown in Figure 8.3. This form gives all of the process design information and should be captured in an enterprise work catalog that collectively defines all work the enterprise is capable of performing on programs. In addition, the responsible functional department must collect all of the organizational RAS line items for which it is responsible and develop a department manual describing the work it must be capable of performing, telling the preferred tools to be applied, and either defining the practices to be employed or defining where they are covered, for example, in a textbook or training video. *System Management: Planning, Enterprise Identity, and Deployment* (Grady, in press) will include a reference to the full common process diagram and organizational RAS provided on a Web site operated by the publisher.

The result of all of this work is a common process design with each F-String identifying an element of work that may be required on a program. During a program planning exercise, these F-Strings can be combined with other program planning strings referred to in this book as P-Strings to directly form the content of the program IMP, as will be shown in Chapter ten. In Chapter ten, we will make this connection forming the program planning data in a seamless process respecting the content of the enterprise work catalog.

9.8   The functional departments

The design of the enterprise operated through a matrix structure must include a functional department structure. Figure 9.3 illustrates an

| Task identification | | FIG A-1 | Department | | Dept | Text |
Task id	Task name	SHEET	NBR	Name	Role	ref
F	Enterprise vision statement	1	D000	Enterprise executive	L	
F	Customer need statement	1	D216	System engineering	L	
F1	Manage Enterprise	1	D000	Enterprise executive	L	
F2	Provide program resources	1	D300	EIT	L	
			D100	Business functions	C	
			D200	Technical functions	C	
F21	Study lessons learned	7	D300	EIT	L	
F22	Functional IRAD	7	D200	Technical functions	L	
F23	Benchmark process	7	D300	EIT	L	
F24	Select improve- ment actions	7	D300	EIT	L	
F25	Maintain knowl- edge and skills	7	D300	EIT	L	
			D100	Business functions	C	
			D200	Technical Functions	C	
F26	Maintain best practices	7	D300	EIT	L	
			D100	Business functions	C	
			D200	Technical functions	C	
F28	Maintain facili- ties and tools	7	D300	EIT	L	
			D100	Business functions	C	
			D200	Technical functions	C	
F29	Release personnel	7	D110	Human resources	L	
F2A	Acquire personnel	7	D110	Human resources	L	

Figure 9.2 Organizational RAS example.

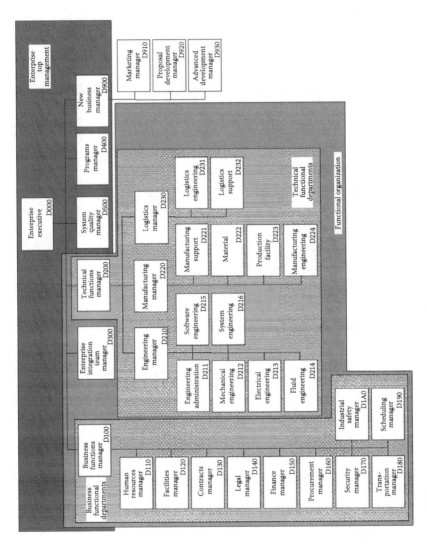

Figure 9.3 Functional department structure.

example of the top-level organizational diagram emphasizing the functional axis. Refer to Figure 6.2 for the two-axis view of the enterprise coordinating functional department connections with program cross-functional teams.

9.9   String levels

Given a function F1, we may allocate it to D300. As we expand the functionality to identify F12 and F122, we may allocate them to D320, a subdepartment of D300. Further decomposition might identify functions F1225 and F12253 that are allocated to D323, a department reporting to D320. This granularity is very important when we come to using these F-Strings in program planning. When planning early program phases where we have not yet developed a lot of information about the program, we should plan the program at a higher F-String level. As details are developed, we should expand the planning depth to a lower level of granularity. The reader is cautioned, however, to be very careful in applying this kind of information in planning real programs to avoid planning an early phase program to a depth greater than the level at which you plan to manage it. Because you have the details there is a bias to use it. Excessive planning data on an early phase program can unnecessarily consume management energy and budget without program benefit.

9.10   Generic program preparation and continuous process improvement

9.10.1   Being prepared is better than not

The history of the world and mankind leads to the encouraging conclusion that we humans strive for the better in the long run. Nature applies the powerful engine of environmental adaptation and survival of the best prepared in a chaotic fashion leading to improvements in the survivors. This process takes a relatively long time and those in business do not have the time or money to wait around for the death of ill-prepared competitors. We have to take conscious, positive action continually to ensure we will always be in the set of survivors. We humans have the advantage of understanding the notion of survival of the fittest and can influence the outcome by constantly improving our capability and readiness. The phrase "continuous process improvement" is an expression of this same notion.

In a business environment in a free market economy, we survive by the quality and value of our products, relative to our competitors, as perceived by our customers. It is therefore of interest to us what the customer thinks of our product. This fact may be lost in the discussion in this chapter so let us emphasize that it must be a part of our understanding. Perception and reality may not be in perfect alignment and we need both pieces of information.

A company needs a feedback mechanism giving it valid information about customer attitudes. This may be the sales force, a field engineering group, a marketing team, a customer relations office, or the program manager. Regardless of how it is implemented, this communication channel must exist and have an influence on changing the way we do business.

While assigned to a new program that in the event of a win we would be returning to a condition of doing business with a prior customer, the author arranged for a responsible former member of that customer organization, at the time employed by a sister division, to speak to the new proposal manager about what that customer's attitude was about us. He told us that when they called Boeing about a problem they would be told that Boeing would look into it right away. They knew that soon a big truck would back up to their loading dock and they would have to fill it with money, but they did not care. He then said that when they called our company with a similar problem they would be told (paraphrased to smooth out some colorful language used by the former colonel), "That is out of scope and we can't respond!"

Probably the best way of keeping our image as a good supplier as high as possible is by supplying good products to our customers that they can afford in a timely fashion. The principle topic in this chapter is the exposition of a method whereby a company may continuously be preparing itself for better work in the future by monitoring its present performance, learning from that performance, and making small adjustments in methods that lead to improved product quality and reduced costs that can be passed on to customers as better value that will encourage customers to return for more.

Where the customer is a large institution like Department of Defense (DoD), NASA, Federal Aviation Administration (FAA), or DoE, the voice of the customer includes understanding their model of system development and product requirements standards described in thousands of documents called applicable documents. This name is derived from customer reference to them in specifications and statements of work. They are referenced in these contract documents as applicable to the product or process, and their content must be complied with under the contract just as the direct content of the system specification and statement of work (SOW) must be complied with. We will see in this chapter that this mass of documentation, for those who must abide it, contains within it a diamond of great value to those who wish to be counted among the survivors.

The coverage in this chapter of customer-applicable documents, and their connection to continuous process improvement, may be lost on those only guided by their own commercial instincts, but the same basic notions apply. Instead of using the applicable documents identified by our military customers for guidance in improving our processes, we may have to rely on more direct feedback from our less well-organized customers or reference to commercial standards. In military procurement, many of the standards that system engineers have to respond to are actually commercial standards today.

9.10.2 Continuous process improvement using metrics

First, let us trace the outlines of a general plan for maintaining readiness to respond to our customer's needs and providing our company with an environment for continuously improving our performance while reducing cost. Figure 9.4 offers a blueprint process applicable to many companies. Commercial enterprises may not have to deal with the proposal process but have to deal with some internalized business acquisition process that can be substituted in place of it.

There are two very important signals we need to be tuned to when trying to improve our process. The first is the voice of the customer, as discussed above. The second is our own internal voice. Very valuable information is available within our own organization and may be currently going down the drain. On-going work on current commitments provides sure knowledge of a company's product quality and cost picture. These signals will be lost, however, if we have no way to capture them. It is necessary to measure our current performance in some intelligent way to acquire the data needed to characterize current performance and the effects of future changes.

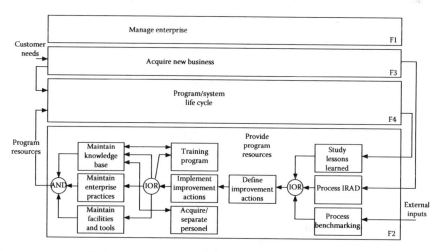

Figure 9.4 Continuous process improvement environment.

Measurement requires a set of metrics that will produce useful numerical figures of merit about our process. These metrics should be automatically produced in the process of doing work. Avoid installing a parasitic metrics manager and staff. This should be a standard practice of good management and managers. The functional axis of the matrix should be responsible for the company methods, tools, and availability of qualified personnel. Therefore, functional management should be charged with implementing appropriate metrics for monitoring program performance of their processes. These metrics provide functional management with lessons learned from current work that must be analyzed to determine what, if any, action should be taken to improve performance. Incidentally, this information can also provide the general manager/Cheif Executive Officer (CEO) with information about the relative performance of the several programs in house provided the enterprise is applying a common process such that information about a task performed on two different programs correlates relative to a common description of a particular piece of work.

Where metrics indicate poor performance, it is suggestive of either a need for improvements in methods, tools, or personnel qualifications or improvements in program execution and management. Functional management should be called upon by programs to participate in design and process reviews as a means of gaining insight into program execution such that all of the information needed to decode lessons learned data is accessible to functional management.

Each functional department should have a metrics list and a means for capturing the data corresponding to these metrics from each program that do not place an excessive burden on programs. Ideally, the computer tools used on programs should be designed to capture and report these metrics automatically. Each metric must be numerical in nature and defined with mathematical precision. For example, suppose we wish to measure the volatility of requirements subsequent to initial release of specifications where low volatility suggests a good requirements definition process. Let's define a metric to measure this.

METRIC NAME: Requirements volatility (V)

DESCRIPTION: A measure of the cumulative number of changes made to a specification subsequent to its initial release.

DEFINITION: $V_i - \dfrac{C_i}{R_i}$,

where

C_i is the cumulative number of changes to requirements made to document D_i since initial release

R_i is the total number of requirements in initial release of document D_i

The figure V_i, as defined above, only applies to an individual specification. We can extend this to all of the specifications on a program by averaging similar figures from all specifications on the program. A company or division figure can be derived by similarly combining the program figures. You can probably imagine that this figure could be automatically computed and tracked by a requirements database. Each time you change the requirements for a particular specification subsequent to an initial release date stored in the system, this metric could be incremented and always available for review at any point in time. The value of the metric over time could be plotted by the system.

Figure 9.5 offers a black-and-white sketch based on the colorful screenshot that the Telelogic Dashboard can generate from data contained in Telelogic DOORS. Note the metric data, not necessarily based on the same logic given above, is presented for the current situation but also provides trends linked to past performance. As in the use of an earned value system, we can project into the future where we might expect to be if we continue doing what we are doing. Given that you start preparing a specification with zero requirements, you would expect some volatility in the beginning. We would want to take action in the case illustrated to bring the volatility alert into a green condition as we approach release of the

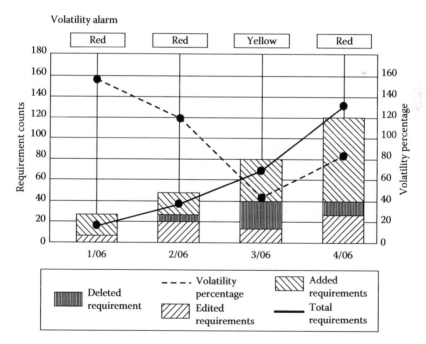

Figure **9.5** Requirements volatility metric generated by Telelogic DOORS dashboard.

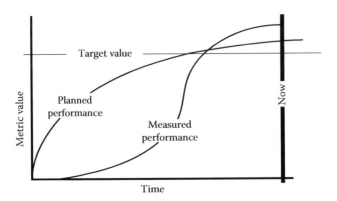

Figure 9.6 Typical metric tracking chart.

specification. As good managers, we would ask ourselves why this condition is not closing and what can we do to make it do so.

Metrics may not have a lot of significance initially because you have no experience upon which to base a conclusion. If you track this metric over time, however, and observe changes that result from specific actions you take to improve the figure on one or more programs, the metric will start to provide value.

These metrics can be tracked in time, as shown in Figure 9.6, exactly like technical performance measurement (TPM) parameters used to monitor program technical performance. In this case, we have set a goal and a projected path predicted for the value of the metric and we have elected some plan to reach that level of capability that looks like it has been slow in improving our performance, but it has eventually moved up to the goal. Like TPM parameters, you want to select a limited number of functional metrics each of which is useful in detecting the quality of performance in specific activities and together give you a comprehensive view of performance. It is almost as bad to have too many metrics as to have none. To manage is to focus on the important. Too many metrics makes it hard to focus on the important.

To the extent that you can cause your tools to generate the metrics automatically in the process of people focusing on doing their work, as in the case of using Telelogic DOORS and Dashboard in your requirements analysis and management work, you will be well served.

9.10.3   Generic preparation

Figure 9.7 illustrates an overall process for the request for proposal (RFP)/ proposal cycle. We may enter this activity following identification of a new need by the customer or as a result of encouragement of the customer by a contractor based on new insights into the customer's problems. We

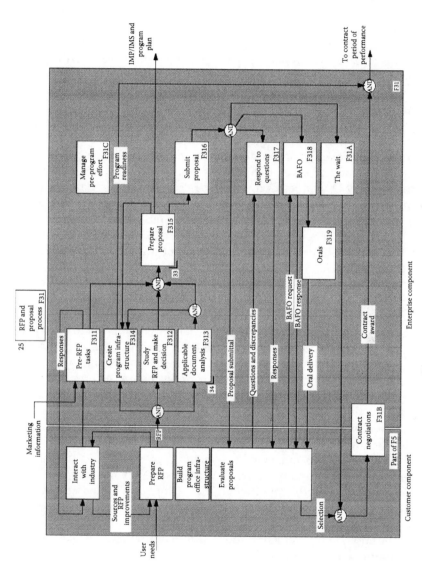

Figure 9.7 RFP-proposal cycle process diagram.

may also enter through interphase cycling on a contract won for a current or prior phase. In the latter case, we may already have performed one or more prior phases and have a considerable database upon which to build. The intensity with which we pursue some tasks identified on Figures 9.7 through 9.9 will depend on this past history as well as the content of the customer's RFP.

The scenario expressed in this section may not relate closely to the problems faced by a commercial enterprise that sells directly to the public in that there may be no RFP or proposal involved. Many of the other components can prove useful, however, as a means to continuously prepare for and implement programs producing greater value for customers. For those interested in DoD or NASA customers, you should take special note of the encouragement to create the program infrastructure concurrently with preparation of the proposal such that you are ready to execute when you win. The author is reminded of the brief euphoric feeling felt by all on the General Dynamics Convair Advanced Cruise Missile team when the program manager announced at an all hands meeting that GD had won the competition. It went something like this, **"Wow! We won!**.... *Oh my God, we won."* The euphoria peaked and receded quickly as we realized that we had built a winning proposal but might not yet be ready to implement.

The preliminary steps in our scenario involve customer preparation of the RFP, which may include issuance of draft copies to prospective contractors for review and comment. Following receipt of the formal RFP, the contractor must study it and reach a decision whether or not to bid. Given an intent to bid, the contractor must strike out on three parallel paths: perform an applicable documents analysis, begin creating an initial program infrastructure in anticipation of the win, and prepare the proposal itself. Some of this work may have to be performed in order to reach a sound decision on the question of whether or not to bid.

We will study applicable documents analysis later in Section 9.10.5, but the central problem in this activity is to match our methods of doing business with the customer's requirements expressed in documents the customer is familiar and comfortable with (commonly military standards and specifications with DoD though increasingly including commercial standards). In Section 9.10.4, we will also explore more fully the connection between tailoring (adjusting content) of customer compliance documents and continuous improvement of our process on the road to acquiring or maintaining world class capability.

For those not familiar with DoD proposal evaluation language, BAFO means the best and final offer that commonly involves some kind of cost adjustment, generally downward. Some customer agencies and some contracts require an oral presentation in person by the program manager-to-be or by video tape subsequent to proposal submission. THE WAIT is the author's term for the period between submission of the proposal and selection when there is little the contractor can legally do to influence the outcome other than answer questions, submit a BAFO, and present an oral presentation. All three of these tasks entail high energy integration work because they do offer the contractor the only window of communication to the customer and the time constraints are very confining. The questions asked suggest customer concerns that you can now answer in voluminous detail that could not have been included in the original proposal due to page limitations. BAFO offers one last attempt to fashion a winning cost figure based on new information provided by questions asked by the customer and the answers provided by all of your competitors. And the oral presentation provides an opportunity to give the customer a very brief and crystal-clear message why you are the best choice.

We are primarily interested in tasks F315, F314, and F313 in this book and their connection to continuous process improvement of our enterprise. The task numbers used in the blocks of these three diagrams are, like those found in many other process diagrams in the book, the author's way of coding all tasks in a computer database to explore useful information relationship concepts.

Figure 9.8 expands task F315, Prepare Proposal from Figure 9.7, and includes several functions intended to prepare the program for a win as well as support the development of a winning proposal. We could simply prepare a proposal and hope for the best. However, we will help our chances for overall success if we also take the opportunity presented by the RFP to create the infrastructure we will need in order to perform in accordance with our proposal after we are selected as the winner. Generally, this will not cost very much money above and beyond proposal costs if we are clever in taking advantage of our continuous process improvement work and work done and materials prepared for this specific proposal. Task F314 on Figure 9.7 provides a collecting point for the work products of value in actually managing the program in the event of a win to the extent they are distinct from feeding the proposal development process.

Note the generic inputs in the left margin of Figure 9.8. Each department has a toolbox they bring to the proposal effort (generally computer tools) and a product standards library containing boilerplates for the documentation products they are commonly required to produce for customers within their customer base. During the proposal period, each department may have to adjust these tools and standards for peculiar customer requirements, but they should be trying to minimize the need by adjusting customer expectations through contract data item descriptions (DID) and applicable document tailoring for equivalence to internal practices.

The customer may provide a partial or complete SOW and work breakdown structure (WBS), but we should carefully study how far we can go

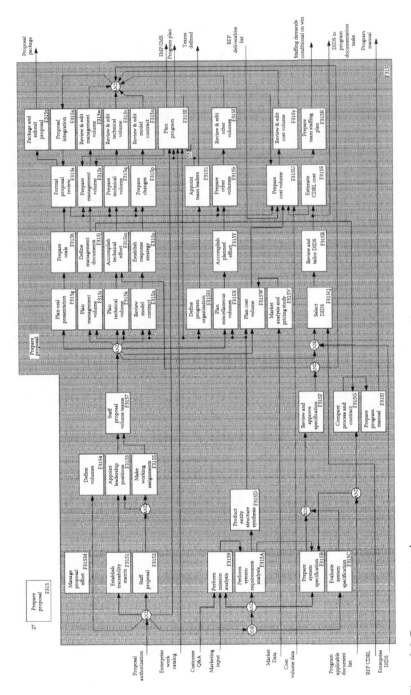

Figure 9.8 Prepare proposal.

in adjusting those structures to our planned development effort, practices, plant, and personnel base. The requirements in the system specification should drive the WBS product component and the SOW should be crafted to identify work needed in the context of our enterprise.

The WBS is used to organize and collect the cost estimate and later in program management. The SOW feeds all other program planning activity including selection of generic tasks defined by our functional organizations through continuous process improvement, linked to our training program, internal practices manuals, and department tool-boxes. The integrated management plan or program plan captures the detailed planning required to integrate the management of all tasks that must be performed to satisfy SOW requirements. The events and work defined in the WBS, SOW, and plan are combined with time to produce the integrated program schedule or the integrated master schedule (IMS).

9.10.4   Tuning our process to customer needs

Applicable documents are specifications, standards, and other documents referred to in specifications, statements of work, standards, and other documents as an economical way of importing large tracts of requirements or other data without the need to duplicate the detailed contents of them. These documents provide standard solutions to problems that people have found troublesome in the past. They describe what have become standard solutions to certain problems in manufacturing, engineering, logistics, material, and many other fields. Reference to an applicable document in a calling document, such as a system specification or SOW, applies it for the purpose and scope defined in the calling document.

The DoD, as well as other customers for large-scale systems, has recognized that if it wants to encourage improved productivity and quality in American industry, it must encourage contractors to develop their own procedures that are tuned to their own particular situations. This is a reversal of past policies that earlier led to contracts calling for full compliance with many applicable documents offering guidance on both product design and process controls. Many companies fell into the habit of redesigning their internal workings for each new contract in order to comply with these many applicable documents, that might be different from contract to contract whether it would lead to optimum customer value and contractor efficiency or not.

Some companies have independently concluded, or been encouraged by government interest in streamlining and total quality management (TQM) efforts, to recognize that they have to develop their own identity defined by their own internal facilities, personnel, methods, history, and procedures. At the same time, there is a great deal of wisdom captured within the many thousands of applicable documents that can usefully be applied to specific situations. As a result, DoD program offices will, for many years to come, find a great deal of comfort in appeals to some of these documents and rightly so. Obviously, there is a condition of balance between these extremes that we need to seek out.

It will likely require several years to fully implement the government's sincere interest in streamlining the use of applicable documents on programs. Contractors can encourage government success in this transformation toward a better condition of balance by development of quality internal procedures that are well matched to their situation and then mapping them to the government-applicable documents most commonly applied by their customer base to programs in their product line. In the process, they will find that this mechanism also provides them with a super highway to a condition of world class capability in their product line through continuous process improvement.

The general pattern involves development of a set of internal practices for each process or task you find it necessary to apply on programs. Then, you develop a list of applicable documents your customer base commonly applies to contracts or defines their value system in accordance with. We then map these documents to our internal practices and tailor them to agree perfectly with our internal practices. Tailoring is an editing process to change the content of a document. It can either be done using a legalistic or in-context method.

In the legalistic tailoring method, we make a list of changes to the document such as

(1) Delete paragraph 3.2.3.5.
(2) Change paragraph 3.4 to read "3.4 The pressure vessel design shall be proof tested for 2.5 times their normal maximum operating pressure."
(3) Add paragraph 3.6.5 as follows: "3.6.5 Limit the pre-shipment test profile for a production pressure vessel manufactured in accordance with approved procedures to 3 excursions to a pressure level no more than 2 times normal operating pressure for 15 minutes each."

In the in-context method, an electronic copy of the document is marked up with a word processor to line out deleted text and highlight added text. While this method is much easier to use because you can see the context of changes, very few of these kinds of documents are available in electronic media unless you scan and edit them locally. An additional problem is that many of these documents are now owned by societies that have them under copyright control. Thus, they cannot be copied and used in this fashion.

9.10.5 Applicable documents analysis

When we bid on a contract for which the customer has offered a particular set of applicable documents, we feed back to the customer, to the maximum extent possible, our tailoring that makes their documents equivalent to our internal practices documentation. The customer can manage the program in a familiar context and our employees can follow the same set of internal practices they have been trained on and become experienced with. There was a time when this action would be referred to in DoD as nonresponsive or arrogant, but DoD has begun to understand the need for contractors to perfect their own processes.

That is the advantage of an internally defined identity in terms of internal practices; it allows the employees to always follow the same process becoming experienced in that process. It supports an internal training program focused on the generic processes defined in company practices that become the textbooks for that training program. It supports the practice-practice-practice notion through which world class athletes become that way. It will also provide the customer with the best possible value that your combination of plant and personnel will permit. These advantages will be short-lived unless you also implement a continuous process improvement program as suggested earlier in this chapter. We have expressed many ideas in this chapter that you may have thought of as separate entities previously. It requires effective cross-process and cross-function integration to mold these entities into a dynamic whole.

Figure 9.9 offers a process for structured tailoring that fits into task F413 in Figure 9.7. This process is repeated for each proposal. From these tailoring exercises, you will from time-to-time gain insights into how to improve your internal practices. Likewise, during actual program performance, you will find better ways to do things than described in your internal practices. These are lessons-learned signals that should stimulate generic process improvements available to all new programs.

The process illustrated in Figure 9.9 passes all applicable documents through a filter and assigns them to one of six classes with the following responses:

Class 1   The content of the applicable document is in agreement with the enterprise practices.

Class 2   There is a difference between the applicable document content and enterprise practices, but the difference can be accepted.

Class 3   Tailor the applicable document for agreement with the enterprise preference.

Class 4   Change preferred enterprise practices for agreement with the applicable document.

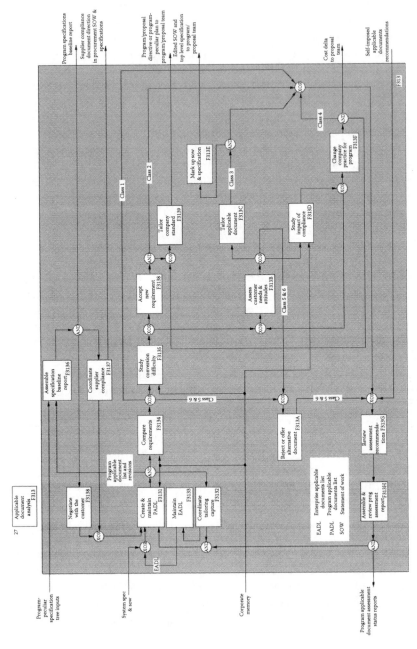

Figure 9.9 Structured applicable document analysis.

Class 5   Reject compliance with the applicable document.
Class 6   Replace the listed applicable document with an alternative
document.

9.10.6   Program audits by functional departments

Given that our functional departments have developed internal practices to guide our personnel in performance of program tasks, it will be helpful to both program and functional management for functional management to audit program performance to those practices. Functional management is thus able to find out what is going on and to provide a useful service to programs. The audit should be conducted to find out the degree of faithfulness in the program application of a given practice using the functional practice description as the baseline.

The audit could be carried out in a very subjective fashion where the functional manager simply gives a pass/fail or numerical grade in a 0–10 or 0–100 range. A better way to do it would be to ask each functional department to prepare an audit checklist for each of their practice descriptions. The answers can then be used to assign an objective score that can be tracked over time providing a useful metric. As a part of this checklist, we may find it useful to include the metrics we discussed earlier. One of the checklist items could, for example, be "Requirements volatility less than or equal to 0.5% one week prior to planned release." Clearly, these checklists will have to be built recognizing the different interests in different program phases.

If your company allows programs to tailor or adjust functional practices for programs, then this audit may have to be made based on a program manual description of the differences. If your company allows programs to develop their own practices independently of the functional departments, then auditing will be money wasted and it will take your company a very long time to be counted among the great system houses. You will not be able to take advantage of the practice-practice-practice and continuous process improvement notions encouraged here.

9.10.7   Benchmarking

The final continuous process improvement process offered is benchmarking. If we only rely on our own experience in our search for excellence, there is a danger of the development of an incestuous group-think mentality that will shield your organization from thinking new thoughts and seeing the value of potentially valuable alternatives. We need to assure that we have access to ideas from outside our immediate organization and we need to consciously recognize the need to compare our performance against these ideas. We should only seek out what we consider to be excellent examples with which to compare our performance in the form of other companies we respect, especially if they are our competitors. This is called benchmarking.

The unfortunate thing is that this kind of information is not always easy to obtain. Your competitors will be reluctant to allow you direct access to it. But, it is not really necessary to engage in covert industrial intelligence operations to find these things out. One source in a large multidivision organization is your other divisions. But, a source that is open to all is provided by the many trade and professional organizations that touch on practically every facet of your business. Examples are the International Council on Systems Engineering (INCOSE), the American Institute of Aeronautics and Astronautics (AIAA), the Society of Automotive Engineers (SAE), and the Institute of Electrical and Electronics Engineers (IEEE). There are members of these organizations that come from all of the best and most successful companies in your field. These members will offer papers covering their best ideas most often derived from their experience in those companies. Personal interaction between your employees and other members will also provide additional insights into alternative approaches of potential use to your company.

The functional audit process and benchmarking can be combined to produce a two-figure metric for your processes. The first figure could be derived as noted under functional audits. The second figure would be derived by an audit of your in-house practice against a benchmarking standard. This pair of figures gives your management added information about how best to apply scarce resources to improve your company's performance.

9.10.8   Where is your process description?

You are encouraged to seek out your company's written description of your system engineering process. Does it exist? If it exists, is it up to date and being followed? You will likely find that the answers to these questions are no. In *System Engineering Planning and Enterprise Identity* (Grady, 1995), you will find a generic system engineering manual (SEM)/system engineering management plan (SEMP). This document may offer you a beginning toward the goal of defining your system engineering process. Once defined, you will find a need to continually improve it as discussed in this chapter based on your experiences on programs. Soon, *System Management: Planning, Enterprise Identity, and Deployment* (Grady, in press) will be available with updated versions of these documents.

chapter ten

Integrated program planning*

10.1 Goals

Many readers will have participated in proposal work in enterprises where they have worked a lot of unpaid overtime amidst a lot of chaos and remained on the staff to begin the program implementation. This often describes the situation even in enterprises with a fairly well-organized proposal development process. If we started from scratch to design a sound program planning process what goals should we seek? Ideally, these goals would be directed at aggregate simplicity for simplicity encourages better management. Simplicity of process encourages more time to focus on real problems where our intellect can be applied to best effect. The mundane will generally take care of itself. Some suggested goals are

1. Clearly state the responsibilities of the functional departments in terms of providing coordinated quality resources to programs (people, tools, and practices) that are continuously improved in a coordinated way within the enterprise to maintain.
2. Clearly state the work necessary to perform on a program so as to encourage understanding of the work at all levels on the program and minimize the management difficulty with the result that management intensity can be placed on smoothing any current discontinuities and the identification and mitigation of future potential risks.
3. Minimize the program planning transform complexity to the end that everyone can contribute from a position of knowledge in pouring generic planning data into the program framework to create the minimum cost, most effective program plan that results in customer best value and enterprise profit.

10.2 Traditional planning methods

For many years, many firms have planned programs from an internal program perspective without considering a need for a common process applied to all programs. Many enterprises continue to plan their programs

---

* The material in this chapter is from Grady, J., *System Integration*, CRC Press, Boca Raton, FL, Chapter 6, 1994. Used with permission.

in this fashion. Generally, this takes place in enterprises that are either projectized or have a very weak functional department axis in their matrix relative to the strength of the program axis. The author is a believer in strong program matrix management balance, but there are valid functions that should be expected from the functional department side including process ownership. The series of program documents needed for this planning process are illustrated in Figure 10.1, and the reader will note that there is no connection to a common process definition; it is strictly program centric.

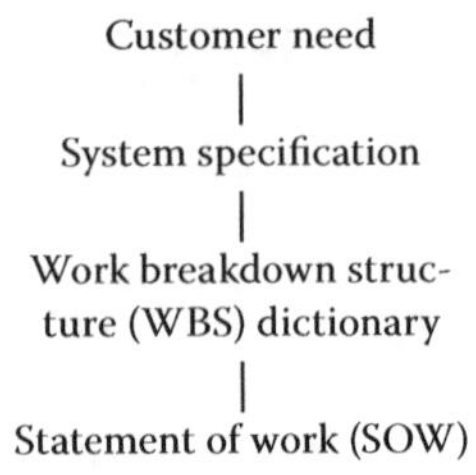

Customer need

|

System specification

|

Work breakdown structure (WBS) dictionary

|

Statement of work (SOW)

Figure 10.1 Program-centric definition documents. (From Grady, J., *System Integration*, CRC Press, Boca Raton, FL, pp. 84–99, 1994. With permission.)

10.2.1   *The need statement*

The U.S. Department of Defense has changed the name of this document to Initial Capabilities Document with the unfortunate acronym of ICD that for many years has provided good service for the interface control document. However, the ICD is still downstream to some extent from the original recognition that the current capabilities are not sufficient and something new is needed. The author continues to cling to the need statement no matter the military or commercial nature of the system and efforts to invent alternative terms for it.

It often happens that the customer's original need has become lost by the time they are ready to let a contract for the development of their system. Sometimes a need statement is never phrased. But even when it is prepared, it is not uncommon for it to have been so thoroughly digested in the process of developing the system requirements that it simply passes from view. Some people would maintain that once you have the system requirements defined in a specification, you do not need one anyway. The author disagrees. The need is the ultimate requirement and, if it truly represents the customer's need and it does clearly state what is needed, it offers useful guidance throughout the development process.

If you cannot find a need statement in materials provided by the customer, you should ask them for it. If they cannot or will not produce one, write one and get customer acceptance. This may be very difficult, especially where the customer has many faces including one or more users and a procurement agency. But because it is difficult, you should not conclude that it is not needed. Quite the contrary, the harder it is to phrase and gain acceptance of the customer's need, the greater the need to press on to a conclusion. A lot of good work will result that will eliminate many false starts during later program execution. This same policy should be pursued in the definition of system requirements—the harder it is to gain acceptance, the harder you should work to understand the customer's needs and achieve agreement. Ideally, in the author's view, the need statement would be captured in Section 1 of the system specification.

10.2.2   The system specification

Depending on the program phase we are discussing, the system specification may be available to the contractor or may not yet exist. Preparing the system specification may be one of the program tasks if the contract involves a very early program phase. More likely, the contractor will receive with a request for proposal (RFP) some kind of requirements document, such as a system requirements document, operational requirements document, or draft system specification from the customer. This input will have to be completed or transformed into a final system specification to the mutual satisfaction of the customer and contractor after contract award. Commonly, the majority of this transform happens during the proposal process with a system requirements review (SRR) scheduled very early in the program execution to reach final agreement on the content of the document. Refer to *System Requirements Analysis* (Grady, 2005) for a description of a process that will result in a quality system specification.

The system specification should be developed based on the results of a structured analysis of the customer need as discussed in the book noted above. In this process, the specification tree becomes an overlay of the product entity structure diagram for which the system specification is the top element. All of the requirements in all of the lower-tier product requirements documents should be traceable from and to this document. All of the requirements in the system specification should be traceable to the customer's need through a logical process of functional decomposition, allocation, and requirements analysis.

The early structured analysis work that results in the system specification content defines what the system must consist of and how it will be organized. We should use the functionally derived product entity structure as the basis for a work breakdown structure (WBS) developed as the basis for management of the program. This is done by deciding at what level the program should be managed and using the product entity structure identification notation or a WBS notation of choice to identify all of the management significant items in the product entity structure. We can then determine the cost and schedule estimates for these items and determine appropriate performance characteristics through a continuation of the structured analysis process applied at the system level. Ideally, our program planning would coordinate integrated product and process teams (IPPT) with these entities selected for inclusion in the WBS.

10.2.3   The work breakdown structure dictionary

The former two documents define the product system the customer is seeking to acquire. The next two define the program they will enter through a contract with a contractor to acquire that system, but the system specification provides the contractually binding technical definition of the system they seek. The WBS has, in the past in the DoD market, all too often been crafted by finance people in industry and government based on one of several MIL-STD-881 appendices reflecting different kinds of systems. Even though MIL-STD-881 encouraged that its appendices be used as a guide only, the mindless way it has been applied has often had a chilling effect on the application of a sound system functional decomposition approach to the development of product systems. Despite the evolution of a conflicting functional system product entity structure on such a program, the inflexible computer cost management tools used by the customer and contractor finance communities has often inhibited alignment of the WBS to the needed functionally derived architecture. Since the conversion of the military standard into a military handbook and the unsuccessful campaign for MIL-STD-499B that would have made WBS a system engineering responsibility, there is a more flexible attitude now than in the past.

The WBS dictionary provides a hierarchical organization of product material and services needed to satisfy the customer's need. The WBS must span the complete system allowing everything in the system to be placed in some WBS category. The content of the WBS must reflect what is needed to satisfy product system functionality and should not be chosen rigidly and arbitrarily based on some financial model still included in the appendices of MIL-STD-881. Figure 10.2 lists the partial content of Appendix C of MIL-STD-881 for missile systems. Only the Air Vehicle, a second-level WBS element, is expanded to the third level.

MIL-STD-881 does not prescribe any particular alphanumeric coding system for the structure, but one is always assigned. For example, in Figure 10.2, the Air Vehicle might be assigned WBS 1000, and 2000, 3000, and 4000 to some other second-tier items. If the air vehicle includes two stages, one might be 11 and the other 1200. The propulsion system for stage 1 would therefore be 1110. Prefixes are sometimes used to denote recurring and nonrecurring cost. The WBS 012-1110 might be assigned to stage 1 propulsion nonrecurring and 014-1110 is assigned to recurring manufacturing of the stage 1 propulsion system. Costs can be accumulated in WBS 1110 across all prefixes for the complete Air Vehicle and within prefix 012 for all development cost.

We have chosen a four-digit code so far, but there is nothing to prevent the selection of a five or six digit code. In a very complex system, a decimal expansion could be added, such as 1110.05 for an engine turbo pump. In this fashion, the WBS can identify everything in the product system in a

Missile system
> Air vehicle
>> Propulsion (Stages 1...*n*, as required)
>> Payload
>> Airframe
>> Reentry system
>> Guidance & control equipment
>> Ordnance initiation set
>> Airborne test equipment
>> Integration, assembly, test, and checkout
> Command and launch equipment
> System engineering/management
> Systems test and evaluation
> Training
> Data
> Peculiar support equipment
> Operational/site activation
> Industrial facilities
> Initial spares and repair parts

Figure 10.2 Example of a MIL-STD-881 WBS. (From Grady, J., *System Integration,* CRC Press, Boca Raton, FL, pp. 55–66, 1994. With permission.)

very organized fashion to any level of indenture desired. It is not necessary to apply a unique WBS number to each product item throughout the system hierarchy, however. The WBS is a management tool and should give managers insight into the system structure. It should be assigned down to the level at which you intend to manage the program.

The DoD pattern for WBS assignment unfortunately collected all system engineering work in a process-oriented partition often joined with program management as observed in Figure 10.2. The author maintains that the WBS should be product-only based. Much of the program management and system engineering work is incidentally associated with the whole system for which a WBS is seldom assigned. MIL-STD-881 properly states that the WBS should be composed of product entities and should not include the work to create them but does still list system engineering and management even though it is work related rather than product oriented. The problem appears to be that the whole system is not commonly assigned a WBS number to which some if not all of the system engineering and management work could easily be attached.

The author employs a base 60 product entity code using the numerals, capital letters less "O" and lowercase letters less "l" so decimal delimitation is not often required, it being uncommon for there to be more than 60 entities immediately subordinate to any other entity. These strings always begin with the letter "A" reflecting a pattern once employed by the author to refer to the system architecture, what he now refers to as the product entity structure. The author would apply the same designators used on the product entity diagram for WBS numbering reflecting a belief that any entity should have a single name or designation if possible. In any case, the WBS should be a financial and management overlay of the functionally derived product entity structure just as the specification tree should be an overlay to identify the specifications that must be developed.

10.2.4   The statement of work

The WBS identifies system entities that program work must be focused on. The next step in the programmatic requirements development process is to determine what work must be performed to develop, design, manufacture, test, and deploy every product element depicted in the WBS. Every bit of work we perform on a development contract should be included in the statement of work (SOW) at some level of detail and should be traceable to the product requirements in the system specification. The SOW, therefore, will tell what work must be accomplished for each product WBS element at some level of indenture. Only then can the program work be said to flow from the system requirements.

In the earliest phase of a program, commonly a study task of some kind, the product entity structure may have no more detailed identification than "the system." In this case, every SOW element coordinates with a WBS entity that is the system, but DoD practice does not apply a WBS number to the system. The author would identify the system with the functional modeling ID (MID) "A" so SOW elements might be A-S01, A-S02, etc., where S0X is a paragraph in the SOW. In that phase, the study might identify a particular system concept with lower-tier entities A1, A2, and A3. The next phase could develop SOW strings including A1-S01, A2-S05, etc. Subsequent phases will have identified sufficient depth of the product entity structure to require a possibly long list of product entities to which work statements would be attached. In any case, the SOW content can be listed in alphanumeric order, and if the program applies integrated teams and those teams are coordinated with the product entity structure that is in alignment with the WBS, then team leaders can be called upon to be responsible for a WBS item at some level of indenture and the top-level specification coordinated with that responsibility.

Note that there appears to be a chick and egg problem here in that we wish to use the functionally derived product entity structure as part of the work-identifying string (as a WBS identifier) and at the same time we wish to form the IPPT around the product entities aligned with the WBS. So, one might ask, how do we propose to know what the product entities are so as to assign the IPPT to do the work that will identify what the product entities should be? The answer is that this work is done in phases. In the first phase, the work may not be broken down any lower than system (A), and during that phase the only team on the study, a system team, will identify lower-tier functionality and an appropriate product entity structure. Now in planning the next phase, we can use the lower-tier product entity structure derived in the first phase as the basis for next phase planning. This process keeps repeating until we have unraveled the product entity structure to include all of the entities around which we will wish to form IPPTs.

In the past, the SOW has often been prepared by someone in the customer's program office copying customer-created boilerplate SOW material from a similar past program into the new program SOW. Today, DoD customers commonly call upon the contractor to write the SOW based on a WBS derived from the same work that produced the definition of the system contained in system specification. The contractor must decide what work must be done within the context of its plant, personnel base, and product history in order to create a product system that satisfies the provided requirements and organize the work around the breakdown in the WBS.

Two competing contractors may very well offer the customer two significantly different statements of work with their proposals because they have different plants, product histories and experiences, and personnel mixes. Neither may necessarily be better than the other, only different for these reasons.

If we have properly identified program work, it should be possible to establish traceability between the paragraphs of the SOW and the paragraphs of the system specification. Where data item description (DID) CMAN 80008A is called by a DoD customer to define the system specification format, the product-oriented SOW paragraphs can be traced to paragraphs under system specification paragraph 3.7, which captures requirements for major items in the system structure or WBS. Paragraphs under 3.7 should have been initially conceived from an orderly functional decomposition of the customer need and the WBS developed from that same analytical process.

We can picture all of the supplier statements of work strung out from the system SOW in a tree structure branching from the SOW block just as we picture all program specifications in a specification tree diagram. Certainly, traceability should exist between the process requirements we accept in the applicable documents referenced in the SOW with our customer and the process requirements we lay upon our suppliers in procurement SOWs. Otherwise, some elements of the complete product delivered by us (containing supplier elements) may not be compliant with our customer's requirements. Should traceability exist not only through the specification tree but through the SOW tree as well and between a SOW and its companion specification? This means that the complete product and process definition for a system can be unfolded from the customer's need in a structured top-down development effort.

Commonly, the system-level SOW is written to cover all of the work in several WBS indentures. We prepare supplier statements of work for major suppliers, but the system SOW commonly provides the only work definition coverage for the prime contractor. The IPPT, or concurrent engineering, paradigm suggests an interesting alternative to this arrangement. We could prepare the system-level SOW to cover only the system-level work under the responsibility of the PIT and write internal SOWs for each major system element identified in the WBS and to which an IPPT will be assigned. These internal SOWs would each define the work that must be accomplished by one of the teams. The product element would be covered by a specification defining the product requirements and the team SOW would define the process or work requirements that must be accomplished to satisfy the corresponding product requirements.

This arrangement results in product requirements and work definition documents aligned perfectly with the product system elements and the development responsibility definition and should result in great precision in management of the program. In order to be successful in this approach, you have to have thoroughly studied the customer's need and requirements and decomposed their need into a stable entity structure that can be assigned to product teams. Instability in this whole structure can result in a tremendous amount of parasitic work. You can begin this planning work too soon as well as too late.

The SOW, as the name implies, identifies work that must be accomplished at a high level to provide the items identified in the WBS. Now, how do we identify or describe the work that must be done in detail? A three-step process is suggested.

Step 1 in transforming generic planning data into the specific program plan for a given program is to map the generic tasks to the WBS. Figure 9.3 illustrates this process. It can be accomplished in a top-down or bottom-up fashion and at any level of management desired either by the program team or functional management. Two alternatives for these three planning process components are listed below under headings in the form Development Direction/Planning Level/Planner and the reader can imagine other possible combinations:

1. Bottom-Up/Department Chief Level/Chief—Each functional department Chief, or their representative, lists under each WBS, their department tasks that must be performed.
2. Top-Down/Upper Management Level/Program Planning Team—A program planning team, composed of people from functional departments, accomplishes the map between the functional department charters and the WBS. The planners shop the functional charters for tasks that satisfy the customer's needs expressed by the WBS dictionary.

Step 2 is to organize the functional task inputs into a program context within the WBS-driven SOW hierarchy. In the author's preferred scenario, no matter how we orchestrated the first step above, the program should form an integrated planning team composed of people from the functional departments assigned to work on the proposal or program initiation work. Ideally, these same people will later be assigned to work on the actual program from those departments.

10.2.5 Work coding and detailed work planning

The program work in a WBS–SOW planned program could be coded using strings composed of the WBS and SOW identifiers. Using the WBS numbers in Figure 10.3, each WBS number, representing a particular system product entity will require work in several categories that can be generically identified with higher-order SOW numbers. These might be

01 Define requirements
02 Design
03 Procure
04 Manufacture
05 Verify requirements

The SOW then becomes an expansion of the WBS by combining each of the WBS entities with each of the top-level SOW elements. This much of the work can be done from the top down. Below is listed the top-level WBS 1100 SOW elements.

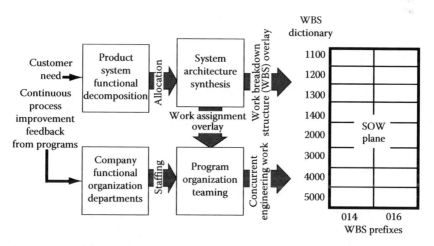

Figure 10.3 SOW building. (From Grady, J., *System Integration*, CRC Press, Boca Raton, FL, pp. 84–99, 1994. With permission.)

1100-01 Define item 1100 requirements
1100-02 Design item 1100
1100-03 Acquire item 1100 material
1100-04 Manufacture item 1100
1100-05 Verify item 1100

This high-level SOW can then be fed to a program planning team composed of experts from the several functional departments that will have work on the program. Each functional participant must identify work that his or her department should perform on each top-level SOW element thus identifying the detailed SOW content in the form XX00-YY-ZZ, where ZZ components are the detailed planning elements contributed by the functional planners. In that the inputs come from several people, the aggregate input may include redundancies and voids and must be integrated to cause the final data to reflect the least cost, most effective plan. Finally, a matrix is commonly assembled with WBS and functional departments on the axes showing how the functional department work and budget is partitioned between the functional departments and the program WBS. If the teams are assigned based on the WBS (and functionally derived product entity structure), this can make for an effective program planning structure, but there is no clear link to an enterprise common process.

10.3   U.S. Air Force integrated management system

In the early 1990s, the U.S. Air Force developed a truly great program planning framework called the integrated management system. It involved an expansion from the WBS–SOW planning scheme discussed above to provide additional program planning details. Table 10.1 lists all of the components in the planning strings applied in this method using the nonrecurring design for WBS 1200, the launch vehicle upper stage. This

Table 10.1 Single Planning String Example

Work identifier	Example	Planning content
WBS WITH PREFIX	016-1100	Launch vehicle upper stage development
SOW TASK	02	Accomplish design of the entity
PROGRAM EVENT	05	Item CDR
SIGNIFICANT ACCOMPLISHMENT	03	Design complete
ACCOMPLISHMENT CRITERIA	02	95% of drawings released

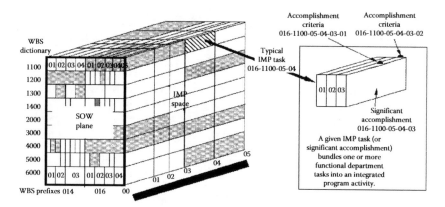

Figure 10.4 IMP content. (From Grady, J., *System Integration*, CRC Press, Boca Raton, FL, pp. 55–66, 1994. With permission.)

method links the work to a program event that can be tied to a particular date or left floating if using an event-driven schedule. In this case, the event is the critical design review (CDR) for which the significant accomplishment is the design being essentially complete, and a manager will know that it is complete because 95% or more of the planned drawings will have been formally released. The full planning string example for this element of work would then be 014-1200-02-05-03-02.

Figure 10.4 extends the WBS–SOW pair shown in Figure 10.3 in time to form an integrated master plan (IMP) space embracing a series of program events that partition the IMP space into particular time spans. All of the planning strings are assembled into a list with explanations forming what is called an IMP. The events are strung into time and the tasks identified by the planning strings can be published as an integrated master schedule (IMS) in a tabular or bar chart format.

During the planning process, the contractor can link manpower estimates from the functional departments, which will have to provide the program with resources, to the planning strings. The problem with this planning approach in common with the previous approach discussed is that it cannot easily be linked to the tasks in an enterprise common process discussed in Chapter nine. One could cross-reference the significant accomplishments to common process definition, but it is fairly complicated. Simple is better than complex in these matters.

10.4 Introduction to the JOG system engineering planning model

The author has developed a method of identifying work using six identifiers as listed and explained in Table 10.2. The model is named after the

Table 10.2 JOGSE Model Structure

Work identifier	Example	Planning content
PROGRAM	P05	The fifth program in house using this method
STAGE OR PHASE	S02	The second stage or phase of the program as defined by the customer
PROGRAM EVENT	E04	Major program milestone as in the USAF model
PRODUCT ENTITY/WBS	A11	The product entity MID (WBS equivalent)
GENERIC FUNCTION	F411	Generic work identifier from the common process model that in this example could be requirements analysis
FUNCTIONAL DEPARTMENT	D262	The Requirements Analysis department will provide the personnel to accomplish this work on the program

consulting company operated by the author for lack of a better term. The elements of this system are consistent with the content of this book and involves a minimum of mapping between different identification systems. It retains the WBS and SOW tradition as intermediate planning structures. As in the USAF model, the IMP would be created from alphanumeric ordered work identification strings like the one described in Table 10.2.

A full identifier then would be P05-S02-E04-A11-F411-D262. The added program identifier permits all program, planning data to be contained in a single database, if desired, with each record program linked by this field. Within the context of a single program, this identifier could be dropped, of course, simplifying the work identification. The prefix letters could also be dropped but are applied here to avoid any more confusion than necessary with these long strings. A stage identifier is also used to link to customer phasing definitions permitting the system to coordinate with any customer's program organizing mechanism. The term "phase" is more common but the author wanted to use the word "program" for the lead identifier thus denying the use of the letter P. Program event and product entity structure identifiers are used in the same way they are used in the United States Air Force (USAF) method.

The reader will note that the identifier string does link to the functional department providing the personnel to do the work on the program, and this linkage offers the functional department managers to filter all of the records in an enterprise database linked to his or her department generating a labor demand curve in time looking into the future. Thus, the manager could know what the personnel demands are going to be for his or her discipline on all enterprise programs. In some cases, the functional manager may be able to work with one or more programs through the PIT to adjust work scheduling to smooth the demand peaks. In other cases, it may be necessary to hire more staff in a timely way or consider ways to trim staff during some periods.

The system also links to the generic work definition as in F411 (Requirements Analysis) around which the enterprise can have constructed estimating models based on past performance. Every program will not have a degree of uniqueness that inhibits development of these vital statistics related to current performance metrics and continuous process improvement.

All of the elements of the planning strings fall into two subsets. Using the example in Table 10.2, the string F411-D262 can be referred to as an F-String that links common process steps (F411) to the responsible department (D282) as explained in Chapter nine. The string P05-S02-E04-A11 can be referred to as a P-String the elements of which are all program related. The set of F-Strings for an enterprise should be fairly stable over time and could be contained in a database as discussed in Chapter nine along with task cost and execution time estimates. In a new proposal, the P-Strings would be formed as the product entity structure is expanded through functional analysis and linked to program definition terms. The top-down planning process simply links P-Strings to F-Strings. This can be done in two steps mirroring the WBS and SOW. First, the program identifies a WBS dictionary by choosing the product entities through which it intends to manage the program. Next, the program links top-level common process functions to these WBS elements. The result is a SOW. Now the bottom-up process would have to be entered where people from the functional departments are brought in to expand the SOW content into detailed planning strings forming the IMP. The IMP is placed in schedule sequence aided by the event terms and one has an IMS.

The JOGSE approach was covered in considerable detail in *System Engineering Planning and Enterprise Identity* (Grady, 1995) and *System Engineering Deployment* (Grady, 2000), and this description will be updated in *System Management: Planning, Enterprise Identity, and Deployment* (Grady, in press).

10.5   Progressive planning granularity

All three of the planning approaches covered in prior paragraphs of this chapter can be applied at different levels of product entity structure because programs do not all start with full knowledge of the product entity structure. In the very beginning of a program, we may only know the system need (now captured in an initial capabilities document in DoD) and the whole IMP content may be as simple as P3-S1-E1-A-F4-D930 for phase 1 of enterprise program 3 where event 1 is submission of the phase end report or briefing. This planning string covers department D930 (a pre-design group perhaps) accomplishing all system development work (F4) relative to the system (A). During this phase of work, the study team will define the system more precisely perhaps identifying the preferred top-level product entities consisting of A1, Launch Vehicle; A2, Launch Site; and A3, Range Interface. In the planning for the next phase, the IMP strings should recognize the entities illustrated in Figure 10.5 and in that phase the product entity structure will be further defined, and in the planning for the phase to follow, the planning should once again be expanded to recognize the additional lower-tier entities.

It is very important to recognize that the powerful program planning capabilities represented by the last two planning methods discussed can be applied prematurely deep during a proposal. Because you have the ability to define work at a very low level there is a magnetic attraction to do so. One must plan a program at the level they plan to manage the program and planning at greater depth results in unnecessary busy work in implementation.

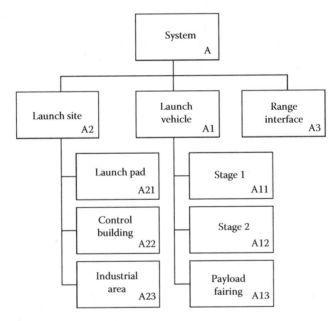

Figure 10.5 System definition expansion.

10.6   Program plans

We would all doubtless accept that a specification tree is necessary on a large program to introduce order into the product requirements development effort. What is not as universally accepted is a similar tree for program plans that capture the design of the development and production process. All too often, programs are implemented allowing planning documentation to be autonomously prepared by the several functional departments (engineering, manufacturing, finance, etc.) contributing work to a program. These plans may be generic company plans or procedure manuals applied to the program or specifically written for the program. Many companies get caught in the trap of trying to be totally responsive to every customer's initially stated process requirements to the extreme that they redesign themselves for each customer in terms of these plans. The future-looking company will apply a continuous process improvement concept to their generic procedures in combination with a rigorous customer procedures tailoring effort on each program to take advantage of the practice-practice-practice notion that world-class athletes use on the road to greatness.

Autonomous, functional department planning, disconnected from program requirements, is the target in this section. Program process plans should be architected just as the product systems requirements and components should be. Plans should exhibit traceability from the top-level plan down through the lower tier plans. A structured, top-down planning process will ensure mutual consistency of all of the program plans with a minimum of surprises during program execution. In the process of preparing program plans, it should not be necessary to come up with a new program design for each proposal and program. It should not be necessary to redesign your company for each customer. We need to find out how to apply the practice-practice-practice technique to our work through careful program planning for specific programs and apply continuous process improvement to our methods with a long-term view. Company personnel should be applying the same proven process, incrementally improved in time, to each program and in the process become expert in their specialized disciplines.

Very little of the business that a company tries to gain through proposal or marketing efforts involves radically new initiatives. Most of our energy is applied to prospects that are close to our historical product line. Therefore, most of a company's procedures and plans should apply in any new program. This is especially true if your company already does apply an energetic continuous process improvement program.

Given that we are organized in a matrix structure, our functional departments should have procedures covering how they perform their function on programs as members of cross-functional teams. The program must knit the functional methods into a coherent process appropriate for the particular product system under development as appropriate to the development phase. In this process, programs should not be allowed to substitute alternative processes creatively without acceptance by functional management because the functional departments should be deploying the very best methods they have developed over time based on continuous improvements fed by lessons learned from prior program experiences.

There are, however, two sound reasons for permitting programs to deviate from the current best practices. First, it is through program implementation that improvements can be developed and tested. A particular program may be asked to experiment with a particular technique in defining interfaces, for example. Perhaps the company's history is to use schematic block diagrams and the proposition is that program XYZ will use n-square diagrams instead in combination with a new and promising computer tool that the customer will make available at no cost.

The second reason is that the customer may have a valid need for a job to be done differently than our current best practices cover. Perhaps the company uses a particular computer program and related procedures to capture logistics support analysis data. The customer may have a big investment in capturing the data in a different computer data structure and have a perfectly valid reason why they need to collect different data than your system will address. You will simply have to adjust your practices to the customer's in this case. In the process of doing so, you may find improvements that can be woven into your preferred practice. But, the suggestion is that this should be the exception and the rule should be to follow internal procedures while incrementally improving them for reasons supportable based on prior or current work.

Given that we all accept that programs should be conducted in accordance with prepared plans for each activity, what plans are needed? At the top of this set of program plans rests the program plan. The program plan can be very simple giving the overall schedule in very broad terms, the customer's need, ground rules and policy, top-level program organization and responsibilities, and reference to other documentation. Figure 10.6 is based on the planning for a program applying the USAF integrated management system but includes a system engineering management plan (SEMP) and subordinate plans to satisfy the IMP requirement for narrative material covering certain planning areas. The Figure 10.6 plan set will satisfy any system engineering standard requirements as well as the integrated management approach. The indicated plans provide requirements for the process that will result in the product system. They should be mutually consistent and this can be demonstrated by establishing traceability between the plans in the patterns suggested by the plan tree.

These plans, in combination with program schedules, tell the humans populating the program what to do, when to do it, and who should do it.

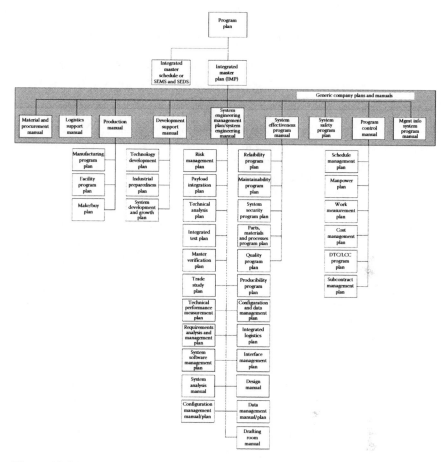

Figure 10.6 Program plan tree example. (From Grady, J., *System Integration*, CRC Press, Boca Raton, FL, pp. 84–99, 1994. With permission.)

Where do these plans come from? What are the right plans to write? Who should prepare them? As mentioned above, we can allow our functional departments to autonomously develop program plans and then try to fit them together during program execution. This is a bottom-up or grassroots approach to planning. Alternatively, we could approach program implementation planning as we should approach product system development—systematically from the top down.

The fundamental difference between the MIL-STD-499 and other standards and the integrated management system is that the IMP defines the work to be accomplished very rigorously linked to the program SOW and WBS by a common work identification coding system. The functional plans become detailed narrative descriptions of how the work defined in the IMP will be accomplished. Also, the integrated management system calls for an IMS rather than the system engineering master schedule (SEMS) called for in MIL-STD-499B. The advance made by the integrated management system is to absolutely link all work defined in the IMP with the schedules in the IMS using a common work identification system.

If you are forced by a DoD customer's requirements to plan a program with an approach that excludes the integrated management system, the IMP is simply collapsed into a WBS–SOW combination. You may conclude from this that the IMP is an extraneous layer of planning, but you are encouraged to finish the chapter before making that conclusion a permanent part of your belief system.

In the integrated management system approach, you would not normally prepare all of the plans subordinate to the IMP. The principal approach offered in this chapter recognizes a planning approach that makes best use of the limited time available during proposals by using the generic planning data as narratives referenced in the IMP; focuses proposal work on how these data can be blended into a program-specific IMP; and encourages company use of standard procedures, incrementally improved over time, that provide customers best value. You cannot develop the latter unless you are allowed to practice-practice-practice. If you redesign your company for each new contract, you will never realize a single identity and your work force will not be able to benefit from repetition.

Please understand that all of the generic plans subordinate to the IMP in Figure 10.6 are intended to be available as a function of on-going functional department planning that defines generically how the various tasks in department charters are to be accomplished. On a given proposal and subsequent program work, that basis should already be in place. The proposal work should focus on fashioning a program plan, IMP and IMS appropriate to the customer's needs drawing on your standard planning data as narratives referenced in the IMP. If you have 10 programs in house, each of the 10 programs would have its own program plan and IMP but each IMP would reference the same set of functional manuals that tell the people on the programs how to do their tasks.

You may find writing a generic integrated system engineering manual (SEM)/SEMP difficult if you do not have a model. *System Engineering Planning and Enterprise Identity* (Grady, 1995) and *System Engineering Deployment* (Grady, 2000) provide a model for this document including a copy on a computer disk. It is structured using the principles of this chapter such that it may be used as a program SEMP or the system engineering process narrative referenced in a program IMP. The JOG System Engineering course Grand Systems Management also offers a process specification with a complete process flow diagram and SEM as well.

It is true that some customers will be uncomfortable with your generic planning in that you can change it without their approval after a contract has been awarded. Some customers will want review authority on your internal plans and that can be a nightmare when you have multiple customers with very different interests. In such cases, you may have to run a copy of a generic plan and reidentify it for specific use on a program. This program-specific copy could thereafter be changed with customer review without causing chaos in your internal documentation. This will force you to deviate to some extent from the practice-practice-practice notion but is not a bad compromise when forced to respect your customer's requirements for review authority on all program procedures. The program adjusted documents should contain the maximum generic content.

The party line in your proposals and in conversations with your customers should be that you are motivated to continuously improve generic procedures that will provide all of your customers with the best possible value. In cases where the needs to satisfy two or more customers are in conflict, you need to consult with those customers as part of the process improvement activity. You should also offer your customers access to your generic internal planning data on the basis that it is proprietary. Let them act as one of the pathways through which you can detect incompatibility using management by exception. One excellent way to do this is to provide internal read-only access to all planning data via computer network throughout your facility and extend this access to your customer base either at their facilities or only at your own.

In any case, the specific work required for each customer's program would be clearly defined in their program-unique and customer-approved program plan, WBS, SOW, IMP, and IMS. These documents would reference the generic planning data that only tells how to do these tasks. The principal obstacle in implementing this approach is, of course, that every company has not done a fine job of documenting their current practices. In the author's opinion, this is not a valid basis for rejection of the offered planning method, nor does it represent an impossible barrier. It means that you must begin developing these generic practices now and keep improving them over time. Your competition may yet give you time to improve your performance for they are likely every bit as fouled up as you are (if that be the case) in this respect.

In the case where a company is organized in a projectized fashion, or only deals with a single product line, you still have to acquire your personnel from the same source as everyone else, from the human race. We are all knowledge limited and are forced to specialize. So, even if you have no matrix and only a project structure in a company with one or more projects, you will have to accomplish program work using specialists whose work patterns should be standardized, one program to another. This standardization can be captured in the kinds of documents shown in Figure 10.6 under the IMP.

chapter eleven

Transition to implementation

11.1   Awaiting contract award

After submission of the proposal, the enterprise proposal team and the program-to-be must await the customer selection process to arrive at a conclusion and announce that conclusion. The proposal team will have been stripped to the bone to avoid unnecessary cost but balanced by the degree of difficulty of reassembling the core team members if a win occurs and the probability of a win. That is, a proposal manager will commonly be allowed to continue with a higher staff level during this waiting period if there is a good change of success and there will be considerable difficulty reassembling the key people if they are assigned to other tasks. This period of time should be used as effectively as possible to prepare for the possibility that you win. There may be continuing proposal tasks to deal with during this period, such as a possible requirement for the intended program manager to offer an aural briefing to the selection team and respond to selection team questions, but remaining tasks required to attain a desired program entry condition should have been listed and those should be integrated into the work priorities as well.

11.1.1   Generic identity

If an enterprise followed the prescription offered in Chapter nine, it will have captured a healthy definition of its internal capabilities for doing work for customers prior to having started to work on this proposal. The enterprise will have a complete work catalog listing all work it can perform on programs in terms of enterprise functionality linked clearly to functional department responsibilities. This pairing will be further linked to resources to be provided to all programs from the functional departments when those programs pull their work strings from the catalog by including the department budget in their program plan. All of the content of the IMP will be traceable to this generic common process and work catalog.

11.1.2   Program work definition

Also, if you have followed the prescription in Section 10.4 during the proposal or in prior study phase work, you will have planned a program based on the product entity structure functionally derived. The items at some level of indenture will have teams associated with them and there will have been two system-level teams established, one for technical control in the form of a program integration team (PIT) and the other a business-oriented team called a business integration team (BIT) or a single integration team supporting the program management staff embracing both components. The other teams, integrated product and process teams (IPPT), will be established by the PIT at contract award based on the program plan. Ideally, the remaining proposal staff will have already been reassigned into the BIT, PIT, and top-level IPPT, at least in terms of leaders in the case of IPPT.

The work will have been planned based on the product entity structure so that when each of the teams becomes fully staffed, they each inherit a contiguous collection of work that is defined in a specification developed by the PIT or proposal team, a coherent work breakdown structure (WBS)-oriented section of the integrated master plan (IMP) and integrated master schedule (IMS), and a budget matching the time and technical requirements difficulty. Each team manager should be able to set the filter on a database containing all program planning strings so as to view only those IMP strings that correspond with the WBS for which the manager's team was made responsible. It is perfectly clear how the planned program budget links to the functional departments because each IMP string includes an F-String that defines the functional department that will contribute the people to do the work. The functional department managers will have been consulting with the proposed program team leaders regarding team staffing in the event of a program win and the movement of these people will have been coordinated with current assignments. Further, the functional department headcounts necessary to support all programs, with those awaiting selection assigned a probability of win influencing the credibility of the totals, will be taken into account by the functional managers to assess whether they have to start planning a hiring cycle or a layoff cycle. All of these planned movements must support the needs of all effected programs in terms of security clearance level, skills, and other factors.

11.1.3   Work assignment and implementation

Each team leader at the top level (PIT, BIT, and top-level IPPTs) will report to the program manager who in turn reports to the enterprise executive or an enterprise programs manager. The selected team leader for a given team must complete the staffing of his or her team, cause them to understand the job at hand, and begin planned team near-term work. Throughout the period covered by the IMP/IMS for this team's existence, the team leader must manage the team's performance so as to meet all cost, schedule, and product performance targets while producing a product design that fully satisfies requirements.

11.1.4   Functional management responsibilities

This book assumes that the enterprise is organized as a matrix with program and functional structures where the functional structure is responsible for program resources and the program structures are responsible for efficiently and effectively using these resources to perform the enterprise's business on programs to satisfy customer needs. Each functional manager is, therefore, responsible for building and maintaining a resource base clearly defined in concert with the enterprise integration team (EIT), that is, the overall enterprise common process owner. The PIT on each program should interact with functional managers and, if necessary, the EIT so as to bring to its program the resources it needs in a timely way. Those resources will have been clearly defined in program planning strings as discussed in Chapters nine and ten. They consist of trained and experienced personnel, good practices, and good tools all coordinated toward specific goals.

This coordination between the PIT and EIT should take place as soon as possible in the new program cycle. It can start during the proposal or commercial product planning cycle when it becomes clear what the program product entity structure is and a general business plan emerges. This information can then be refined as program work progresses toward a full program go-ahead. The source of the resources needed are the functional managers. The ultimate customer for these resources are the PIT, BIT, and IPPT leaders. If each team on a program negotiates with each functional manager independently and there is more than one program forming at one time, there is a good chance for chaos. Therefore, the author recommends that the emerging PIT collect all of the resource demands from their program and negotiate through the EIT to acquire those resources.

There should be no surprises for the functional managers because the enterprise cost schedule control system (CSCS) or earned value system (EVS) should have all of the information loaded from each existing real program and programs-to-be (some of which will not come to fruition, of course). These entries should be prioritized and identified relative to probability of a win/go-ahead so that the functional managers can determine to what extent they should staff for the aggregate of enterprise demands for their personnel. Now, many enterprises are probably still using CSCS/EVS that do not serve the needs of lower-tier functional managers, and these enterprises will have continuing difficulty in the future matching program demands and functional department availability of personnel. Only if the enterprise employs a common process supported by an effective database system will one be able to appraise functional managers efficiently of their near-term personnel needs. When a manager at General Dynamics (GD) Space Systems division, the author often had to spend 2 or 3 days to review several programs to determine his headcount needs for 3 months. This was repeated every few months. The finance systems were intended to serve program manager needs at the time. At the time he left GD Space Systems division, they were beginning to bring online systems that would serve the needs of the functional management better.

11.2   Populating the teams

In order to bring in the people, there must be some work done to prepare a facility in terms of space, office partitions and space improvements, utilities, computing and communications equipment and wiring installations, team space allocations made, and other improvements made. Those will be discussed shortly. For the moment, we will focus on the personnel acquisition side of the issue.

This book encourages the implementation of a PIT on all large programs. On small programs this may be one person or the project manager doing double duty. The PIT should be staffed as early as possible to perform the system-level technical work to define the system and accomplish program planning. Generally, on programs dealing with large acquisition agents like Department of Defence (DoD) employs, this team can be formed during the proposal and be responsible for the technical volume of the proposal and any required system specification work. It is commonly not feasible to populate the product-oriented program teams before the system development process begins in the form of a commercial formal go-ahead decision or contract funding from an acquisition agent. Enterprises must function in accordance with a sound financial plan and cannot staff up programs without some hope of recouping the costs expended unless the enterprise is doing so in accordance with a consciously evaluated risk.

As the PIT identifies product entities and associates them with IPPT, those teams can be staffed starting first with selection of a team leader who should then collect the related team defining information from the PIT including the following.

Item specification: At the top team level, the PIT should have developed a specification for the item for which the IPPT is responsible. The IPPT may have to develop lower-tier specifications.

Interfaces: The PIT must make it clear to the IPPT what their external interface responsibilities are and with what other teams they must interact to develop each of them.

Funding: Ideally, during the program planning process, the WBS will have been overlaid upon the product entity structure and not include a services component as discussed in Chapter ten. If this work is done well, then some single WBS element at some level of indenture will apply exclusively to each IPPT.

Planning data: Assuming the program has been planned using the integrated management system or JOGSE approach discussed in Chapter ten resulting in an IMP/IMS, there is a contiguous section of program planning and scheduling that connects to each IPPT.

11.3   Bringing in the resources

Throughout the early existence of a program, its PIT must be working toward the potential for a full go-ahead decision. This must include providing its teams with the resources they will need to do the program work. This preparatory work should be coordinated through the EIT with enterprise facilities, the functional departments, and other organizations responsible for providing the resources needed.

11.3.1   Facilities

Clearly, the teams will need space within which to work. These facilities can be very bare if absolutely necessary, but the finer they are, the better the performance of the teams, probably. The author has worked in factory spaces on small projects and has known engineers whose desks were at the mercy of the pidgins in factory spaces as well as leaks in the roof during rainy periods. In all of these cases in the author's knowledge, the engineers performed well despite the poor conditions. That was a few years ago and it is not clear that today's crop of engineers would accept these kinds of conditions particularly during a period when jobs are plentiful.

Perhaps the worst facility conditions that the author can recall were at Teledyne Ryan Aeronautical when management failed to have leaks in the roof fixed before the annual rainy season. On one day, the author can recall wearing his raincoat while sitting at his desk holding an umbrella in one hand and writing with the other also catching water in a couple of trash cans located under other leaks. The author was never too adversely affected by these conditions because he had spent 10 years in field engineering, three of those in country or afloat off Vietnam, and considered it good working conditions when there were no rocket or mortar rounds incoming. Most people will not be so understanding.

A program should do its very best to provide clean, well-appointed space within which good work can be performed. The space available during the proposal or early program work may not be optimum and should be improved as soon as possible by improving the space or moving to another space when it is necessary to staff up for the program. Ideally, all

of the people on a program would be located in one facility with the easi-
est possible physical movement of people in and around each other. The
teams should be physically collocated, of course, but the relative locations
and distances between teams should also be determined by the strength
of the interfaces between the products for which they are responsible. If
IPPT 1 and IPPT 2 are responsible for two product entities that have an
intense interface between them, then teams 1 and 2 should be in close
proximity in the facility.

Where it is necessary to physically disperse the team members, perhaps as widely as in different countries, it puts an increased demand
on good inter-team communication and computer connectivity possibly
including videoconferencing capabilities.

11.3.2 Space improvements and team space allocation

In the dying days of the twentieth century and early days of millennium
2000, there was a lot of support for the use of modular cubical structures that give each person an office. This is an easy way to turn a large
clear space into fully equipped individual spaces configured for specific
program needs. Most of this equipment is reusable and the panels and
attached office equipment can be broken down from a terminating program and reerected for another program in another part of the plant.

The author found that the most productive system engineering environment he had ever worked in during his employment in industry was
in large open spaces without bulkheads that seem to be so popular today
in the interest of privacy and status apparently. The cubical structures
constrain communication between people in a situation where you are
depending on human communication to weld together the specialized
minds of all of the people on a program into the equivalent of one great
all knowing mind. It is true that some people find this work environment
a very disturbing experience.

One of the author's system engineering heroes once told him as they
prepared for staffing a big program that he had considered locating people on the program randomly because it might trigger some interaction
between the people relative to product development that seemed to have
evaded him on other programs despite a carefully planned desk location strategy. We did not do that on that program but it was a thought-
provoking possibility.

The teams will need space, of course, for their team members to sit
and do work but also within which to meet. This space need not be a
meeting room but that is probably preferred. The author has seen hallway
space used temporarily where there is wall space available for mounting
large charts. Wall space is part of the communications picture. We need it
to mount the big picture around which we can gather and jointly discuss

evolving design features. But the teams also need joint access to the details available on computer networks. Individuals can be connected in their cubicles, yes, but this is not quite the same experience as when the team is assembled in one space all exposed to the same image and the same conversations. This can be accomplished through computer projection onto an available wall or projection screen but it does require thoughtful work with those people responsible for computer network availability.

Because we must bring many people together to solve complex problems, we need work spaces where these many people can interact in the most efficient ways possible. Despite the increasing capabilities available through networked computers, there has not yet been developed any improvement over a team of people gathering in the same space and interacting appealing to the maximum number of senses in communication attempts, audible and visual in particular.

11.3.3 Computing and communication

The space within which the team will work must be networked today and it is hard to imagine that anyone would debate this. While the author worked for GD Space Systems Division, the Systems Engineering Director set a goal that we would outfit at least one meeting room on every program during a particular year using Systems Engineering money. One program chief engineer did not want a computer in his meeting room so the program system engineering lead had to build a mobile computer cart that could be brought into the room when the chief engineer was not using it and the room had to be wired while the chief engineer was on vacation.

There is no technological development that has more effectively supported the skillful implementation of specialty engineering work than networked computer information. We need to bring together people into meeting venues to gain the full benefits of synergism it is true, but we also must provide opportunities for people to do their own individual work reasonably free of interruption. Where the specifications or computer databases containing this data, computer-aided design data, and the results of detailed analytical work are all available to all members of the program teams, these engineers can carefully study the work of other team members contributing their own improvements in program information stimulated by a common monotonically increasing store of information.

These enabling communications resources include good computer networking, telecommunications, audio-video conferencing, meeting space computer projection, fax, and agreement on computer software applications to the extent possible. Programs using five different word processor applications may have problems sharing information, for example.

11.4   Firing up the program relationships

Programs do not exist as separate entities rather are living organisms with internal and external interactions of some complexity. The external relationships must be brought to full life at the earliest opportunity.

11.4.1   The customer

On programs where there is a customer acquisition agent as on DoD programs, the contractor should not delay its efforts to build a sound business relationship. Most of these customer program managers would not sit still while being ignored by their contractors anyway. The customer and contractor program managers should encourage the exchange of contact information between key people on the two sides of the contract and encourage that these contacts be exercised early and often. The customer will likely insist on an early beginning of program management reviews initially at a high frequency such as on a monthly basis possibly using videoconferencing.

11.4.2   The procurement process

Preliminary work on defining the product entity structure should have also included identification of source possibilities. First in terms establishing the source category: in-house build, purchasing, commercial off the shelf, or customer furnished. Each supplier should provide its product through a contract or purchase order in accordance with a clear set of expectations on the buyer's part. Where the item will require supplier development, this should include a procurement specification done well as early as possible. Where the item will be an in-house design with production purchased to engineering drawings, the responsible team should first define the requirements in some kind of in-house requirements document and then proceed with a sound design responsive to those requirements and any sensible encouragements from potential suppliers. Commercial off-the-shelf equipment and software contacts should also be implemented with every reasonable effort made to control the configuration acquired and coordinate it with the plans for the other system elements.

Customer-furnished equipment and software should, of course, be listed in the contract. Because it is so listed does not, however, mean that it will be available when the contractor first needs it. In some cases, DoD may overextend itself with regard to supplying customer-furnished equipment. In one case, the author can recall, the customer had to pay the contractor to write a specification for some aerospace ground equipment yellow gear, develop the engineering drawings, and to produce it. This was a common item of USAF equipment but not only did they not have enough to supply the contract agreement, they did not have the specifications or engineering drawings on it. The contractor had to reverse engineer the items from the one of each supplied by the customer.

11.4.3   Associate relations

A DoD contract involving the development of large systems may entail procurement of major elements of the system from different major suppliers. These contractors will not typically have any kind of contractual relationship between themselves as they would in a prime-supplier arrangement. Rather, the common customer with whom each has a contract will require them to cooperate in accordance with contract provisions governing associate contractors. These contractors will be required to sign letters of agreement or understanding saying that they will abide by a common interface development plan and jointly staff an interface control working group (ICWG) that develops and applies an interface control document to control the design of interfaces between the system elements under the control of the two or more associate contractors.

Each contractor must identify people who will support the ICWG and one of the contractors may be identified as a prime integrating contractor responsible for setting up the ICWG, interface development plan, and initial ICD. Alternatively, the customer program office may bring in a system engineering and integration contractor to provide that function or elect to do this work itself. The customer may even neglect to do anything until urged by one of the contractors.

11.5   Winding down the beginning

At the earliest possible time, the program should settle down to work in accordance with the program plan and schedule under the full control of the team and program managers. One way to make this transition is to apply an action item approach to attaining readiness. In this approach, the program maintains a list of open actions linked to the person or organization responsible for bringing about closure. The program manager or surrogate holds periodic meetings with the cast of characters responsible for closure determining status on open items, closing ones as appropriate, and opening new ones possibly. This list would link each item to a team as the recipient of some service or material and a supplier responsible for bringing about closure. This can also be moved to closure by dealing with it at each program manager's meeting with the team leaders reporting on their status. But, it is important that the program staff get in sync with the program plan as soon as possible.

part three

Product system definition

chapter twelve

System modeling and requirements identification

12.1 Overview

This chapter offers structured modeling methods for requirements analysis that will work for systems, hardware entities, and software entities. It also explores the gap that commonly exists on development programs bounded by software and system requirements analysis. Older system engineers tend to use some form of traditional structured analysis (TSA) to decompose a system need into lower-tier functionality, performance requirements, and system entities so long as they do not encounter any entity that could best be implemented in computer software. When that happens, many of them are helpless to continue the analysis in that product entity branch because TSA is not effective in modeling the problem space where the solution is a software entity and they are commonly not effective software analysts.

At that point, the work must transition to software analysts who can apply modern structured analysis, early object-oriented analysis (OOA), or unified modeling language (UML) to continue the analysis while not having any way to clearly link the results of their work to the lower fringe of the system analysis work. It is intended that this gap be filled using a combined model base composed of UML and TSA with the understanding that the development organization should be working to replace TSA with system modeling language (SysML) as the latter matures to fully cover the TSA territory and as the workforce matures in the transition. Figure 12.1 suggests that there is a collection of modeling artifacts that are formed by the union of UML and TSA, as well as between UML and SysML, that will provide us with a complete set of modeling artifacts to use in analyzing any problem space no matter how it is intended to be implemented. Figure 12.1 also suggests that the Department of Defense Architecture Framework (DoDAF) can be introduced into this same pool of modeling artifacts. In many system development efforts, DoDAF can be used to initiate the system analysis with a subsequent resort to TSA as elements of the system are identified that must be developed as hardware entities. Also, we will see that DoDAF can rely upon UML, SysML, and TSA for the modeling artifacts

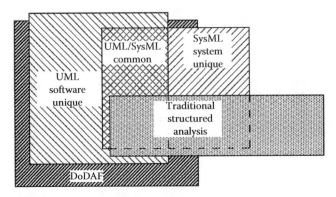

Figure 12.1 Modeling artifact pool.

it employs. The message is that today the system engineer needs to be more than just familiar with a pool of modeling artifacts as suggested in Figure 12.1. System engineers have to be ready to apply the complete model set that provides them with comprehensive coverage of the problem space.

The models considered in this chapter were all developed by different collections of people in different time frames. They were each developed for an intended purpose and with one exception: no planned coordination with the others. However, our intent in this chapter is to use the aggregate of these models in a coordinated way as if they were a part of one large modeling family. We will also see how over time, the family will shift toward a more integrated set approaching what the author likes to refer to as the universal unified modeling language (UUML).

Some recent software modeling methods, like early OOA, rely upon first identifying the static entities that will compose the system followed by then identifying the dynamic behavior of those entities. To system engineers, this is a violation of a fundamental concept long accepted, the Sullivan notion of form follows function. There does seem to be a clear advantage in first knowing what a system must do before deciding what the system shall be composed of to accomplish what it must do. Therefore, the software modeling methods employed should be applied following Sullivan's idea encouraging that a common pattern be applied in the development of all of the elements of an unprecedented system. This is not easy to do with early OOA, but it is entirely possible with UML.

The system engineering approach to system development entails identifying an ultimate definition of the problem space referred to as a need statement. In the most recent language of DoD, this is referred to as a joint or initial capabilities document, but the older term will be used in this book because it is so expressive of the intent. The need is then decomposed into lower-tier functionality. Performance requirements are derived from this exposed functionality and are allocated to physical entities that will have to be purchased or developed respecting a set of requirements derived from the analytical process. During the development process, it will be decided that some of the entities will be developed in hardware and some in software. It is intended that the analyst will reach into the TSA and SysML portions of the modeling pool and extract the modeling artifacts used for the systems and hardware portions of the analysis and into the UML and DoDAF portions for those parts related to software.

As a result, it is necessary for the program system engineer or system engineering team (referred to as the program integration team or PIT) to have some means to make decisions as the analysis unfolds to allocate functionality to specific entities of one of these two types with a possible additional need to allocate functionality to people accomplishing some form of work, thinking, or action. The opening model should provide this service for the top levels of the analysis. Lower-tier analyses can then apply the appropriate model set as the analysis unfolds. One might ask which set might best be used at the beginning? One can use either pair in the beginning but the author believes that, in general, that SysML or TSA would be the right place to start because there is no such thing as a software system entirely separate from some collection of hardware upon which the software must run. But, there are cases where the primary area of interest is computer networking for which DoDAF might offer the best starting point.

The author is suggesting a top-down development direction that is preferred where the problem space is new to the team, that is, the problem space is largely unprecedented for the team. The direction may be different in other situations. Where the problem space is highly precedented and physical product (including software) already exists that will accomplish much of the system functionality, the entry modeling may profit from exploring the existing system product entity structure and determining which of these entities can be depended upon to provide good service in the new system with no change, which items can be modified and continue within the system, which items must be discarded, and what new items must be developed. Modeling methods discussed in this paper will be effective against the new and modified entities that comprise the functional differential between the As-Is and the To-Be versions of such a system.

After a brief overview of the two patterns of development, we will first review TSA and then discuss UML. Finally, we will attempt to integrate across the gap. The author believes that this story will eventually be completed using UML teamed with SysML.

12.2   Development pattern overviews

12.2.1   The current system and hardware pattern

Given that the organization in question applies some form of TSA using a simple functional flow diagram, IDEF-0, enhanced functional flow block diagram (EFFBD), or behavioral diagram, performance requirements are derived from the identified functions and allocated to product entities. Other models are applied to identify interfaces and their requirements in the form of n-square or schematic block diagrams, specialty engineering requirements using a specialty engineering scoping matrix supported by specialty modeling efforts, and environmental requirements applying a three-layer model (system, end item, and component) addressing four kinds of environmental stresses—natural, hostile, noncooperative, and self-induced.

This process can be applied in layers as encouraged by Bernard Morais, President, Synergistic Applications, Sunnyvale, California and the late Dr. Brian Mar in their use of another alternative structured approach called FRAT. The analysis of the behavior of a layer N product entity is based on having previously defined the level $N-1$ product entities by applying a functional analysis of the product entities at level $N-1$ as suggested by Figure 12.2 (Morais and Mar, 2004). This is an expanding process as the pyramidal structure suggests. A product entity at level $N-1$ becomes one or more subordinate product entities at level N. The analysts for all of the branches apply functional analysis to decompose the product entities for which they are responsible at level $N-1$ and when the functional analysis for level $N-1$ has been fully exhausted to identify level $N-1$ entities, the process may be extended to an analysis of level N functionality.

The layer-by-layer staging approach is only one of four approaches that could be applied. Two faulted approaches are instant allocation of functionality, to which the layer-by-layer approach is similar but better organized, and delayed allocation where one does not start allocating functionality until the functional analysis is complete. It is fairly difficult

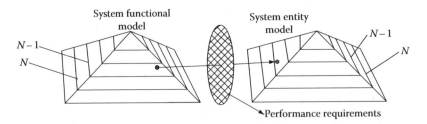

Figure 12.2 Modeling layers. (From Morais, B. and Mar, B.W., Bridging the Gulf—Systems engineering and program management. Presentation at the *2004 INCOSE Symposium*, Toulouse, France, June 2004. With permission.)

to determine when you are at the end of the functional analysis in the latter case because that decision can best be made in the product entity pyramid. The author prefers a progressive allocation process involving more than one layer between functional analysis and product entity identification but we will see that the layer-by-layer approach will make it easier to compare the hardware and software processes.

12.2.2   The current software pattern

Many software engineers have entered the field since OOA entered the modeling set and have been intellectually damaged by a fundamental flaw that the author believes early OOA embraced. All of the early OOA book authors with whom the author is familiar encouraged the identification of objects as the first analytical step. This was to be followed by applying data flow diagramming to determine the interior functionality and data needs of the objects and state diagramming to determine the interior behavior of objects. This pattern reverses the sequence that system engineers take at faith crafted by the architect Sullivan as form follows function. While UML permits an approach reflecting Sullivan, many software engineers continue to apply it in the sequence encouraged in early OOA. It is possible to apply UML in essentially the same pattern that system and hardware engineers apply in TSA and in the process encourage more effective hardware–software integration while not damaging their analytical process. These same ideas can be extended to DoDAF to the extent that one employs UML constructs within DoDAF work. Unfortunately, DoDAF does not appear to recognize the concept of product entities other than as terminals of interfaces. Systems consist of things and relationships. DoDAF is strong on the relationships and weak on the concrete things that compose them.

12.3   System definition

It is an essential step in creating a program plan to understand what the system shall consist of in that program planning is focused on this important point. This work is best accomplished by modeling the problem space functionally and deriving the product entity structure from the exposed functionality. The program planning is then grounded in the product work breakdown structure (WBS) overlaid upon the product entity structure.

12.3.1   Initial system architecting

Early in the development of a new system, the development team must determine the higher-tier product entities that will be employed perhaps using TSA, and in the process determine where computer processors shall be employed in the system and what functions those processors will be responsible for accomplishing. At this point, it is possible to set up the product teams that in each case will involve some subset of the engineering domains needed for the aggregate program. The program should initiate the requirements analysis work in a system team, referred to in this book as a PIT. The PIT should have developed the system specification and the next tier of specifications that will end up being the responsibility of the top layer of integrated product and process teams (IPPTs). Given that the program has applied the integrated management system or the JOG System Engineering variation and produced an IMP/IMS, the highest-tier IPPT oriented in each case about a top-level product entity subordinate to the system can be brought on board and given a specification defining the problem they must solve and their portions of the IMP-IMS that has been organized around the WBS that was overlaid upon the entity for which the team is responsible.

We will have to choose whether to organize the software development responsibility as one or more software teams coordinated with the processors or make each of the hardware dominated teams responsible for the software that is related to the processors included within the product entity for which the team is responsible, that is, in a collected or distributed pattern. In the latter case, one realizes well-defined interface relationships between the hardware–software pairs but will often realize poor software-to-software interface relationships within the processors. The relationships problem reverses, of course, if the computer software is developed by one or more dedicated software teams. This book does not recommend either particular software team concept as it can be successfully done either way and there may be other factors that would mitigate against one selection or the other on a particular program. As the software teams or engineering positions are identified and staffed, they begin work with knowledge of the computer architecture and languages to be used and the functions for which they are responsible for developing the software because of the previous work by a parent team. Ideally, the parent team would not attempt to identify the machine in any detail or the languages that will run upon it until the related software team or engineers have determined a reasonable set of requirements for the software and the machine upon which it will run. These decisions should be handled as an interface discussion working toward the aggregate best solution from both hardware and software perspectives.

12.3.2   *Traditional structured analysis*

In TSA, the ultimate function, the need, is represented by a single function block identified as function F. This function is expanded into a life cycle model that is essentially the same for every system an enterprise may be called upon to develop. The author's view of this diagram is shown in Figure 12.3. The life cycle is split between development and employment. It is the employment overlay functions that must be understood to

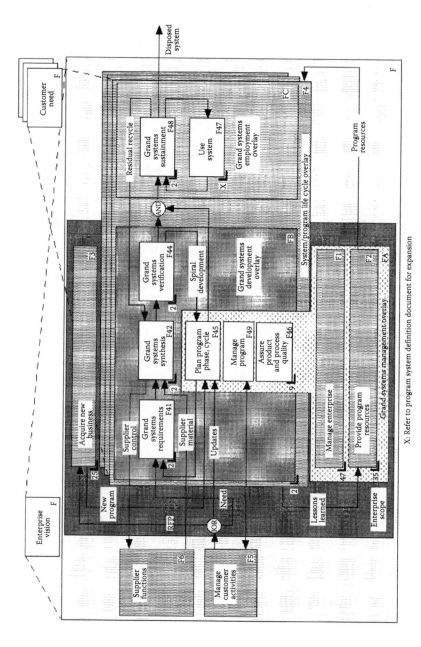

Figure 12.3 Life cycle model.

identify the best choice of system entities and their requirements. In particular, the Use System function is decomposed exposing more detailed functionality. Each function is translated into one or more performance requirements and those requirements allocated to something in the system product entity structure and thus appear in the specification for the item to which they were allocated.

Figure 12.4 is advanced as an integrated view of the definition and synthesis processes shown in Figure 12.3. The system definition process defines the system specification and identifies the top-level entities in the product entity structure. The PIT applies TSA to build the specifications for these top-level entities that can be handed off to the top-level IPPT as they are formed. In the event that one of the very top-level entities is software then the PIT will apply UML to develop the specification for that entity. No team should begin work on the program without being handed a copy of the specification for the entity for which that team is responsible plus the related component of the program IMP/IMS. Each team leader is intended to be held responsible for the cost, schedule, and performance of the entity for which his or her team is responsible.

Top-level hardware development teams can continue to apply TSA down to the lowest-tier hardware entities possibly forming lower-tier teams in one or more layers, and to the extent that multiple teams form, the parent team must take over as the system agent for its child teams. The performance requirements derived from the functions are entered in a requirements analysis sheet (RAS) ideally implemented in a computer database application. The functional decomposition process continues so long as the allocations are to hardware entities. As requirements are allocated to computer software, the analysis has to shift gears to UML for each separate software entity identified. The top-level software entity in each case may be treated as a node or component and the analysis accomplished as described in Section 12.3.3.

The steps discussed so far cover steps 1 through 7 of Figure 12.5. As the product entity structure matures, the analyst can evaluate the allocation of functionality to product entity pairs using an n-square diagram to identify interfaces needed (step 8). For system and hardware entities design constraints analysis is applied to identify needed interface, specialty engineering, and environmental requirements. A three-layered model is useful in identifying system, end item, and component environmental requirements (step 9) featuring tailored standards at the system level, a three-dimensional service use profile arrangement for end items, and an end item zoning technique for component level. A technique called design constraints scoping matrix applied in the *Air Force Systems Command Manual 375* series many years ago can be resurrected to coordinate the entities illustrated in the evolving expansion of the product entity structure with demands for specialty engineering requirements analysis

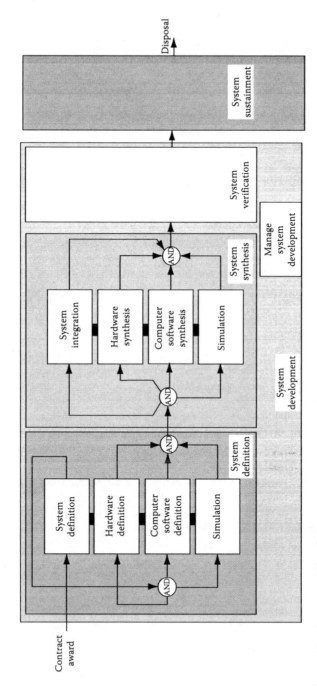

Figure 12.4 Integrated development.

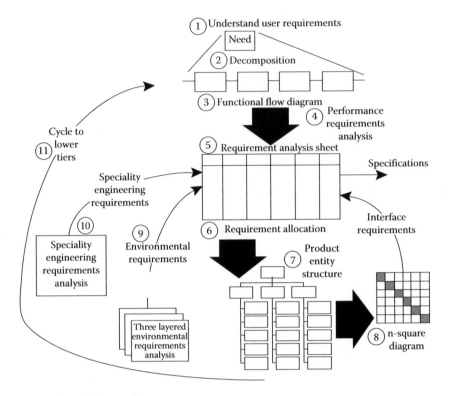

Figure 12.5 TSA overview.

work (step 10). Thus, every kind of requirement that appears in a system or hardware specification can have associated with it a structured analysis model within TSA. Also, the requirements that flow out of these analyses can be introduced into the common RAS for the program that should be implemented on a computer database within which traceability can be maintained and from which specifications can be printed. As suggested in Figure 12.5, the whole process circles back around to identify the next layer of functionality and product entities as suggested in the pyramidal model of Morais and Mar (2004).

It should be very clear to system engineers that traceability between the requirements and the models from which they are derived should be a simple matter given that they follow the encouragement to derive all requirements from models and in this case, TSA components.

12.3.3   *Top-down software development*

The goal in this chapter is to describe a top-down development approach using UML similar to that employed by system engineers using TSA.

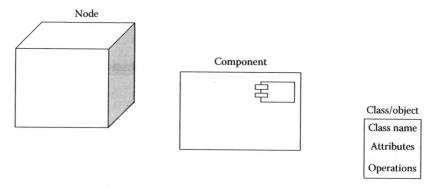

Figure 12.6 The static structure of software.

The intent is that as SysML comes online and starts being applied on programs and is coordinated with UML application, the development gap between these communities will continue to move toward closure.

Figure 12.6 illustrates the three UML layers of static product that software development should be trying to characterize. Since software must operate on a hardware device, it is reasonable to expect that there shall never be a system totally implemented in software, but using a little imagination, we can refer to the top-level software entity as a software system. That system will satisfy some predetermined function or provide some needed capability, more likely a collection of capabilities not all necessarily functionally related. UML provides for partitioning the software system into a hierarchy of software entities noted in Figure 12.6. The node is a specific deliverable collection of software entities that operate on a specific platform and it consists of one or more software components. A component consists of one or more components (multiple layers of components is possible) or classes or objects where an object is an instance of a class. In very complex systems, the notion of subsystem might be added between system and node, referred to as a subject in some early OOA models. The question is, in what direction will we identify these static entities—from the bottom up or top down.

We will refer to the entities depicted in Figure 12.6 generically as classifiers, and it is our purpose in applying UML to determine what these shall be and, eventually, what the lines of code should be to comprise each of them. The intent is to identify these static entities through an exploration of their functionality in a top-down fashion when developing a largely unprecedented system. It is granted that bottom up may be preferred when developing a highly precedented system looking for functional differential. Ideally, models should be bidirectional with functional problem space entry for heavily unprecedented problem spaces and product entity problem space entry for heavily precedented problem spaces. UML can be applied in this fashion as can be TSA.

Figure 12.7 illustrates an overview of UML as it will be covered in this paper. We enter Figure 12.7 with the understanding that we must develop the software for a particular computer processor or set of processors with particular functions possibly extended into a clear set of performance requirements provided by the preceding TSA work that identified the software entity. True, once the problem is turned over to one or more software IPPT by the system team, they may individually or collectively conclude that the system and software could have been organized more advantageously leading to a possible iteration. Figure 12.7 captures a process for applying UML in a top-down pattern first identifying the Nodes that will be required. Each of these Nodes may have a separate specification developed for it and we should apply UML to first identify the requirements corresponding to the problem space and to feed information into the parallel simulation work. To analyze the classifier of interest (system at the top level, node at the next tier, and component below that) we can start with a context diagram consisting of a single bubble representing the classifier of interest with lines connecting the bubble with each of the external terminators representing entities in the system environment that will interact with the classifier. For each terminator, the analyst should build a set of use cases corresponding to external agents deriving benefits from the system. The context diagram, derived from modern structured analysis, is not part of UML but can be used to help bound the use cases.

Each use case should then be represented by a set of scenarios that can be represented by UML sequence or activity diagrams. Some large programs have applied use cases teamed with sequence diagrams (step 5 of Figure 12.7). Alternatively, one could apply activity diagrams that a system engineer would understand as functional flow diagrams (step 6 of Figure 12.7). Software people draw the latter in the vertical (a heritage from mainframe computers printing flowcharts on line printers using ASCII symbols) and system engineers draw them in the horizontal (probably driven by an engineering drawing focus of left to right and top to bottom) but that is not an impediment to communication. Thus we have used context diagrams and two UML artifacts to get us to the equivalent of the system engineer's functional flow diagram. System engineers might benefit from the use of a context diagram and a set of scenarios as a preparation for the functional flow diagramming as well.

We may enter the dynamic analysis using a sequence diagram, a communication diagram, activity diagram, or a state diagram depending on the nature of the problem space. The suggested approach for analyzing any classifier dynamically can be explained in the hierarchical structure illustrated in Figure 12.8. For each context diagram terminator, we recognize one or more use cases that may have related extended or included use cases. For each use case, we can build one or more text scenarios describing the use case what some people call use case specifications.

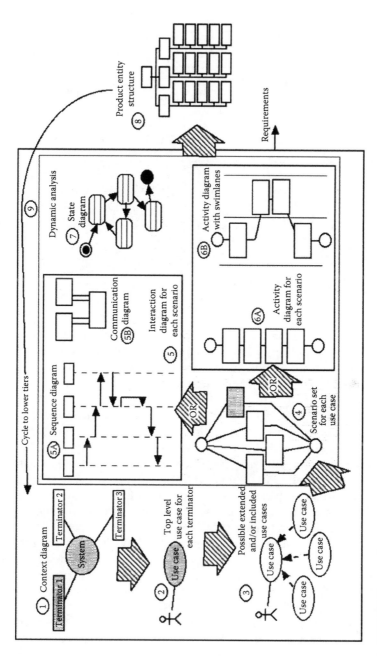

Figure 12.7 UML overview.

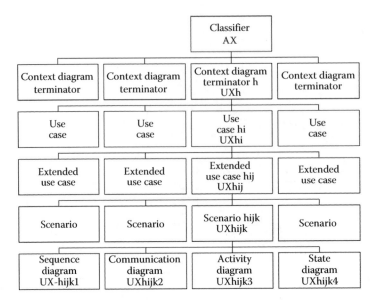

Figure 12.8 Hierarchical structure of UML analyzes.

As a result of this analysis and the requirements previously developed for the entity being analyzed, we can select which one or more of the four dynamic modeling artifacts to use for the scenario to be analyzed. This process may entail many use case analyses each of which is dynamically analyzed using some combination of the dynamic artifacts.

Requirements flow from the dynamic analyses and are captured in the RAS. In the process of building the sequence, communication, and/or activity diagrams it is necessary to identify lower-tier SW classifiers in the form of activity diagram swim lanes, sequence diagram lifelines, or communication diagram entities. These become the next lower-tier classifiers expanding laterally and in depth as the analysis proceeds.

The classifiers identified in this work are added to the system product entity diagram. Concurrently, the system- and hardware-oriented teams are expanding the functional analysis from a hardware perspective with both hardware and software entities being added to the same physical hierarchy. Communication and sequence diagrams help identify interfaces between software entities. Hardware and software engineers on the several IPPTs on the program cooperate to identify hardware–software interfaces that are audited by system engineers.

The fact that we are partitioning the problem space in a hierarchical fashion in the work exposed in Figure 12.7 suggests that the results will have to be integrated and optimized. Other alternatives are investigated in search of an optimum definition of the problem space. The analysts may then determine how the software will be organized into lower-tier classifiers by rebuilding the dynamic modeling diagram set respecting activity diagram swim lanes or sequence diagram life lines corresponding to the intended next lower-tier classifiers. Essentially, this action parallels the work that a system engineer does to allocate functionality or the performance requirements derived from the functions to physical entities that will comprise the system at a particular hierarchical level. This is the key adaptation of the UML approach to cause closer hardware–software development methods alignment.

This pattern can be repeated for each classifier. If in the analysis of a node we identify three components, we can apply this approach to each component thus identifying the next lower-tier entities (lower-tier components or classes). Having identified the next lower-tier classifiers, we still have a task to analyze the entry classifier for dynamic behavior that is done using the diagrammatic treatments through which the lower-tier classifiers were identified with. These modeling entities should give insight into performance requirements of merit for the physical entities identified in the static analysis. A computer software specification data item description developed by JOG System Engineering is available with *System Requirements Analysis* (Grady, 2005) that coordinates the use of UML with the paragraphing structure of the specification.

12.4   Integration at the gap

A program must make a decision about the modeling techniques it will apply as it builds the program proposal. This may include some mix of TSA, UML, SysML, DoDAF, and IDEF-0. A program that is going to be primarily focused on development of computer software or networked assets could enter the program with UML or DoDAF. If the product is a database, it could enter with IDEF-1 or UML. But, generally, modeling entry should involve the use of TSA or SysML at the system level because software, an intellectual entity, must run on hardware entities that provide the product with real substance. As noted earlier, at the time this was written, SysML was not yet fully ready to completely replace TSA in the author's opinion, so TSA would be the author's preference at the time this was written. But, an enterprise should continue to follow the development of SysML and work toward replacing TSA with SysML. In any case, there is a need to recognize that for some period of time, there will be a need for a model that works well for systems and hardware and another model that works well for SW. For now, also, there is not one great computer application within which one can model that one can apply to HW and SW requirements work, permit easy cross-model traceability, and provide specification publication capability so it may be necessary to use two or three applications to cover the needed tool suite.

Work can be accomplished well for systems and HW as covered under Section 12.3.2 and for SW using the approach covered under Section 12.3.3. When applying both, problems will tend to arise when transitions have to be made between these two approaches. It is not possible for SW to include HW but the opposite case is perfectly normal. So, the transitions will only be a problem as the analysis shifts from HW to SW moving from the use of TSA or SysML to UML. There are two concerns at this point: one in the models applied and the other in the computer applications employed.

The transition point will occur when the highest-tier software entities are identified. There may, of course, be several of these transitions distributed about the expanding product entity structure. The program has the option of pooling all of the software into an integrated entity or permitting it to be distributed within multiple processors that may still all be under the responsibility of one team or distributed among teams with both hardware and software responsibilities. If we can solve one of these hardware–software handoffs, we will have solved the general problem of requirements traceability across these gaps.

It should be clear that requirements traceability to models is assured in the approach covered in this discussion because all of the requirements are to be derived from a model. Vertical or hierarchical requirements traceability is very simple in specialty engineering areas in hardware, software, and across the gap. The environmental requirements are vitally different between hardware and software, and one can make a case that lower-tier software environmental requirements should not have to respect traceability across the gap to higher-tier hardware or system environmental requirements that are largely environmentally related. Precisely the same method of identifying hardware interface requirements can be used to identify software interfaces as well as hardware–software interfaces because we identify them between entities that appear in the joint product entity structure. So, if interface requirements traceability involves lower-tier interface expansion requirements to higher-tier interface requirements, traceability is assured. This leaves only the performance requirements a remaining problem from a vertical or hierarchical traceability perspective.

Given that the system entry analysis was accomplished in TSA using some form of functional analysis and the lower-tier software analysis is going to be done using UML, there is a temptation to employ activity diagrams in UML to analyze software entities from a dynamic perspective because it is very similar to functional flow diagramming and might give us some interesting opportunities to link up hierarchical traceability. However, for a given software entity, there may have been 10 performance requirements derived from 8 functions allocated to the software entity in question. There is no clear way to link up the UML activity or sequence analysis and requirements derived from it with the several functional analysis strings and the performance requirements derived from them that can easily be automated.

So, let us pursue another tack in an attempt to coordinate the traceability relative to the sequence-oriented dynamic analysis approach described previously. If requirement $R\%_v$ is one of a set of requirements $R\%_1$ through $R\%_{10}$ where $R\%_v$ is derived from function $F\#_q$ of a set of functions $F\#_1$ through $F\#_8$ and requirement $R\%_v$ is allocated to product entity AX. Further, it is decided that AX is going to be developed as a software entity. Then one of the scenarios to be analyzed will be UXhijk as suggested in Figure 12.8. Assume that we accomplish the dynamic analysis using sequence diagram UXhijk1 from which we derive requirement $R@_u$. What we are looking for is a way to establish hierarchical traceability between requirement $R@_u$ and some requirement in the set $R\%_1$ through $R\%_{10}$. The X, %, and # characters are being used to designate base 60 strings in this discussion that the author prefers to use to identify specific entities explained in previous chapters. We know that requirement $R@_u$ must be traceable to one of the 10 performance requirements allocated to classifier AX through the use of TSA and we can look at that list of requirements and select the one most closely related.

To make this selection more organized, we can form a u by v matrix, in this case a 10 by 12 matrix, and pair-wise compare the sets $R@$ and $R\%$. In Figure 12.9, you can see this whole process taking place. The 10 functionally derived requirements, are captured in the RAS mapped to the set of functions $F\#_1$ through $F\#_8$ and allocated to product entity AX. Based on these requirements, we build a context diagram for entity AX and analyze AX from the perspective of each of the three terminators shown. As an example, Use Case UX3 is extended to three use cases and we build three scenarios one of which, UX3111, is analyzed from a dynamic perspective

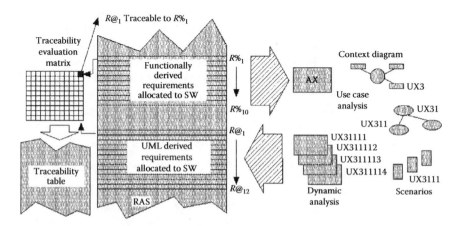

Figure 12.9 Requirements traceability across the gap.

along with the possible use of some combination of sequence, communication, activity, and state diagrams as well. Requirements $R@_1$ through $R@_{12}$ are derived from these analyses and captured in the RAS (possibly linked to the RAS database from a UML modeling application).

There are 10 requirements ($R\%_1$ through $R\%_{10}$) to which the requirements in the set $R@_1$ through $R@_{12}$ will have to hierarchically trace. We can build a 10 by 12 matrix and pair-wise analyze the relationships between the two sets of requirements, perhaps concluding that one of the matches is $R@_1$ traces to $R\%_1$. All of the matches are marked in the requirements management database table for traceability relationships. There are no known databases that provide the traceability evaluation matrix discussed here so it may have to be accomplished as a pencil-and-paper aid. However, we should be able to set the requirements management database filter for the two sets of requirements of interest aiding in the identification of the sets of interest for a particular case. In this example, there might be 10 or more sets of requirements like $R@$ derived from 10 or more dynamic analyses. In each case, an u by v matrix would be needed to pair-wise analyze the traceability relationships. In any case, it should be clear that we can have good requirements to modeling traceability and even good hierarchical traceability across the gap between performance requirements.

In both TSA and UML, we have discussed a decomposition process that partitions the problem space into parts in which the analysis is accomplished. Whenever we partition any whole we have an obligation to integrate and optimize across the boundary conditions thus created. The PIT must accomplish this integration work relative to the top-level IPPT and each IPPT with lower-tier teams must accomplish this work relative to its own immediately subordinate teams. Much of this integration work will take place at the interfaces ensuring that requirements on one end of an interface are compatible with those for the other terminal. Each team with subordinate teams, however, should also integrate across its immediately subordinate teams relative to the requirements derived at the subordinate team level relative to those at the parent team level. Part of this work can be accomplished by simply establishing the traceability between the requirements at the two levels. Another approach of value is to accomplish higher-tier function effects across the lower-tier team responsibilities. For example, one can inquire collaboratively into lower-tier performance of higher-tier functions like turning the system or entity on or off, moving from one major mode to another, accomplishing some kind of transfer function, or physical separation or joining of two entities.

Another kind of traceability can also be used to stimulate integrating results. This was pointed out to the author by an engineer at Puget Sound Naval Base in Bremerton, Washington. Given a requirement at level m, we can inquire if the intent of the requirement was fully implemented in the requirements for the n entities at level $m+1$ (downward). This kind of traceability inspection must await the development of the subordinate specifications, of course.

12.5 Tools integration in the near term

As the organization becomes more proficient in using SysML, it may phase out its use of TSA. The resultant changes in the integration and optimization work should be favorable and the aggregate work force should move closer to understanding the opposite side of the hardware–software divide with the integration-inhibiting peaks diminishing in altitude. Over time, the author believes that the Object Modeling Group and International Council on Systems Engineering will evolve UML and SysML into an even closer relationship moving toward what could be called UUML formed from these two model sets rather than the several that the author has cobbled together in this chapter. Ideally, this merger would also provide all of the artifacts needed to implement DoDAF suggesting that DoDAF may not be necessary as a separate modeling language only a particular perspective on a suite of models.

The enterprise should adopt a set of computer applications that collectively support all of the modeling work they apply while providing good traceability across the gaps. The latter is probably the most serious failing in available tools. Figure 12.10 suggests one near-term combination of applications using DOORS as the RAS within which traceability is established. DOORS is playing the role of what the author refers to as the big dumb database surrounded by some manually implemented analysis applications and computer modeling tools. The tool makers are working

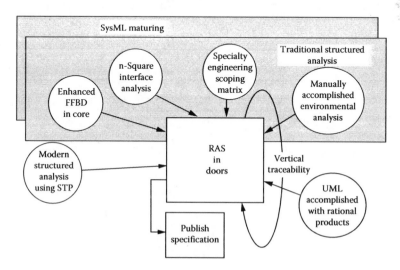

Figure 12.10 Tools integration.

closely with the evolution of SysML and will likely evolve some more effective integrated tool sets.

12.6   Next generation integrated development

Today, a development organization can conduct the software modeling activity shown in Figure 12.7 in coordination with a parallel system and hardware development activity applying TSA as shown in Figure 12.5. In this parallel development, the system team must integrate and optimize during lower-tier IPPT development of their assigned elements. At the system level, those who perform the analysis are reaching into a pool of modeling methods that the enterprise has selected as its norm and applied such parts of the whole set as they find useful for the particular modeling challenge offered by the program.

The author believes that the UML–SysML combination shown in Figure 12.11 still lacks the artifacts to cover specialty engineering and environmental requirements analysis. Figure 12.12 suggests an extension that might make the combination complete. The three-layer environmental model was briefly discussed previously as was the specialty engineering scoping matrix. For complete details, refer to *System Requirements Analysis* (Grady, 2005).

At one time, in the 1950s when software was a very young discipline, it happened that hardware and software analyses, to the extent that they were done, were accomplished using exactly the same model, flowcharting. Over time, probably encouraged by the ease with which flowcharts could be outputted onto line printers using ASCII symbols, computer software people got into the habit of building flowcharts in the vertical

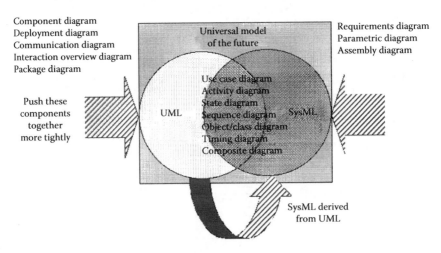

Figure 12.11 Modeling over the years.

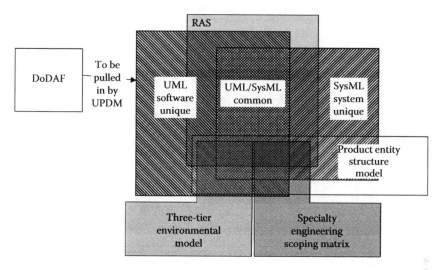

Figure 12.12 Augmented modeling suite.

rather than the horizontal axis still used by system engineering people in their functional flow diagrams. The activity diagrams of UML still reflect the vertical orientation, but it is really of little significance which orientation is used. The absolutely fascinating approaching reality is that system and software people will rejoin the same house in the near future. As UML and SysML become more fully integrated and truly comprehensive as suggested in Figure 12.13, we will achieve a tremendous milestone of universal unified modeling capability.

As we pass through this door into a world of integrated modeling and supporting computer applications, we will find it a more reasonably affordable task to integrate across the hardware–software boundary than has been the case for many years. But then as now, integration takes place in the minds of the system engineers working on the program. These engineers must be ever vigilant for inconsistencies between information

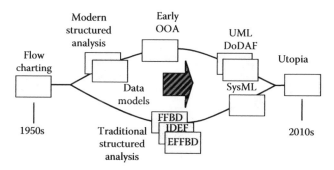

Figure 12.13 The approaching merge.

sets that signal that two different domains are not working from a common understanding of the problem and solution spaces.

12.7   The outputs

In the process of accomplishing the problem space modeling work we will have developed insight into three things of interest: (1) knowledge of the entities of which the system should consist, (2) knowledge of the relationships (interfaces) between these entities, and (3) knowledge of the requirements that apply to the entities and the relationships that should flow into specifications for the former and interface documents for the latter. Chapter thirteen picks up the discussion for item 1 and Chapter fourteen for item 2. For a complete story on the content of this chapter refer to *System Requirements Analysis* (Grady, 2005).

chapter thirteen

Product entity definition*

13.1 Structured analysis

The work described in this chapter is properly accomplished early in a program or the proposal leading a program to define and organize the product system and associate the development organization with responsibilities for the development of product entities. This work results from having accomplished problem space modeling as discussed in Chapter twelve. There is a strong flavor of process integration embedded in product entity definition, but the principal focus is on determining what shall be in the system in terms of hardware, software, people, and facilities. It is an example of cross function and product integration. Specialists from many functional organizations must cooperate in the task to develop the complete product structure of the system and the several different views of the system we will overlay upon the product entity structure.

The product entity structure flows from the structured analysis modeling discussed in Chapter twelve. It is expressed as a hierarchical relationship between everything that is in the system. The very top of the hierarchy is the system itself. The next layer down will consist of some number of major elements in the system. Each of these is further broken down into their components and this pattern continued to the lowest tiers of the system in each branch. When we begin to develop a new and unprecedented system, using the top-down structured approach recommended in this book, the only element of this hierarchy we know is the system that will accomplish the ultimate function, the need. During the early development phases we need to completely define the product entity structure by some organized means based on what we know the system must do.

Figure 13.1 illustrates a variation of the view offered in Figure 12.3 intended to emphasize the relationship between the requirements analysis and concept development processes shown in bold. As the problem space modeling work explained in Chapter twelve is accomplished, hardware and software entities are identified and added to the expanding product entity structure. Subsequently, work should be accomplished to define a preferred design concept responsive to the requirements also derived from the structured analysis. This may require a trade study to select a most advantageous

---

* The material in this chapter is from Grady, J., *System Integration*, CRC Press, Boca Raton, FL, Chapter 11, 1994. Used with permission.

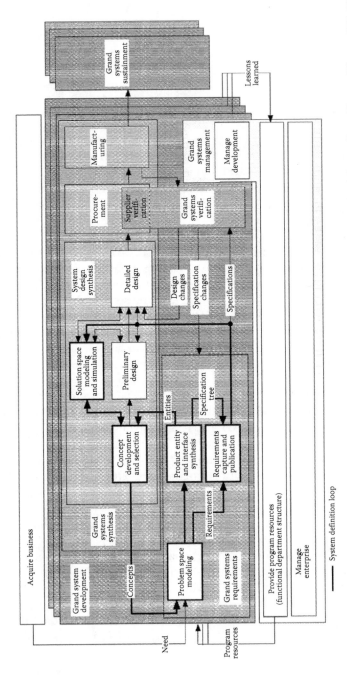

Figure 13.1 Concept development for product entities. (From Grady, J., *System Integration*, CRC Press, Boca Raton, FL, pp. 174–183, 1994. With permission.)

candidate as discussed in Chapter eighteen, but in any case, a preferred concept will then influence the lower-tier functional analysis. This study and concept work may require support from solution space modeling and simulation that hopefully reflects the higher-tier requirements already identified.

Cross-functional teams should be established, as discussed in Chapters six and nineteen, based on the evolving product entity structure starting with just the program integration team (PIT) that should start appointing IPPT leaders as the PIT completes the specifications for the top-tier entities subordinate to the system. The team that will accomplish the concept development work should be the team assigned responsibility for the parent item unless the new entity has been established as a new team on its own. Each IPPT should come into being at a time such that the team leader can be presented a specification for the item the team is focused on as well as the planning data for that item in the form of a contiguous portion of the WBS, SOW, and IMP/IMS. If the WBS is simply buttered onto the evolving product entity identifications, there will be perfect alignment between them, and the interface relationships will also align with the inter-team communications patterns needed to develop those interfaces. Thus, there should exist a closed-loop process for expanding the program knowledge of system composition linked to the continuing identification for the requirements appropriate to the system entities and their relationships.

The author would give an exception to the formation of a team to develop a software entity that is immediately subordinate to a hardware entity immediately upon identification that the item exists in the product entity structure rather than awaiting the full development of the item specification, unless the parent team includes people qualified to do the software analysis.

If the system has been previously developed and our purpose is to significantly modify or reengineer it, our work path should be a little different in the beginning, but the same end game is valid for those parts of the system that must be changed. We must understand what the new need (goal or purpose) is and then determine what in the existing system product entity structure remains useful in satisfying that new need. New or significantly changed functionality that cannot be satisfied by existing or modified resources will have to be satisfied by new resources that we can gain insight into from the methods discussed in this chapter and Chapter twelve. If, in the unlikely event the original analytical data is available for the system, we can begin by determining impacts based on the changes in the customer's need and then ripple those changes down through the analysis to find product entity impacts. In the more likely case, we will have no information from a prior system development cycle for the existing system even if an organized systems approach had been applied during its development. In this case, we should quickly recreate the analytical basis for the system using methods discussed in this chapter and Chapter twelve being alert to shut-off analysis on strings where we conclude no changes should be made.

One computer software development model called for the analyst to understand the existing physical model and then the existing functional model as a basis for then creating a functional and physical model for the modified system. One criticism of this approach was that no customer could restrain their interest in the new development long enough for the analyst to understand the old system from which they were trying to escape. This is a valid concern that the development team must be conscious of. Any attempt to recreate a proper functional system description should be carefully monitored against predefined goals for which there is an iron clad definition of completeness criteria.

Prior to discussing product entity definition, let us first introduce a rigorous top-down method for determining an appropriate product entity structure for a system. For more expansive coverage of this matter, refer to *System Requirements Analysis* (Grady, 1993, 2005). The structured analysis method common to the system requirements analysis process developed by major Department of Defense (DoD) weapons system program offices and aerospace contractors involves some kind of diagrammatic aid to understand needed system functionality followed by assignment or allocation of that functionality to things, elements, components, or items in the system. This approach can be applied effectively to any kind of system whether for commercial sale or DoD contract. It is effective for systems and hardware entities, but there are better models where it is known that the product will be completely composed of computer software. Figure 13.2 illustrates this structured decomposition process.

Perhaps the most traditional approach is to use a functional flow diagram as a tool to explore what exactly the system must do. Alternative methods (behavior diagramming as used with the Ascent Logic tool RDD-100, enhanced functional flow block diagramming as used in the Vitech tool CORE, or IDEF-0) are available that result in a richer understanding of the problem space at the expense of using a more complex model. Hierarchical functional analysis is a simple function outlining technique but it leaves much to be desired. As the analyst gains understanding about needed system functions, this functionality is transformed into performance requirements and allocated (assigned) to particular system elements responsible for satisfying that functionality. Where some doubt exists about the most appropriate allocation, a trade study can be performed to clearly reveal the best approach of two or more possible ones (hardware versus software implementation, for example). Please note that we are using the word function to denote an activity or capability of a system. Previously, we have used the word function to refer to the organizational elements on one axis of a matrix-managed company.

If the item is software, functional flow diagramming is not the ideal modeling approach even though that is where software requirements

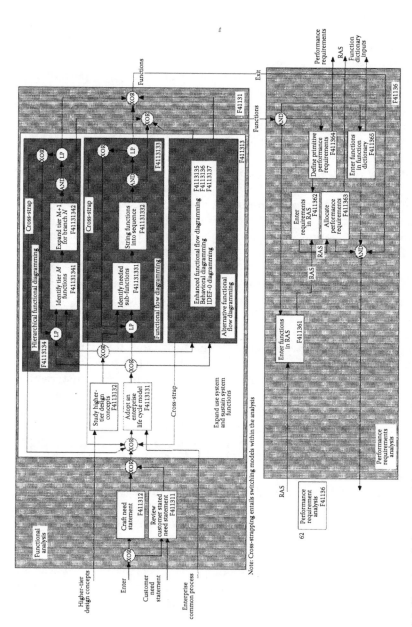

Figure 13.2 Structured decomposition process. (From Grady, J., *System Integration,* CRC Press, Boca Raton, FL, pp. 174–183, 1994. With permission.)

analysis began. There are several effective software development techniques principally unified modeling language (UML).

The functional allocations that result from this process become the basis for both the identification of the system product entity structure, through allocation, and precursors of performance requirements appropriate to the elements of that product entity structure. This is the heart of the structured development process. The act of allocating a performance requirement derived from a function to a system element places a demand on the responsible development team to transform that allocation into a design feature that is responsive.

The resultant system elements are assembled into a hierarchical block diagram, the top item of which is the complete system to which the top-level function, the need, was allocated. The method for assembling the evolving product entity structure into related family trees suggested here is to group them based on minimizing the cross-organizational interface relationships that are defined as interfaces between product elements where the organizations responsible for designing the terminal elements are different. It is exactly these kinds of interfaces that will result in the principal system development difficulty, so we can reduce the development difficulty by reducing these kinds of interfaces. We cannot eliminate them since they are responsible for the richness of the system, but those that remain should be very clearly known to the development team and responsibility very clearly established.

In Figure 13.2, you will note that the functional flow diagramming approach expands to provide increasing detail. Three major steps are noted, each further expanded on the diagram: (1) first we must identify needed functionality and link it into a sequence of blocks representing the functionality, (2) we expand each function into one or more fully quantified performance requirements, and (3) the identified performance must be allocated to something in the physical model, the product entity structure.

Three alternative flow-oriented decomposition approaches are noted in the figure but not expanded. Note that this process must be cyclically applied to lower and lower levels to fully define the problem space. As we allocate functionality to the solution space, the physical model, the product entity structure, the concepts developed for those items should be fed back to the decomposition process to refine the lower tiers of the functional analysis with higher-level solutions. This acts to tune the lower-tier functionality to the evolving solution and can greatly speed up the process. If done too zealously, this can also encourage a rush to point designs of the past. Pacing this process is an art form learned by doing.

When a function is allocated to computer software, the team responsible for developing that software will prefer to apply one of the many computer-aided software engineering approaches and tool sets to the further analysis of needed functionality as pointed out in Chapter twelve. That should, of course be permitted. In fact, while some system engineers will be shocked, the author believes that a team should be permitted to use any of the decomposition techniques discussed in Chapter twelve based on the kind of product and the experience of the team members unless the enterprise has standardized on a particular set. As a result, the complete decomposition and analysis may be composed of a mixture of techniques. Over time, enterprises should be trying to refine their system development modeling capability to narrow the models they find necessary or useful to the evolving combination of SysML and UML that hopefully will merge into a universal modeling capability.

In this chapter, we will explore only how to establish the system product entity structure and map it to other hierarchical structures of interest by the development team. In Chapter fourteen, we will pick up on the interface analysis process that is useful in refining the product entity structure along the lines of minimized cross-organizational interfaces.

13.2   Product entity structure synthesis overview

One of the principal products of product entity synthesis is the product entity block diagram, an example of which is illustrated in Figure 13.3. The top item is the complete system. This block can be allocated mindlessly from the top function, the customer's need statement. The next tier includes major elements of the total system. This pattern keeps breaking down to the lower levels of system composition. As noted, structured analysis generates allocations of needed system functionality to specific physical entities through trade studies, an appeal to historical precedent, good engineering judgment, and respect for customer direction on use of specific resources in the system. Customer furnished property is an example of the latter and this occurs when a customer has residual property from other systems they wish to make continued use of in a new or updated system.

On any given project, a PIT should be made responsible for the complete product entity diagram. The PIT should conduct the structured decomposition process down to the point where it yields allocations to elements that can be assigned to product-oriented integrated product and process team (IPPT). Once that assignment has been made, the responsible IPT should continue the structured decomposition process within the confines of the element(s) allocated to them.

The engineering community, in concert with the production and logistics leadership on the PIT and IPPT must decide in team efforts how to organize these allocations into families of things. They must map the evolving product entities to (1) organizations responsible for the development of each item (thereby giving birth to IPPTs), (2) the planned

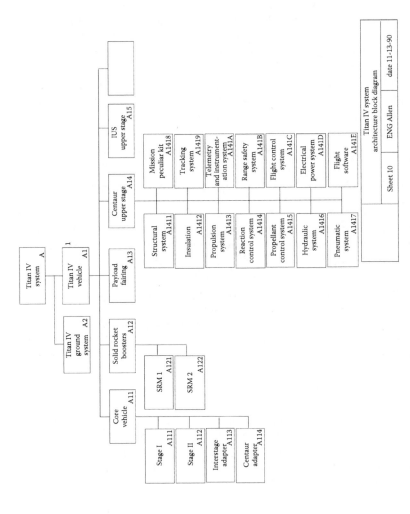

Figure 13.3 Typical product entity block diagram. (From Grady, J., *System Integration*, CRC Press, Boca Raton, FL, pp. 174–183, 1994. With permission.)

manufacturing process flow, (3) the customer's interests in managing the development process through configuration items, (4) make or buy considerations, (5) specification needs, and (6) the customer's WBS that will be used to organize and report contractor cost and schedule performance. All of these views of the system should have an influence on the final product entity structure synthesis.

Mistakes can be made in the process of allocating functionality to product entities. The system engineering component of PIT must continually evaluate the evolving product entity structure for suboptimum structure and interface relationships. Suboptimal structures will commonly result in reduced product system efficiency and development difficulties. The most difficult system development problems occur at the interfaces where functionality differences coincide with organizational responsibility differences. Optimum interface is defined as that interface condition characterized by a minimized interface count between elements under development responsibility by different design agents (different teams or contractors). Chapter fourteen covers this area of cross-organizational interface in detail.

The results of the interface analysis should be fed back into the product entity definition process to close the loop on this system optimizing activity. In the process, you may find many other changes will be triggered, but this is part of the iteration, or churning activity, that must occur as we become smarter about the system we are creating. We proceed from the simple to the complex, from the general to the specific. On the route, we may find that ideas we had when we were less knowledgeable were not the best ones. We must have the courage in this event to change those ideas as early as we possibly can since it is much less costly to do so early in a program than later.

The same people on the PIT responsible for functional allocation at the system level should also be responsible for product entity synthesis and system interface optimization. These three activities are closely related and offer synergistic and serendipitous opportunities where the same people are doing all three. If people with different reporting paths are responsible for parts of this work, they must be collocated physically and encouraged to interact to the degree that they effectively form one team.

Figure 13.4 illustrates the process we will be discussing in this chapter. The discussion opens with product entity block diagramming and follows up with mapping of this structure to several important hierarchical views of the system including WBS, manufacturing breakdown structure (MBS), drawing breakdown structure (DBS), customer configuration items list, and specification tree. Configuration items are identified by the DoD customer as the major elements through which a DoD customer would prefer to manage a development program. Other customers may refer to them as principal elements, end items, or contract end items (NASA). A commercial

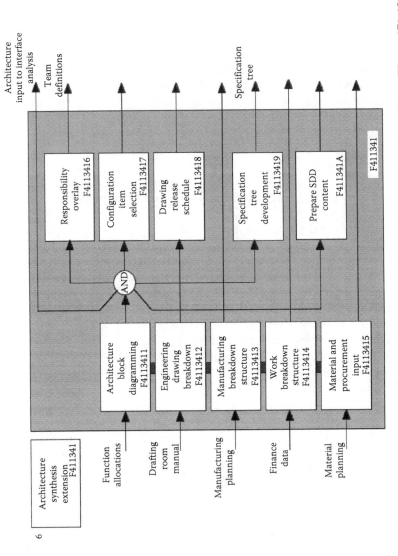

Figure 13.4 Product entity synthesis process. (From Grady, J., *System Integration*, CRC Press, Boca Raton, FL, pp. 174–183, 1994. With permission.)

products organization is, of course, unconstrained in deciding through which items it shall manage the evolving program but should pick some that allow them to focus on managing the important, driving aspects of the evolving system. During early program phases, the customer may ask the contractor to supply a candidate list of configuration items. We will, later in this chapter, describe a process for assembling that list.

The interface analysis block on Figure 13.4 will be covered in Chapter fourteen. But, note that we are suggesting that as the product entity structure unfolds, it should be reviewed (by the PIT) for interface appropriateness. This is the activity mentioned above where we are trying to minimize the number of cross-organizational interfaces. The product of the product entity synthesis process can be captured into a single baseline definition document for record purposes as illustrated on Figure 13.4 in the form of a system definition document (SDD).

The cross-organizational interfaces are predefined by two actions we take: (1) all interfaces are predetermined by the way we allocate functionality to the product entities and (2) entity responsibility is a function of the way we assign teams to the product entities. Together, these two actions predetermine where the cross-organizational interfaces will occur.

13.3   Product entity block diagramming

A product entity diagram is a simple hierarchical diagram consisting of blocks that represent the things composing the system and interconnecting lines that show parent and child relationships. Commonly, one item is composed of two or more subordinate items and this pattern is repeated all the way from the system block to the lower tier composed of things that will surrender to detailed design by small teams or procurement at that level.

In placing the things, identified through functional decomposition, on the diagram, we must consider many views and identify diagram overlays based on these different views as we proceed. The next section describes some of these different views that must be evaluated concurrently as we try to reach a conclusion on the best system organization. These different views will not always be in agreement on the best product arrangement, so the team must provide for a decision-making process that may entail trade studies where very difficult decisions are involved.

The PIT should be responsible for top-level product entity definition and this responsibility should be passed on to the IPPTs for items for which those teams are responsible. PIT should monitor lower-tier product entity and interface development and integrate those lower-tier components into the system product entity structure diagram.

Figure 13.3 uses a base 60 product entity modeling ID (MID) notation to avoid use of decimal points between levels. The decimal-delimited codes have the advantage of handling many items at any one level, but the decimal points do not sort well in simple database systems some people may be forced to use. The base 60 system uses the 10 Arabic numerals plus 25 capital English letters (capital "O" not used to avoid confusion with the numeral zero) and 25 lowercase English letters (lowercase "l" not used to avoid confusion with the numeral one). This system permits identification of up to 60 items at any one level that will accommodate most system needs.

13.4   Product entity overlays

13.4.1   WBS overlay

The WBS is a cost-oriented view of system structure defined for the DoD in MIL-STD-881. Large system acquisition customers manage program cost and schedule by associating all program work with a WBS. WBS elements are related to those responsible for performing the work through work authorization numbers and plans. The WBS includes both product and service elements. The only portion of that tree that we are immediately interested in here is the product portion. Figure 13.5 illustrates a typical WBS.

A complete WBS may include multiple structures differentiated by a prefix code. We might have one prefix for recurring work and another for nonrecurring, for example. One prefix might be established for the nonrecurring design activity and another for nonrecurring flight test activity. The prefix codes essentially establish a layered structure such that a given WBS number may appear in several prefixes. Cost can be collected in two axes this way. For example, a WBS 1300 might be for the structure of a launch vehicle. All cost corresponding to development of the design for the structure might be collected in WBS 123-1300, where 123 is a prefix for nonrecurring design development work. The work corresponding to structure testing might be collected in WBS 342-1300, where 342 is a prefix for nonrecurring test and evaluation. All work and cost related to the structure can be compiled by adding all of the WBS 1300 figures from all of the prefixes. At the same time, all nonrecurring work can be accumulated by adding all nonrecurring prefixes.

One of the most contentious problems in the early phases of system development is the imposition of a cost-oriented structure of the system onto the engineering organization that is trying to understand in an orderly way what the system should consist of. It is all very well for the finance world to develop an initial WBS, but they should recognize that the structure of the system should be defined in terms of its functionality not some arbitrary finance standard for which the finance computers happen to be programmed. The WBS should be adjusted concurrently

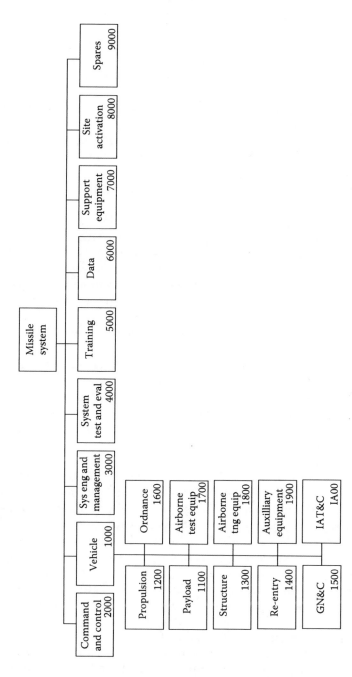

Figure 13.5 Typical WBS diagram. (From Grady, J., *System Integration*, CRC Press, Boca Raton, FL, pp. 174–183, 1994. With permission.)

based on the results of the process to build a product entity block diagram jointly by finance and the engineering community through membership on the PIT.

The reason that this is not always possible is that customer and contractor finance groups generally have access to cost models that are insufficiently flexible to change and that the inflexibility of the models is matched or exceeded by the people who staff these groups. This problem is an example of past foolishness in allowing functional organizations to build walls between themselves rather than cooperating in a teamwork environment. MIL-STD-881 was converted into a military handbook so that it could be called for guidance on government contracts but it could not be used to force contractors to employ the WBS structures contained in its appendices. Over a period of several years, DoD concluded that this was a bad change and in 2009, the document was returned to a standard status.

The engineering community should, nonetheless, assertively press its case for a functionally organized view of the system product entity structure and make an effort to force adjustments in the WBS to reflect this structure. If it is not possible to move the heavy weight of an inflexible WBS, be prepared to simply map the product entities to the WBS in the product entity dictionary. The Finance and Program Office people are right in that everything in the system must belong to some WBS in order to properly manage program cost and schedule, but a common identification scheme will result in simplicity that will encourage better management.

MIL-STD-881 includes several appendices that give suggestions for a WBS arrangement for different kinds of systems. These include deliverable product system elements as well as contractor service elements. The WBS seeks to embrace every possible program cost element. It is best if the WBS can do this so as to reflect the physical and functional organization of the system. Better still if the WBS can be used directly as the MIDs. One way to facilitate this result is to overlay the functionally derived product entity structure with the WBS assignments. Whenever you observe a crossing of the lines between the physical, functional, and cost (WBS) organizations of the system, you have a potential system organizational conflict. What can happen if this situation prevails?

A common problem in system development is a discontinuity between the functional and physical organization of the system. If you follow the pattern expressed in this book, the system will be decomposed into functional subsystems each the responsibility of a design group. Let us say by way of example that the WBS for a system matches the functional breakdown because it is clear that the functional subsystems line up with the way the contractor is organized and therefore the WBS will mate up well with reports generated by the contractor's cost schedule control system.

These functional subsystems must be integrated into one or more physical entities. If the design evolves to allow one or more subsystems to cross over the boundaries of system elements that form obviously separate physical components that are respected in the engineering drawing tree and manufacturing plans, it can result in great difficulty in determining the cost of the physical entities. If one or more of the physical entities is a configuration item (see next section) through which the customer manages the program, the difficulty can be increased for the customer.

For example, picture a space transport launch vehicle upper stage that is composed of two major physical end items, a main body and an equipment section. Now lay on eight functional subsystems with two of them wholly contained on the main body section, three wholly contained on the equipment section, and the remaining three including components spread between the two items. If the WBS is organized about the functional structure of the system, it will be very difficult to determine the cost of either one of these physical entities. It is important to avoid crossing of these physical, functional, and cost boundaries in elements that have a useful separate physical existence. The General Dynamics Atlas Centaur was actually built in this fashion.

Space transport launch facilities, oil refineries, and computer network installations provide other examples of this problem. These complexes may consist of facilities interwoven with functional systems for fluid handling, electrical control, and communications. The best way to avoid conflict in these cases is to recognize that the customer is taking delivery of two kinds of things: (1) facilities and (2) operating systems that are installed within these facilities in some pattern. The WBS can reflect these two sets with little difficulty even when the functional systems cross facility boundaries. In this case, the complete complex is an integrated whole that will always be used together.

13.4.2   Configuration item overlay

Customers like DoD and NASA will wish to especially identify certain items through which the complete program will be managed. In DoD, these items are called configuration items and in NASA, they are referred to as end items. The intent is to select items in a band across the product entity structure such that everything in the system is contained in one of the configuration items. Figure 13.6 illustrates this concept. The items corresponding to the shaded portions would be the subject of major reviews held by the customer that occur at major program milestones. The IPPT structure should be coordinated with this arrangement to simplify and avoid program control problems.

The customer may select high-level items where they wish to manage the program at a high level or detailed items where there are many new development items and they have the program management resources to

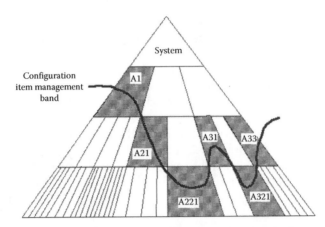

Figure 13.6 Configuration item management band. (From Grady, J., *System Integration*, CRC Press, Boca Raton, FL, pp. 174–183, 1994. With permission.)

cover the details. A very complex system involving new technology may have up to three layers of configuration items. Refer to *System Requirements Analysis* (Grady, 1993, 2005) for detailed coverage of a comprehensive item selection logic.

13.4.3   Specification tree overlay

The specification tree defines the elements in the product entity structure that require formal documentation of their requirements in a customer specification format, procurement specification format, or in-house format. The identification of configuration items carries with it the need to write specifications that will be contractually binding and require formal verification of product compliance with them. Specifications are the repository for the output of the requirements analysis process for a given item. The requirements content of specifications is what holds the whole structured development process together. They ensure that the solutions to the many small problems into which we decompose the ultimate function, the customer's need, will fit together to solve the system problem. This is attained through traceability of requirements and the specification tree indicates the principal traceability paths that must exist.

Figure 13.7 illustrates a typical diagram. In this example, the blocks are shaded to tell what kind of document and the block contains the specification number in the lower right corner of each block. It is useful to show the top-level specification tree in diagrammatic form, especially down through the level where all customer specifications are included. At lower levels, it is better to use a tabular list to avoid the cost of maintaining the graphics. If you have a product entity database, you can include specification numbers in another field and arrange for an indentured output report.

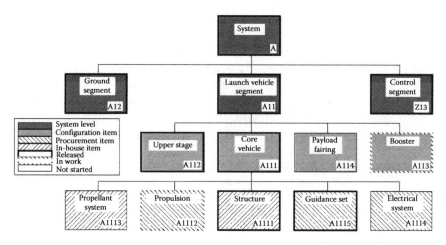

Figure 13.7 Typical specification tree. (From Grady, J., *System Integration*, CRC Press, Boca Raton, FL, pp. 174–183, 1994. With permission.)

The specification tree will show three different kinds of specifications. First, and commonly at the top of the product entity structure, we will find the customer specifications (system and configuration or end item specifications). We will also wish to clearly define the requirements for any items that we will procure where those items will require some development and testing. These requirements should be captured in procurement specifications. Finally, there will be some items that we will accept responsibility for the development of and also conclude that we should define their requirements prior to performing design work. These we might call in-house requirements documents.

You may have to have two versions of a specification tree. A DoD customer is mainly interested in the tree showing only system, segment, and configuration item specifications. The contractor may want an augmented tree with the other specification types indicated as well. In any case, the specification tree should be an overlay of the product entity diagram. There should be nothing on the specification that is not on the product entity diagram.

13.4.4   Manufacturing, procurement, and engineering breakdown structure overlays

The manufacturing, procurement, and DBS activities feed off from the product entity block diagramming activity. In the manufacturing and procurement cases, we seek to influence the organization of the system into major components that neatly align with our manufacturing plant and capabilities and our procurement plans.

Manufacturing will wish to be able to identify the assemblies that are worked upon at its workstations in terms of specific engineering drawing numbers in the simplest possible way. We want our production line to have to deal only with physically coherent wholes where possible, and avoid situations where a manufacturing workstation has to work with items only partially covered on two or more drawings. Ideally, the engineering drawings should map to the workstation activities view of the product in a simple one-to-one relationship. The more complex the relationship between these two views of the product entity structure, the more difficult the integration task during production.

The PIT manufacturing engineer can achieve this desired simple relationship by planning the manufacturing facilitization concurrently with product design development and influencing the product entity structuring accordingly concurrently with engineering efforts to map the product entity structure to the engineering drawing structure. Manufacturing membership on the PIT must define exactly how the system will be manufactured. What facilities will be used? How will these facilities be allocated to particular production processes? What workstations will be included within the process? Answers to these questions will encourage a particular physical organization of the product.

The product entity structure should also evolve with conscious identification of the results of the make-buy decision process. Each item to be procured should have a separate engineering drawing at the top level and appear on the product entity diagram as a single entity at some level of indenture. When this condition exists, procurement can contract with a supplier with a very simple drawing. All of the notes and data on the drawing should apply to the procurement. When a mismatch is permitted to develop in this area, procurement must fabricate special instructions that may even countermand engineering drawing instructions. This leads to complexities in the manufacturing process because a part may not comply completely with the engineering when received from the supplier. Special manufacturing planning may be required to perform steps on purchased parts before they are installed.

The manufacturing and procurement breakdown structuring of the product should be two driving forces balanced with a third being minimizing cross-organizational interfaces. Once the product entity structure is screened for manufacturing breakdown, it can then be used as the basis for the engineering DBS. If you come out of this activity with minimized cross-organizational interface and consistency between the engineering drawing structure and the manufacturing process, you are well on the road to a money-making program that produces an item that will be highly valued by your customer.

Clearly, it requires procurement, manufacturing, and design knowledge concurrently applied in a team environment to properly structure the system product entity structure. Finance and configuration management also must cooperate for WBS, configuration item, and specification tree overlays. With all of these people from specialized groups involved, it also makes sense to insist on a system engineer to coordinate the overall product entity structuring ensuring that all of these views are respected and conflicts are resolved based on sound logic derived from the conflicting inputs from the several specialized system views. It is precisely this conversation, possibly very intense in nature, that must take place in block F4113411 of Figure 13.4.

13.4.5 IPPT assignment

We have saved the best until last. The point is made throughout this book that the teaming arrangement should align with the product entity structure. Task F4113416 in Figure 13.4 is exactly where this happens. The reason this is important is that it causes perfect alignment between the IPPTs and the planning data collected in the WBS, SO, IMP, and IMS assuming the recommendations included in Part two are complied with. But there is another important result that will be touched upon in Chapter fourteen. We will have perfect alignment between the cross-organizational interfaces that have the potential for causing great problems and the communication paths that must be working effectively between the teams.

chapter fourteen

Interface definition*

14.1 Interface analysis

Interface development is a core system engineering activity because of the fundamental development technique employed for decomposing large problems into many related smaller ones. The small problem solutions will have interfaces between them and these interfaces must have solutions on both terminals that are compatible as well as being consistent with the functionality allocated to the terminal entities. All product entities should have only a single party assigned responsibility. Interfaces commonly will have two or possibly even three parties responsible for their design. The difficult part of interface development is that each of the two terminals may be the responsibility of a different person or team. Clearly, the work of these people or teams must be integrated in an application of cross-product integration. This chapter was derived from *System Requirements Analysis* (Grady, 1993) and updated in a book by the same title (Grady, 2005).

Interface analysis is one of the methods by which we arrive at an optimum product entity structure for a system before we commit to its design. We define precisely where the interface planes are in the system, who is responsible for them, and ensure that the principal engineers responsible for the interface terminals and the media all have precisely the same understanding of the requirements and design concept for each one. In the process, we may affect changes to the system product entity structure derived from allocation of system functionality and influenced by the several overlay views discussed earlier. The desired goal is to refine the match between how the design team is organized and how the product system is organized in order to simplify the system development process and assure that we have accounted for all needed system functionality.

This book encourages the use of n-square diagramming techniques augmented by specialized analytical techniques corresponding to the several engineering disciplines (avionics, fluid dynamics, and so forth) to determine the optimum interface relations between system elements. Generally, the analyst should seek to maximize the capacity for interaction between system elements while minimizing the system need to interact. Generally, this results in minimizing the number and complexity of system interfaces.

---

* The material in this chapter is from Grady, J., *System Integration*, CRC Press, Boca Raton, FL, Chapter 12, 1994. Used with permission.

Given that we have identified an optimum interface condition within the system through interface analysis, we need to communicate the results of that analysis very effectively to all development team personnel to ensure they are all working to this model across their own portions of the model. Schematic block diagrams (SBD) and interface dictionaries can be very effective in satisfying the communications objective. Schematic block diagramming is a graphical methodology for reporting upon the preferred interfaces defined through interface analysis. An interface dictionary simply lists every interface and correlates them with other system information.

14.2   Interface defined

An interface is a plane or place at which independent systems or components thereof meet and act or communicate with each other. An interface is characterized by two terminals, each touching one element in the system product entity structure and possessing a medium of interaction. An interface is completed between these terminals via an interface media such as physical contact, electrical signals in wiring, fluid flow in plumbing, or a radio signal in space. The interface media is provided by either an element of the system or the system environment. The interface is not the media itself, rather the functionality facilitated by the media.

For example, as a function of being bolted together, two physically interfacing components will remain attached. As a function of being connected by a wire harness, an aircraft on-board computer may send a command signal to the aileron actuator of an aircraft that causes the aileron to move interacting with the air mass of the environment, in relative motion with respect to the aircraft, to roll and turn the aircraft (swept wing aircraft with high degree of yaw-roll coupling assumed) such that the command signal is nulled out in the on-board computer as a function of the guidance set detecting approach to the direction commanded.

It is common practice to associate an interface with one of three types: functional, physical, or environmental. A physical interface involves the form and fit of mating parts. Examples of physical interfaces include mounting bolt patterns, drive shaft flange connections, mating wire harness connector physical attachment, and the tires of a fighter plane resting on the tarmac.

Some examples of functional interfaces are a 28 V DC signal passing from a solenoid driver in an on-board computer to a valve solenoid in a pressurization control unit, a digital data stream flowing from an instrumentation control unit to a flight data recorder input port, and the flow of liquid oxygen from a rocket propellant system to the engine.

Environmental interfaces can exist between two items when the natural environment communicates environmental stresses between them. A reconnaissance camera company once argued that one of their cameras could take excellent pictures through the engine exhaust of a Teledyne Ryan Aeronautical unmanned reconnaissance aircraft. If this were true it would have made it possible to outfit a very inexpensive unmanned photo reconnaissance bird. Unfortunately, no one questioned that the environmental interface provided obstruction offered by the jet engine exhaust between the desired imagery and the camera system would preclude good imagery before money had been spent to prove the existence of an irreconcilable environmental interface incompatibility. In general, however, environmental stresses are treated as a separate kind of relationship in a system referred to as an environmental relationship even though they can be perfectly well described as an interface.

14.3   The interface dilemma

Included in the definition of a system is the existence of two or more elements of a system that cooperate via interfaces. It is through these interfaces that a system attains its superiority over an unorganized collection of things. We must conclude that a system must have some interfaces. The richer the interface complexity, the greater the potential for synergism between the components.

We know from Chapter thirteen that we must decompose system functionality into system elements that can be designed by teams of specialists. We commonly organize our engineering departments according to how we apply these specialized engineers to designing elements in our company's product line. It is a fact that the majority of system problems will occur at the planes in a system where different specialized organizations are responsible for the opposite terminals of an interface. The way to minimize these problems is to minimize the number of such interfaces to the degree that it is possible. It also helps a great deal to disconnect the functional organization from the program organization by organizing the program by product-oriented teams we have called IPPT.

We must have interfaces between elements designed by different agents yet we desire to minimize these interfaces. This is perhaps the most important task of a system engineer or the PIT, to control the development of the interfaces in a system to satisfy these conflicting demands. Obviously, the solution will be a compromise between two extremes, a rich interface in a system that will drive you to the poor house to develop and an uncoordinated collection of independent things that does not satisfy the definition of the word system.

14.4   The solution

The principal technique for arriving at a reasonable solution to the interface dilemma is careful development of the product entity structure to respect the known design team organizational boundaries. With some careful thought we may be able to allocate the functionality and aggregate responsibility for the product entities so as to minimize the need for cross-organizational interfaces. As the product entity structure is created we must continue to analyze its expanding interface implications within the system itself and between the elements for which the design teams are responsible in order to assure that we remain in a condition of optimum interface definition.

We will describe two diagrammatic tools for the development of a clear definition of the interface relationships needed between the system elements defined on the product entity structure diagram. These tools may be used together with n-square diagrams used as an analytical tool and schematic block diagramming used to publish the results. Either tool can also be used to accomplish the complete interface definition task, the analytical and the exposition portions. The reality is that they present precisely the same information to the viewer. We will also find that an interface dictionary implemented with a computer database can be very useful in controlling the development of interfaces throughout the system.

Our methods will focus on producing some combination of three work products:

1. An n-square diagram that identifies interface relationships between a set of system elements in a way that forces us to consciously consider every possibility.
2. An SBD, that defines system interfaces. It uses simple blocks and interconnecting lines to define the existence of interface elements and their terminal relationships.
3. An interface dictionary that is a tabular listing of system interface elements illustrated on the SBD (one tabular line item for each line on the SBD) containing the element name, a description of the element, its two terminals and media identification from the product entity dictionary, and other reference information too extensive to place on the face of the SBD.

14.5   n-*Square diagramming methods*

An n-square diagram is a square matrix with size N (N cells on a side). The corresponding units of each axis of the matrix are associated with the same set of product entities aggregated into families that are under the same organizational responsibility. Our interest could be between systems in a system of systems, between subsystems within a particular system, or between the components in a subsystem as a function of the situation we find ourselves in.

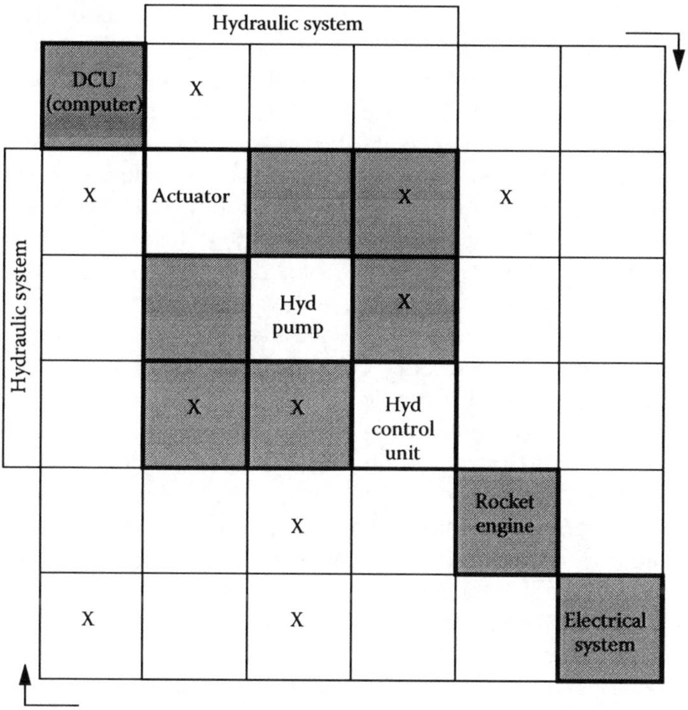

Figure 14.1 Typical n-square diagram. (From Grady, J., *System Integration*, CRC Press, Boca Raton, FL, pp. 185–206, 1994. With permission.)

The diagonal corresponds to internal interfaces for each product entity covered by the diagram. The other squares correspond to interface opportunities between the elements. Figure 14.1 illustrates such a diagram. The squares are marked with an "X" where there is an interface between the elements identified on the two axes. You will note that there are two squares, one above and one below the diagonal, for each pair of elements. We choose the cells above the diagonal for interfaces that have their source from the right face of a diagonal square higher up on the diagonal and those below the diagonal for those that have their source on the left side of a diagonal square appearing lower on the diagonal. There is nothing wrong with making the opposite selection. Just be sure that you use the same convention throughout the analysis. Note the arrow at the top right corner of the matrix indicating clockwise rotation applied in this analysis

As we can see in Figure 14.1, there is an interface requirement between the actuators that are a part of the hydraulic system and the DCU (on-board digital computer unit) that develops the command signal. There are also interfaces between the actuator and the hydraulic control unit that provide hydraulic power to the actuator in response to the control exercised by the DCU. We can use this diagram to evaluate the functional allocation choices we have made against the organizational responsibilities.

Let us assume that each of the subsystems we have identified on the diagram is the responsibility of a separate IPPT. Therefore, the shaded squares on the diagonal represent interface situations between system elements within an IPPT responsibility. There is only one design agent responsible for both terminals of these interfaces and they will tend to be properly developed with little outside integration by the PIT or system engineering community. Any interfaces identified outside of these shaded blocks with an "X" in this case represents a cross-organizational interface of the kind we wish to minimize and the kind that makes our system a system.

As one means of reducing cross-organizational interface, we can reassign a product entity from one subsystem to another based on the off-diagonal interface count. This is done on our n-square diagram by moving a row and the corresponding column to a new location in the matrix such that the component in that row and column are now located within a different subsystem. The interfaces will automatically switch to the new configuration. In this case, we have not changed the functionality allocated to the system elements, we have reordered the elements with respect to the IPPT organization.

We may also change an existing interface condition by revisiting the functional allocations. For example, we could conclude in the case of the actuator that we should have allocated the steering function to a separate steering propulsion system that does not require actuators, only exhaust jets that are turned on and off under the control of the on-board computer. In the process we would eliminate the hydraulic system if it is not required for any other function.

Alternatively, we could change the design concept from a hydraulic actuator solution to an electrically driven actuator solution. Now the complete control system is in the avionics design domain, the computer and the actuator. We may have removed the need for a hydraulic system as well.

These three approaches are effective in refining the system product entity structure to simplify system interfaces while assuring that we are satisfying the needed system functionality. We will expand on the n-square diagramming process in a moment and describe a companion technique called schematic block diagramming. But first, let us explore a troubling but interesting question about our organizational priorities.

Figure 14.2 offers a compound form of the n-square diagram. To make an n-square diagram the analyst makes a square marked off on each side or down the diagonal with a space for each system entity under consideration. These spaces are then annotated with the product entity ID or element names. The analyst then marks the intersections within the square to note the required interfaces between the elements. The diagonal represents the internal interfaces required for each of the entities and need

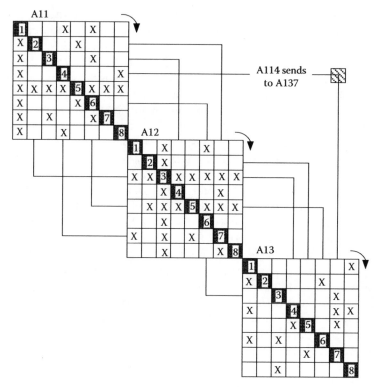

Figure 14.2 Compound SBD. (From Grady, J., *System Integration*, CRC Press, Boca Raton, FL, pp. 185–206, 1994. With permission.)

not be explored in this analysis. We may have to subsequently define the internal interfaces in another n-square diagram that shows the lower-tier entities on a separate n-square diagram.

Figure 14.2 is a compound n-square diagram in that it is illustrating more than a single level of interface relationships. At first glance it is easy to see that there may be a degree of ambiguity in this matrix since it includes two intersections for each pair of product entity elements. However, this allows us to use the matrix to define interface directionality as well as existence as noted earlier. You have to define which half corresponds to which direction. Establish a convention and stick with it throughout any particular analysis. In Figure 14.2, clockwise has been selected by the arrow at the top right corner of the squares. It happens that all three of the subsystems illustrated in Figure 14.2 are composed of eight subordinate entities but that need not be the case, of course. A11 could have 3 entities, A12 could have 5, and A13 could have 11.

The method of noting the existence of an interface in the intersections of the matrix can be as simple as marking the intersection with an "X," using a code with an explanation in accompanying list format, or by writing cryptic notes within the intersections that describe the interfaces needed.

The interfaces within the major entities are broken down into their components on the diagram. The external interfaces between the subsystems are shown as lines such as the one annotated with the remark "A14 SENDS TO A37." Entity A124 could conceivably be expanded to show its five included entities on the same diagonal. You could conceivably expand this to include three or more levels, but the complexity of the diagram may quickly overwhelm the clarity of its message.

Someone might complain that we should use the same convention for defining all interfaces. The interface sending something from A13 to A37 could be indicated by enclosing all of the three squares within the same large square such that we could mark that interface with an X in the cross-hatched square shown. In so doing, however, you would loose a very emphatic message. Assuming that we have assigned A11, A12, and A13 to three different IPPTs, each of which can be held completely responsible for their internal interfaces marked with Xs, then these lines represent the most difficult interfaces of all to develop, they represent cross-organizational interfaces that we will discuss in more detail later. The lines jump out of the page at us screaming, "MANAGE ME BECAUSE I CAN REALLY MESS YOU UP."

14.6   Schematic methods

14.6.1   Schematic symbols

System SBD are structured from very simple symbols, blocks and lines. The blocks are the objects, and only the objects, illustrated on the product entity diagram and the lines that join them represent the interfaces between the objects. Figure 14.3 defines all of the symbols used on the SBD.

Interface lines are either of the bundled or elementary type. At the higher levels bundled lines are used denoting the existence of an interface requirement between two system elements. These lines do not include arrows to denote direction because they very likely include elementary interface elements of many kinds involving both directions. An elementary line, which represents a single specific interface, should show direction with an arrow at one end (unidirectional) or both ends (bidirectional).

At the lower tiers where an interface element represents a single elementary interface, interface element names may be included on the line. Also at the lower tiers, different line types may be used to illustrate different classes of interface such as solid lines for electrical, long dashed lines for pneumatic, and alternating dashed lines for fluids, for example. This diagram should be created by the system engineering community or

Blocks

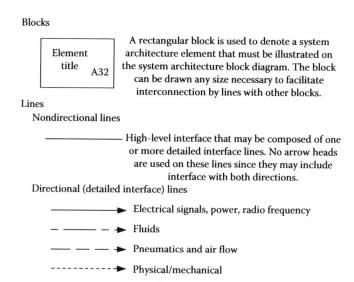

A rectangular block is used to denote a system architecture element that must be illustrated on the system architecture block diagram. The block can be drawn any size necessary to facilitate interconnection by lines with other blocks.

Lines
Nondirectional lines

High-level interface that may be composed of one or more detailed interface lines. No arrow heads are used on these lines since they may include interface with both directions.

Directional (detailed interface) lines

Electrical signals, power, radio frequency

Fluids

Pneumatics and air flow

Physical/mechanical

Figure 14.3 SBD symbols. (From Grady, J., *System Integration*, CRC Press, Boca Raton, FL, pp. 185–206, 1994. With permission.)

PIT in order to ensure continuity. For example, left to their own devices, the specialized design disciplines will each use the solid line to represent the line they have to use most often because it is easier to draw than any other kind. On electrical SBD you will see the solid line for wiring and on hydraulic and pneumatic schematics it will represent plumbing. Each specialized discipline has its own component symbols as well. The system SBD should use simple blocks to represent the system items.

14.6.2   Schematic symbols schematic discipline

It is vital that only those blocks used on the product entity diagram be used on the SBD. If in the functional analysis process, one finds that it is not possible to satisfy system functions currently allocated to system elements as a result of constructing the SBD and the solution is to add a system element, then the analyst should cause the object to be added to the product entity structure before it is used in the SBD. The interface planes on the SBD must accurately reflect the system product entities and this condition will be realized by respecting this discipline.

14.6.3   Schematic symbols interface coding

Each interface can be coded for unique identification in computer databases. The author uses the same codes applied to the product entity ID system. The complete set is identified as "I" that need not be used in cases where the context is clear, but the author retains it for use in computer databases. At the top level the interfaces are numbered I1, I2, I3, ..., In. At the next level each of these interfaces are broken down to the next level as I11, I12, I13, ..., I1n. This process is continued until the component product entity structure level is reached in the SBD.

A base 60 system is used here employing the 10 Arabic numerals, 25 capital English alphabet letters (O excluded), and 25 lowercase English alphabet letters (l excluded). The number of characters in an interface ID indicates the interface level. Within each level and branch, up to 60 different interfaces can be coded using this system. If more than 60 are needed they can be broken up into subclasses each no more than 60 in number.

Some people prefer a decimal-delimited ID system allowing an unlimited number of identifications within any one level. The code 1.10.4.17 is an example of this method. The problem with this ID system is that it does not sort and index properly in simple computer database systems available for microcomputers. The ID 1.10.4.17 will appear before the ID 1.2.5.7 because a simple database implementation sorts character-by-character using the precedence established for ASCII and the 1 in 10 of the first ID has a higher priority than the 2 in the second ID.

The work-around often used to make the decimal delimitation system work is to always add spaces to each level corresponding to the maximum number of characters allowed at any one level. For example, we may conclude that there will never be more than 99 elements at any one level. So we would enter the ID 1.10.4.17 into the computer database as _1.10._4.17, where the "_" symbol denotes a spacebar entry. This will sort properly but it is tedious to keep the books straight.

14.6.4   Schematic symbols interface coding ultimate SBD

The ultimate, or level zero, SBD is exactly the same for every system and is illustrated in Figure 14.4. Every system interacts with its environment to achieve its top-level function (expressed in the system need statement). That environment includes the natural environment and may include hostile systems that the system is intended to destroy or avoid, cooperative systems that provide the system with useful services, and noncooperative systems that may adversely interact with the system unintentionally (by creating electromagnetic interference, for example).

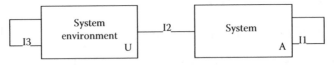

Figure 14.4 Universal ultimate SBD. (From Grady, J., *System Integration*, CRC Press, Boca Raton, FL, pp. 185–206, 1994. With permission.)

The system environment is labeled with a product entity ID of "U" and the system with a product entity ID of "A." The interface between the environment and the system is labeled here with a level one Interface ID "I2" and the system innerface (internal interface) with a level one Interface ID "I1." In order to reduce the number of levels in the Interface ID, the analyst may choose to assign level one Interface ID to the major interfaces within the top-level system elements as well as to the environment interface. The use of the letter "U" for the system environment relates to it consisting of everything in the universe other than the system in question. In a logical sense it is really U-A.

Figure 14.5 illustrates a typical top-level SBD for a space transport system. As you can easily see, this diagram could be fitted into the form of Figure 14.4, which is nothing more than a systems engineering curiosity. The top-level SBD is created by laying out each of the top-level system product entities (as a minimum, those entities immediately subordinate to the system block on the product entity block diagram) on a sheet of paper and connecting pairs of them appropriately with interface lines. This diagram is for the Atlas space transport system ca. 1990.

Lower-tier diagrams must expand the interfaces defined on the top-level SBD by creating an expanded SBD for each interface line on the diagram. This process is carried from the top downward progressively in step with the advancing product entity definition. The product of the interface analysis work should be fed back to help define the expanding product entity responsibility assignments in terms of their interface development responsibilities.

Note the use of the lines that are attached at both ends to the same block. These interface lines signify the internal interface for that item. These lines correspond to the diagonal on an n-square diagram. It is useful to illustrate them on an SBD as a part of the discipline in assigning interface ID codes. The internal interface is coded at the same level as other interfaces at that level. If an SBD is developed for that item, all of the internal interfaces have the prefix of the internal interface ID. This is the gimmick that assures that interface ID codes can be expanded from the top down in step with the expanding product entity structure.

Every interface in a system can be said to have a source, a destination, and a media (wires, pipes, attaching bolts, etc.). The intensity of interface problems in a developing system are directly proportional to the percentage of system interfaces that possess development organization responsibility differences among these three interface aspects of source, destination, and media. Therefore, in years gone by one of the principal objectives in laying out the system product entity structure was to structure the functional elements of the system, driven out by the functional analysis, into subsystems that were cleanly related to the engineering design organization structure. Given that you are using cross-functional

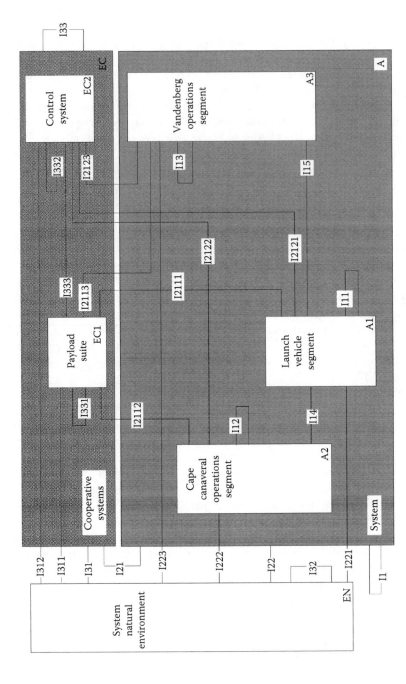

Figure 14.5 Typical system SBD. (From Grady, J., *System Integration*, CRC Press, Boca Raton, FL, pp. 185–206, 1994. With permission.)

teams that are coordinated with the product entity structure, it is not necessary to follow this pattern today.

Interfaces that have different design agencies responsible for source, destination, or media are critical interfaces that the author refers to as cross-organizational. These interfaces must be tracked by system engineers to insure they are properly developed by the two or three agents responsible. Those interfaces that involve different associates or teaming partners should have interface control documents (ICD) prepared to manage the interface development. Interfaces that have all three aspects (source, destination, and media) under a single design agent's responsibility are noncritical and will be the complete design responsibility of that same design agent.

This objective is satisfied by the concurrent development of the product entity structure and the SBD under the responsibility of the PIT. As elements are added to the product entity structure from functional analysis action, they are fitted into the SBD with a watchful eye on alignment between interfaces and any existing team boundaries. This may be done by coloring the SBD blocks different colors for the different design organizations, by overlaying the design responsibilities onto the SBD with dashed boxes, or by simply encircling elements under each team's responsibility with a line of a particular color (color computer graphics and automated identification of critical interfaces could be a great help here).

The analyst then takes advantage of interface insights provided by the annotated SBD to help guide placement of product entities in the product entity structure so as to result in simple interface relationships between the responsible development teams.

It is of crucial importance that the analyst respect the discipline that the SBD uses no blocks that are not already identified on the product entity structure. The reason for this is that product entities should have assigned to them organizational development responsibilities, which, in turn, can be used to define interface development responsibility pairs. If these are not clearly connected, then management of the development process will suffer.

14.6.5 Schematic symbols interface coding ultimate SBD expansion

Later we will refine our definition of system interfaces to identify three subsets—innerface, crossface, and outerface—as a function of the principal engineers responsible for product development through the assigned teams. For the time being let us just say that innerface includes those interface elements that have both their terminals at the same product entity. A crossface interface element has two different product entity terminals. It is useful to have a name for interface elements that are of no interest to a particular development agent, thus the outerface class where neither interface terminal touches a particular entity.

The top-level SBD is expanded in step with the expanding product entity structure diagram. There are two principal expansion approaches: (1) innerface expansion, and (2) crossface expansion. These two subsets of the complete interface set for a system encompass all system interfaces, so it is unnecessary to focus special attention on the third class of interface called outerface. More on this method of classifying interfaces momentarily. Just accept these as names for interface classes for now.

14.6.6   Schematic symbols interface coding ultimate SBD innerface expansion

Innerface is identified on any SBD as a line with both ends on the same block. This line will be coded like any other with an Interface ID. To expand one of these innerface elements, the analyst should prepare an SBD sheet in either the block diagram or triangular matrix form that illustrates each of the system elements subordinate to the block that has both ends of the line attached. The subordinate blocks are derived from the product entity diagram. The analyst will find that initially it may be useful to simply make a circle on a piece paper for each block. Then draw a line joining each pair that must have an interface. Add an innerface line to each block. Add the Interface ID to each line in an arbitrary fashion by appending 1, 2, 3, through z (using our base 60 system) to the Interface ID for the lines expanded in this SBD.

From this preliminary sketch, an example of which is shown in Figure 14.6a, it will be possible to see a pattern of lines and blocks that can be arranged in an artistically pleasing as well as technically accurate way. The final illustration, shown in Figure 14.6b, should have as few line crosses as possible. Prepare the final diagram on a computer using a graphics package or using templates and a pencil.

If the rough sketch becomes too cluttered consider using a triangular matrix format to illustrate the SBD. The triangular matrix format corresponding to the interface shown in Figure 14.6 is illustrated in Figure 14.7. It is formed by simply listing the architecture IDs for the set of elements on both axes. For each intersection corresponding to a needed innerface, mark the square with an Interface ID expanded from the original Interface ID.

14.6.7   Crossface expansion

A slightly different technique is necessary for expansion of crossface lines on an SBD. A crossface line will, of course, always have its two ends on different blocks. To expand one of these lines draw a rough sketch with

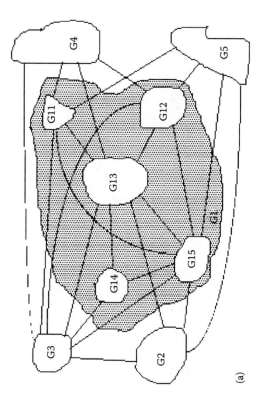

Figure 14.6 Innerface SBD expansion: (a) primitive SBD. (From Grady, J., *System Integration*, CRC Press, Boca Raton, FL, pp. 185–206, 1994. With permission.)

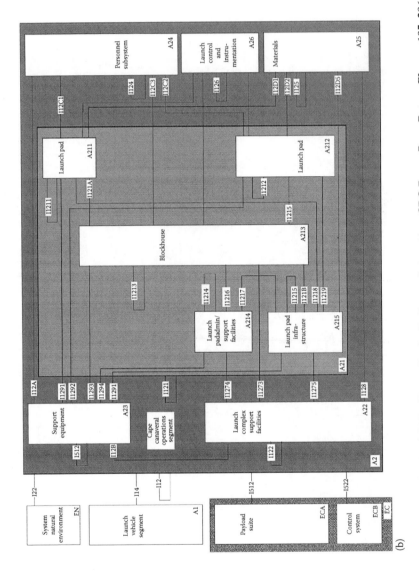

Figure 14.6 (continued) (b) finished SBD. (From Grady, J., *System Integration*, CRC Press, Boca Raton, FL, pp. 185–206, 1994. With permission.)

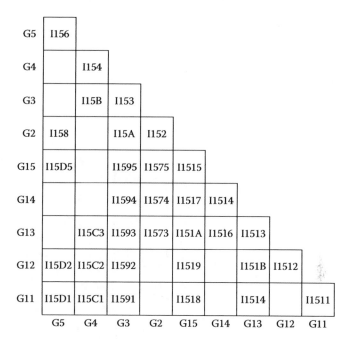

Figure 14.7 Triangular matrix SBD example. (From Grady, J., *System Integration,* CRC Press, Boca Raton, FL, pp. 185–206, 1994. With permission.)

all of the blocks subordinate to each of the two parent blocks illustrated on one side of the sheet. One way to get this sketch is to use the innerface expansions for each of these elements, if they have already been prepared. Next connect pairs of these blocks, one end on a subelement block on one side of the page and the other end on a subelement block in the other side of the page.

When all of the line pairs have been entered, number them all expanding on the Interface ID that is being worked as the root for each of the lower-tier Interface IDs. Use the sketch to see the non-crossing patterns of interest in creating the final drawing. Figure 14.8 illustrates the initial top level, primitive, and final form for a typical diagram expansion.

If a block diagram format becomes too cluttered, use a square matrix format where the blocks of one of the two elements are on one matrix axis and the blocks of the other are on the other matrix axis. Place the Interface ID codes in the appropriate intersections. A dash or a blank space is understood to mean that no interface exists between those two elements.

14.6.8 *Crossface compound expansion*

There may be cases where the analyst finds it advantageous to combine both forms of expansion discussed above. For example, you may wish to

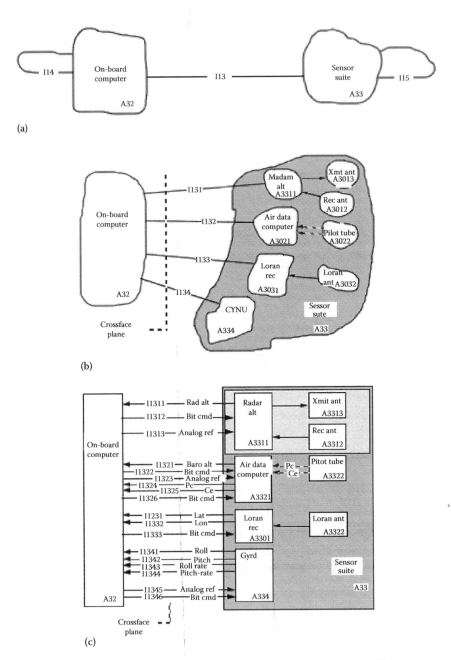

Figure 14.8 Crossface SBD expansion: (a) top-level representation, (b) primitive SBD, and (c) finished SBD. (From Grady, J., *System Integration*, CRC Press, Boca Raton, FL, pp. 185–206, 1994. With permission.)

expand a particular interface joining two blocks and the innerface of each block on the same SBD. The advantage to this approach is fully realized only when there are very few other interfaces between those two blocks and other blocks. Avoid preparing SBD that have so many lines that the value of the diagram is degraded.

14.7 Interface dictionary

It is not possible to use the SBD or n-square diagram to retain all of the information of interest about the interfaces illustrated there. The SBD actually only illustrates the existence of interfaces and identifies unique names for each interface. The analyst should also prepare an interface dictionary that provides an inventory of all system interfaces in the form of an alphanumeric listing by Interface ID corresponding to all of the interface lines on all of the SBD.

Ideally, the dictionary is retained in and printed from a computer database that is a part of a computerized requirements system. It includes information about the source, destination, and media of each interface element plus other data of interest. The database allows system engineers to map all interfaces to design responsibility groups for the purpose of identifying the critical interface subset as well as providing interface definition and statusing data. Figure 14.9 illustrates the interface dictionary columns that define each interface in terms of its source, destination, and media. Other columns can be included for other purposes.

14.8 Three views of interface

Interfaces exist between objects of systems for the purpose of introducing synergism between the objects. A system is said to be a collection of two or more objects that interact in purposeful ways to achieve an objective that no subset of the objects could otherwise achieve and it is through the synergism provided by interfaces between the system objects that this takes place. Interfaces communicate information from one object to another that the other would have no way of obtaining through its own unaided capacity. They provide energy with which to function as in the case of electrical

Interface identity		Interface responsibility		
Mid	Name	Source	Dest.	Media
I5111	Fuel on command	A21	A251	A32
I5112	Fuel flow light	A251	A21	A32

Figure 14.9 Typical interface dictionary listing: (a) top-level representation, (b) primitive SBD, and (c) finished SBD. (From Grady, J., *System Integration*, CRC Press, Boca Raton, FL, pp. 185–206, 1994. With permission.)

or hydraulic power. Interfaces act to physically tie a system together where objects are bolted or otherwise joined together.

An interface element has two fundamental characteristics: (1) a terminal at each end and (2) a media of communication. The terminal will always be an element of the system depicted on the system product entity diagram or an element of the system environment. This includes the elements of the system of interest, possibly some cooperative and hostile systems within the system environment, natural environmental elements, and no other possibilities. Likewise, the media of the interface element will be provided by some element of the system of interest, or the system environment (natural environment or cooperative systems). Wire harnesses, fluid plumbing, and the space through which radio signals pass are examples of interface media. The first two of these media are provided by the system and the last is an example of an element completed by the system environment.

We take three different views of the system interface here as a function of the perspective of the observer. These three views take into account the variable intensity of interest and responsibility for interface by the design community. The basis of these differing views is the relationship between the terminals and media of an interface element and the responsible design organization. It is a great systems management truth that the interfaces that will result in the greatest system problems are those where different organizations are responsible for two or three of the following: source terminal, destination terminal, and media. Relatively little difficulty can be expected where the same organization is responsible for all three. Where differences occur, the interface is a member of the cross-organizational interface set that system engineers must focus their energy in interface integration work.

The traditional view of interface is that an interface exists where two different contractor organizations are responsible for the design of the two terminals of the interface. We take a more general view here where an interface is any means of relating one system element to another at any level of system indenture. Interface thus imagined may be partitioned into subsets of interest as a function of organizational responsibility by establishing three classes as follows:

1. Outerface—Interface elements with neither terminal nor media under the design responsibility of organizational element X are in the organization X Outerface class. Organization X has no immediate interest in this class of interface and is not responsible in any way for its proper design.
2. Innerface—Interface elements with both terminals and the media under the design responsibility of organization X are in the organization X Innerface class. Only organization X is immediately concerned with this class and is completely responsible for its proper design.

3. Crossface—Interface elements with some subset (other than the null set) of two terminals and media under the design responsibility of organization X and the remaining subset (not the null set) under the design responsibility of a different organization are in the organization X Crossface class. Organization X and some other organization are jointly responsible for the design of this class. This is the interface class where system engineering energy and budget must be focused because it is where system problems will develop. These are the cracks between the specialized knowledge and experience of the engineering organizational elements. They are inevitable in any system and are determined by how the organization is structured with respect to the structuring of the system product entity structure. It is important that the system engineering community be capable of sorting all of this class of interface from the whole and assigning responsibility for its development.

14.9   Interface responsibility models

14.9.1   Set theoretic view

How shall we assign responsibility for interface development? It is clear there is a fundamental difference between architecture development responsibility and interface development responsibility. The former can be completely assigned to a single development agent with little difficulty. Interface on the other hand has two, and with the interface media possibly three, agents that should be interested in any one interface element. We need a way to make interface responsibility unambiguous.

The previous discussion of the three classes of interface as a function of your outlook offer a solution. First, let us create a hypothetical SBD in Figure 14.10. It includes seven components in three subsystems. Our challenge now is to determine who, among the seven principal engineers at the component level and the three principal engineers at the subsystem level, is responsible for each of the interfaces illustrated here.

The system illustrated in Figure 14.10 is composed of a set of product entities consisting of A = {A11, A12, A21, A22, A31, A32, A33} arranged into three subsystems (A1, A2, A3) and a set of interfaces defined by I = {I11, I12, I13, I21, I22, I23, I31, I32, I33, I411, I412, I42, I431, I432, I441, I442, I511, I512, I52, I53, I54, I55, I6, I71, I72, I7311, I7312, I7313, I732}. Let us now assign some human beings as principal engineers for these elements.

It is easy to assign product entity responsibility. Let us assume that A1 is a mechanical subsystem with Bob the principal engineer supported by Bill for A11 and John for A12. Similarly, we will assign Adam the principal engineer for the Avionics Subsystem, A2, with Ruth responsible for A21 and Allen for A22. Jane is the Fluids Subsystem principal engineer, A3,

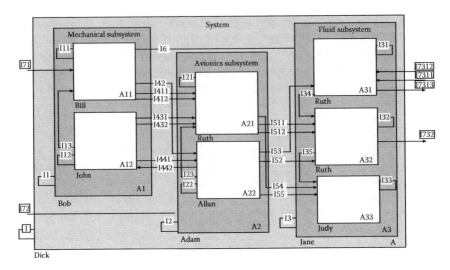

Figure 14.10 Interface responsibility model. (From Grady, J., *System Integration*, CRC Press, Boca Raton, FL, pp. 185–206, 1994. With permission.)

with Blain responsible for A31, Brian for A32, and Judy for A33. We must have a system engineer responsible for the system as a whole, assign Dick to that task. Now, how do these people share development responsibility for the interfaces between the elements for which they are responsible?

To answer this question let us partition the set of interfaces into the three views of innerface, outerface, and crossface from the perspective of each one of the subsystem principal engineers. Figure 14.11 provides a Venn diagram of all of the interface possibilities for the system based on an input partition of the set I, an output partition of the set I, and the superimposition of these two sets. This diagram includes subsets for each of the three subsystems and one for the system environment.

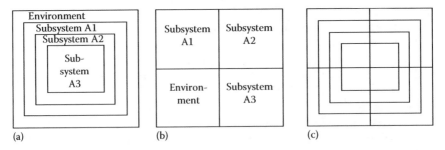

Figure 14.11 Interface partitions: (a) input partition, (b) output partition, and (c) superimposition. (From Grady, J., *System Integration*, CRC Press, Boca Raton, FL, pp. 185–206, 1994. With permission.)

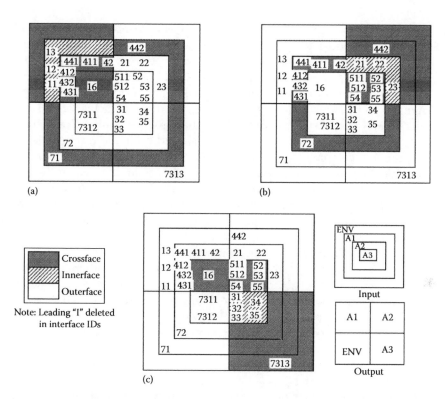

Figure 14.12 Subsystem principal engineer views: (a) subsystem A1 perspective, (b) subsystem A2 perspective, and (c) subsystem A3 perspective. (From Grady, J., *System Integration*, CRC Press, Boca Raton, FL, pp. 185–206, 1994. With permission.)

You will note in Figure 14.12 how the 29 interface elements illustrated on Figure 14.10 map into the 16 subsets defined on Figure 14.11 as a function of subsystem level input/output. Some subsets are voids with none of the 29 interface elements involved. Now let us, in Figure 14.12, see how each of the three subsystem principal engineers views these interfaces from their innerface, outerface, and crossface perspective.

The final step in defining interface responsibility is at hand. Note in Figure 14.12 that it is clear that the subsystem principal engineers can easily be made responsible for all of their innerface. They may wish to flow this responsibility down to their component principal engineers and the same method we are about to complete for the subsystem level will serve the subsystem principal's needs. This same pattern can be continued through all layers of system interface in a top-down development of interface in step with the top-down development of the system product entities.

The problem remains to decide which of the subsystem principals should be responsible for interfaces identified as crossfaces on Figure 14.12. The leading "I" symbols have been dropped from Figure 14.12 interface IDs to provide adequate space for a type font that we humans can actually see. We have three choices: (1) make the product entity principal engineer on the source end responsible, (2) make the principal on the destination end responsible, or (3) let them share responsibility. Clearly they have to work together, but one of them should be made responsible for the proper development of each interface. It is possible to select either rule 1 or rule 2 above to avoid interface responsibility ambiguity. In this era of interest in total quality management, it may be popular to select rule number 2 since the destination principal engineer can be thought of as the customer for that interface. It turns out that this is a good choice for other reasons as well.

Whichever rule you select, you will find many good reasons for exceptions based on (1) product entities that are grandfathered into the system as customer furnished equipment derived from previous systems, (2) existing equipment designs that are only modified for this application, and (3) biased engineering judgment. Since it is so difficult to pick a foolproof interface responsibility rule, you may conclude that it is foolish to try, but try to resist this feeling.

Where a system is being developed from a clean sheet of paper and will use no existing designs (a very uncommon situation), we could pick rule 1 or 2 without fear of compromise. In a real-world situation, where exceptions are possible, the rule could be 1 or 2 as the default with exceptions decided on a case-by-case basis. Given that we have an interface dictionary, we can include interface responsibility information in that dictionary and remove all doubt about who is responsible. Either the default rule applies or the default responsibility rule is reversed as noted in the dictionary.

14.9.2   *Annotated schematic block diagram*

The responsibility approach covered in the previous section is intellectually satisfying but not very easily implemented on programs. A very much simpler way to detect cross-organizational interface is to simply annotate an SBD with team responsibility overlays. Remember, it is cross-organizational interface that is the biggest problem and the principal target for spending limited system engineering resources. Figure 14.13 shows the SBD of a Centaur Upper Stage now produced by Lockheed Martin. This diagram was developed while the author was employed at General Dynamics Space Systems division on a small study focused on converting the vehicle propulsion system from two RL-10 rocket engines to a single extendable skirt version of the RL-10 in the interest of reducing engine cost.

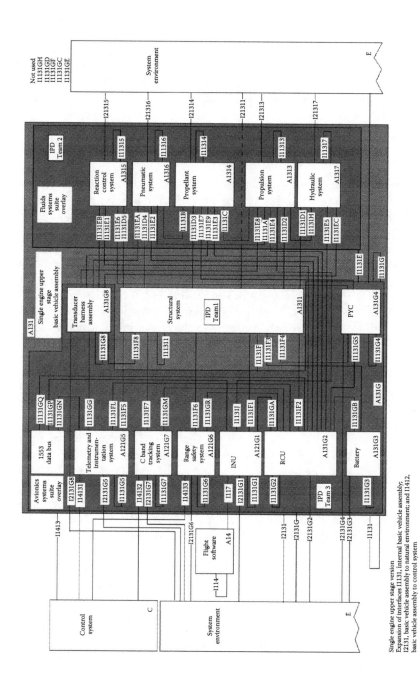

Figure 14.13 Cross-organizational interface identification through responsibility overlays.

Note that three teams are identified on the figure. One for fluid systems, one for structural systems, and a third for avionics. Any interface that crosses one of these team boundaries is a cross-organizational interface and should be very energetically subjected to integration work. The interfaces contained entirely within a team boundary should be developed by the responsible team and should not require a lot of system-level interest to cause them to be developed well.

14.10   The special need for external interface development

Interfaces between elements that are both under the same company's design responsibility should never be a problem in a development activity, but they commonly are even where different design departments of the same company are involved. Interfaces between elements a contractor is developing in-house and those that are procured from an outside vendor should never be a problem since the contractor controls the vendor through a subcontract or purchase order. Interface definition is a common area of dispute between contractors and their suppliers, however. If interfaces that are fully under the control of a contractor are difficult to develop flawlessly, those between two contractors that have no contractual relationship could be nearly impossible. Because it is so difficult to develop these interfaces, we need some special controls at these very important interface planes.

Interface development is an essential engineering management and technical approach to the development of interfaces where the contractors responsible for the design of the elements at the two terminals are different and linked only through the contractual language that may or may not be included in their contract from a common customer. Commonly these contractors are called associates, and each has an independent contract with a common contracting agency. These contracts should include special interface management clauses that seek to introduce discipline and precision into the technical communications between the contractors and to formally resolve problems across the interface between their products.

A customer will commonly require two interfacing associate contractors to reach an agreement on how they will jointly work a mutual interface and to jointly sign a memorandum of agreement stating the means by which they will cooperate. They may also be required to jointly develop an interface management plan to expand on the memorandum of agreement and both sign that plan. One of the associates will be required to prepare an ICD and they and the common customer will manage the development of the interface through that ICD.

An interface control working group (ICWG) will be formed with the customer, principal integrating contractor, or a system integration contractor chairing its meetings. The ICWG will meet periodically and take up interface issues. Over time, the interface definition evolves between the two elements managed in this way. Interface issues will be resolved by joint action and approved as evidenced by approving signatures on the ICD and revisions (interface change notices or ICN) by ICWG members.

Two possibilities exist for this ICD. One possibility is that it will only be used as a means to an end, that being to provide each associate with the same interface definition for use in their terminal item specification and design work. In this case, each associate uses the ICD as the source of the interface requirements for the specification covering the element on their side of the interface plan. When these two specifications are authenticated (signed) by the common customer into the program baseline, the interface can thereafter be managed though engineering change proposals (ECP) against the two specifications. The ICD goes away at this point having served its purpose.

The other alternative is that the ICD remains a living document throughout the program. In this case, the interface requirements contained in it should never be copied over into the specifications at the terminals. Each terminal specification should reference the ICD for the interface requirements and becomes an applicable document in the two terminal specifications. There is a third alternative that is nothing but trouble and that is the case of a living ICD combined with individual contractor attempts to capture the interface requirements in their terminal specifications. This results in a triple set of books that will almost certainly get out of synchronism as changes occur.

14.11   Degree of interface extension

The system engineer or analyst may analyze needed interfaces from a pair-wise perspective, as discussed in this chapter, initially as a means of identifying those interfaces. But, as in all cases where partitioning has been applied, it is necessary to also apply integration. It is therefore also useful to consider interface threads or strings in terms of a pattern of relationships across several product entities. Granted this is of less interest today than in the past because of the movement to central computers and the computer software running in them being responsible for much of the hard functionality in the system.

In years past, it was a useful system engineering approach to develop what were often called one-line diagrams. An example of these would be the range and bearing three-speed synchro loops from radar and sonar to the fire control instruments. Between source and destination these signals would have passed through one or more signal switchboards and bulkhead connectors or stuffing tubes. The unmanned photo reconnaissance aircraft the author worked on while employed by Teledyne Ryan Aeronautical in the 1960s and 1970s used relay diode logic to switch analog signals in their flight control systems. A signal might start out as a 115 V AC 400 Hz voltage from an inverter, be acted upon through a transformer, relay contacts in more than one black box, and arrive at a resistive network for summing at the input to a servo amp for pitch axis control. The 28 V DC control voltage for the coil of a relay might wend its way through the contact of several relays located in different black boxes on its way to that coil.

The reason that this one-lining technique is not so useful at present is that systems are simply built differently now. Systems are composed of sensors providing data to the central computer that internally generates solutions that are applied to loads. Also, the data that flows between items in the system is often converted early in its existence to a digital format leading to relatively few interface formats often completed via a computer network. In years past, one would find 28 V DC discrete, 400 Hz analog, 0–5 V DC analog, one-, two-, and three-speed synchro signals completing the pathway between sources and destinations in the same system. Systems today have the structure of a bicycle wheel without the rim, spokes connecting the hub where the computer is located. There may be interesting physical routes for some of these spokes but there is relatively little of interest happening outside the computer.

This interest in threads can be effectively applied in systems of systems today. A Marine in the boondocks ashore may call in an air strike relayed by an unmanned aircraft overhead to a naval ship in the littoral space. The request might be processed to determine the best way to respond to the request for fire support and the answer relayed to a close air support aircraft armed with laser-guided bombs. The same Marine who started this chain might have to illuminate the target with a laser designator to attract the bomb dropped by the aircraft. While each of these pair-wise interfaces is important and can be identified and defined in isolation, it is important to consider the whole picture as a thread of related actions.

14.12   A rationale in support of IPPT

We may ask ourselves if it would not be possible to forget about how our engineering team is organized at the beginning of the project and organize only after we have created the product entity structure and understand the interfaces. That is, let the product design our development organizational structure. This way we could form our design team organization to minimize cross-organizational interfaces while focusing our system decomposition process on needed system functionality without worrying about the organizational constraint. This is a case of turning what was once the conventional world upside down, but it is an alternative that we should consider. This approach is the basis for concurrent development or integrated product development as expressed in this book.

Another way to say this is to form integrated design teams composed of design disciplines that match the technology associated with the functionality of the element. Many system elements in complex systems require an appeal to several disciplines anyway, so why force yourself to live with a constraining functional organization? The IPPT notion linked to the concurrent development concept are based on exactly this concept of organizing the development team to correspond to the product organization. In years past we got into the habit of functionally organizing ourselves in terms of our specialized knowledge bases. We made it worse by collocating by functions and mutually withdrawing from the cross-organizational interfaces.

In years past, programs often formed functional department extensions on each program. Budget was doled out to each of these islands by WBS managers only responsible for cost and schedule. Technical problems were dealt with by a chief engineer who would coordinate with managers from the several islands as it related to problems and opportunities within the various product entities that the different functionally dominated teams would have primary responsibility for. This is not describing an effective practice of system engineering. It is expressing a structure dominated by functional departments in a company where their longevity has permitted them to dictate the way the enterprise organizes to accomplish program work.

Hopefully, your company has been saved from this unfortunate organizational structure by now through the adoption of the use of IPPTs cross-functionally staffed and program management by the program manager through these IPPT managers rather than the imposition of functional department managers.

Section 6.8 offers some insights into how to manage the evolving interface definition through IPPTs and organizing for interface integration using a lowest common team concept.

chapter fifteen

Requirements integration*

15.1 What is requirements integration?

Requirements are essential attributes for things (in the system product entity structure) and processes in the system that must be respected by designers when synthesizing those requirements into a design solution. There are integration measures that should be applied to the process of developing these requirements for individual items or processes and across the whole system composition. Some of this work requires active interaction during the requirements definition period while other work involves auditing the results of requirements work previously accomplished. So, there is a temporal or program phase–related aspect to requirements integration work.

How we assign responsibility for requirements integration work has a lot to do with how we are organized on programs and the range of requirements analysis strategies and methods allowed or used. We have assumed throughout this book that we were organized within a matrix management structure with personnel supervised on programs within integrated product and process teams (IPPT) defined by program management with the team organization aligned with the product entity structure. The principal discussion in this chapter will focus on this structure.

Requirements integration entails ensuring that each requirement for an item satisfies a set of minimum criterion, that the set of requirements for an item satisfy a minimum criteria, that all of the requirements for an item are traceable to other requirements and related information, that the requirements have been validated meaning that the risk of achieving the requirement in design is of low risk, within budget and schedule limits imposed, and that all mutual conflicts have been removed. It is largely a requirements quality control process. The big question is, who is going to be responsible for this work. Some of it should be the responsibility of those writing requirements while other integration work can only be done at the system level by the program integration team (PIT).

15.2 Requirements integration responsibility

Requirements will flow into system and team item specifications from the many specialists staffing the responsible teams. The aggregate set of

---

* The material in this chapter is from Grady, J., *System Integration*, CRC Press, Boca Raton, FL, Chapter 13, 1994. Used with permission.

requirements for any one item may not be free of conflicts. It is very important that the PIT make it clear who is responsible for requirements integration within the PIT and within the IPPTs. A principal engineer should be assigned for each specification and that principal engineer should be made responsible for development of the document content including proactive identification and integration of conflicts.

In addition to performing requirements integration on system level specifications and interface control documents where one terminal is outside of the system, the PIT should also monitor the requirements integration work done by teams. This commonly cannot be done in total because of budget constraints, so any audits must focus on the most critical items and requirements categories that should include cross-organizational interfaces.

15.3   System level SRA overview

Figure 15.1 offers a view of the generic system requirements analysis (SRA) process, during the phases when development requirements for performance, part I, or Type A and B specifications are being identified. The SRA process should be accomplished by the PIT at the system level including the publication of the specifications for the top-level team items and at the appropriate IPPT level by IPPT members. The process illustrated here fits into each of the several IPPT processes on-going simultaneously via a concurrent engineering bond. Within the IPPT the SRA work entails cross-function, coproduct integration work while between the IPPT and PIT teams the integration work is more focused on cofunction, cross-product integration. All of the engineers and analysts involved in the item and SRA processes should, as we have discussed earlier, have a means of cooperating easily via physical collocation or excellent communications and computer networking of tools and information products.

Refer to *System Requirements Analysis* (Grady, 1993, 2005) for a detailed discussion of the process illustrated in Figure 15.1. We are most interested here in the tasks identified as verification requirements analysis and top-down planning and requirements audit, assure traceability, integrate, and validate requirements.

15.4   Requirements integration activities

Those responsible for requirements integration should perform the five specific audit actions on requirements described in subordinate paragraphs. You will likely not have the resources to complete all of these audits on all of the items discussed, so you must prioritize the work and distribute it such that IPPTs are doing the majority of the work and the PIT does spot checks to identify items for further audit action.

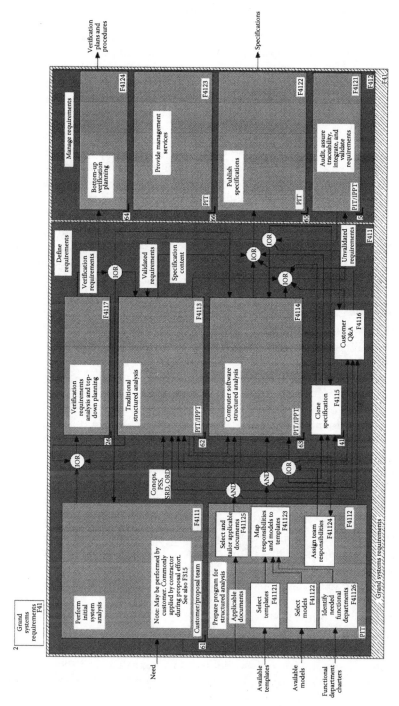

Figure 15.1 SRA process. (From Grady, J., *System Integration*, CRC Press, Boca Raton, FL, pp. 209–222, 1994. With permission.)

15.4.1   Individual requirements audit

Each individual requirement should be checked against at least the following seven criteria: traceability, style, singleness of purpose, quantification, verifiability, unambiguity, and good sense. These criteria are from *Systems Requirements Analysis* (Grady, 1993, 2005) with some refinement.

15.4.1.1   Traceability

Every requirement should be traceable to a driving need in terms of parent requirements, the requirements analysis process through which the need for the requirement was stimulated or a source reference, and the verification process through which it will be proven that the design is compliant. Every requirement should, theoretically, trace all the way up to the customer need. Any requirement that does not upwardly trace to a parent specification requirement is suspect and the rationale for non-traceability should be captured and approved or a condition of traceability established. All requirements should be derived using models so all requirements should be traceable to a modeling artifact from which it was derived. Ideally traceability would be maintained through requirements IDs assigned automatically by a database system rather than through paragraph numbers. Product requirements should also be traceable to the verification requirements and thus to the whole verification documentation suite (plans, procedures, and reports) as discussed in *System Verification* (Grady, 2007).

15.4.1.2   Correctness of style

Every requirement should be written with correct use of the language of choice, spelling, and punctuation. It should reflect the customer's style guide and data item description where those are controlled.

15.4.1.3   Understandability

Every requirement should be easy to understand by a broad cross section of people. Only a single semantic interpretation should be possible. Simple words, simple sentence structures, short sentences, and avoidance of negatives helps to achieve this criterion. Brevity and adherence to the one-paragraph-one-requirement rule also helps to satisfy this criterion, with one possible exception discussed in the next criteria.

15.4.1.4   Singleness of purpose

There seems to be a tremendous attraction to writing requirements in long paragraphs. It is not unusual to find many requirements in one specification paragraph. The only valid argument for this, and the argument is marginal, is that some requirements have to be stated as two or more closely coupled thoughts in one paragraph, and to break these kinds of ideas apart may result in violating the understandability criteria. This should be the exception and will never be a valid argument for paragraphs three pages long. In general, we should strive to include only one requirement in each specification paragraph.

The principal argument for singleness of purpose is that to do otherwise complicates verification and vertical traceability due to the potential for ambiguity. The argument can be made that new requirements analysis tools permit identification of fragments of paragraphs (called parsing) for purposes of verification and traceability. If you are served by such tools, the argument for singleness of purpose may be less valid. Even in this case, however, the author confesses to a bias in favor of simplicity of structure over other considerations. The argument is also made that this rule results in boring specifications but specifications are not intended to entertain, they are intended to inform people of the facts.

There is one other problem raised by the singleness of purpose criteria. MIL-STD-961E, the military standard for specifications, bounds the number of indentures in paragraph numbering to seven levels. Limiting requirements statements to one requirement per paragraph generally will increase the number of levels and commonly beyond the seven-level restriction. Adherence to this criterion may require tailoring of the standard on a military contract. Alternatively, we can use lettered subparagraphs to satisfy both the singleness of purpose criteria and the seven-level restriction. Once again, if your requirements tool can maintain traceability and verification hooks uniquely to the fragments of interest and generate specifications from this structure, it may be a good overall solution. The application of bullets to identify requirements statements should be discouraged in favor of having a unique alphanumeric paragraph identification method employing letters if necessary at the lowest tier.

15.4.1.5 Quantification
Requirements should include not only what is required but the needed value or values. Some kinds of requirements must be stated in qualitative terms, but they should be exceptions. We should seek to quantify our requirements. Otherwise, it is very difficult for the design team to determine just how good does the design have to be and for verification people to determine if the requirement was satisfied in the testing they accomplish.

15.4.1.6 Verifiability
One of the most effective ways to ensure good requirements are written is to require that they be verifiable through some practical process. In the process of trying to determine how to verify a requirement, you are forced to write the product requirement properly. This is why many people believe that the verification requirements (Section 4 of a specification)

should be written by the same person who wrote the Section 3 requirement and that it should be done at the same time. This also encourages that quantified requirements be written because it is very hard to devise a verification process for nonquantified requirements.

15.4.1.7   Good judgment and good sense

We need to check the requirements set for good sense. Requirements must be written against characteristics of the product and not against things that cannot be controlled with all the money in the world, like the weather. We must not violate the laws of physics or any other formal discipline. Once again, simple language is a great assist in making this check. Failures in this criterion are best detected by engineers with a lot of experience.

15.4.2   Requirements set attributes

The complete set of requirements for an item should satisfy the following five criteria: consistency, completeness, minimized, uniqueness, and balance. The specification principal engineer must ensure that these criteria are satisfied during the requirements analysis effort and the management approval process should encourage the principal engineer to give evidence to that effect during reviews.

15.4.2.1   Consistency

The set of requirements is internally consistent if it does not entail self-contradiction. The set is consistent with all other requirements sets if it does not conflict with those other sets. An example of inconsistency in an item requirements set would be where one requirement called for nonuse of strategic materials while another requirement cannot be synthesized without the use of such materials.

15.4.2.2   Completeness

How can we be sure we have identified all of the appropriate requirements for an item? Unfortunately this is a question that cannot easily be answered. But, there is a good chance we will satisfy this criterion if a qualified staff of engineers used a systematic approach to identify the attributes that must be controlled. That is exactly the basis for the structured approach encouraged in Chapter twelve. If we have conducted a thorough structured analysis of what the item has to do, an effective environmental analysis, a systematic interface analysis, and an integrated specialty engineering requirements analysis (or the software equivalent) and expanded the attributes thus derived into quantified requirements statements using sound quantification methods, there is a good chance that we have satisfied the completeness criterion. In addition to the set including every needed requirement, all TBD (to be determined) remarks must be replaced with appropriate values, and all figures and tables referenced in requirements text supplied and complete.

If we have used freestyle or cloning strategies rather than a structured approach, there is less assurance that we have identified all of the appropriate requirements. A boilerplate can be useful in cross-checking specialty engineering constraint categories of a requirements set generated through the structured approach. Cloning approaches can be effective in specialty engineering and environmental constraints, but are not effective in assuring completeness for performance requirements and interface constraints.

15.4.2.3   Minimized

But, we just identified one of the criteria as completeness. Now we have a criterion for the opposite situation. Is not that a contradiction? Strangely enough, it is possible to create too many requirements. Our objective is to decompose the customer's need into a series of smaller problems that will yield to the creative genius of specialized design engineers and their teammates. We need to identify only those requirements that will ensure that the product of the engineer's creativity will work synergistically when integrated into the system so as to satisfy the original driving need.

Requirements have a constraining effect on creativity. We purposely write them for that purpose to ensure that the design solution will have certain important characteristics. Unnecessary requirements constrain unnecessarily and can have the effect of eliminating some potential design solutions that could be better than the remaining options. Requirements do reduce the solution space available to the designer. They should tell the designer the attributes the solution must have in order for the item to function synergistically within the system. They should not identify a particular solution or unnecessarily confine the solution space. They should be design independent.

The way to check the need for a requirement is to ask, "What effect would it have if this requirement were deleted. Could the designer, as a result, select a design that would be unacceptable from a system perspective?" If the answer is no, the requirement should be a candidate for deletion.

15.4.2.4   Uniqueness

Each requirement in the set should be unique in the set with no repetition. Each unique requirement should only appear once in a requirements set.

15.4.2.5   Balance

Some kinds of requirements are invariably in conflict and we need to find a reasonable balance point in such cases. Reliability and maintainability requirements, as well as cost and most everything else are potential examples. One Atlas rocket had to be destroyed as it diverged from the planned launch path over the Atlantic Ocean. Destruct was commanded by a radio setting off an explosive charge that triggered the propellants causing the vehicle to come apart and its pieces to fall harmlessly into the sea.

There was great interest in what had caused the fault that required the use of the destruct command. Luckily the on-board computer was found washed up on the beach and strangely enough was intact. When tested, it disclosed that a memory location contained engine gimbal angle data corresponding to the divergence observed by radar and the cause was traced to a lightning strike during ascent from the launchpad. This computer was engulfed in a tremendous explosion stimulated by a lightning strike, fell several thousand feet to impact in the sea but retained its water-tight integrity, and still worked well enough to determine memory content at the time of the incident. The accident investigation team was very happy that this information was available to it, but is it possible that this computer was overdesigned? Is it possible that this overdesign was driven by requirements values way out of step with real-world needs? While it was very helpful to be able to pinpoint the cause in this case, it would not have been unreasonable to expect that the computer would have been totally destroyed.

15.4.3   Margin check

If you use margins to provide slack requirements values to permit easier solution for difficult design problems, the principal engineer should check to ensure that appropriate margin values have been identified, respected, and preserved to the extent possible for the benefit of any future program technical risk management purposes.

15.4.4   TPM status check

If there are any technical performance measurement (TPM) parameters actively in work for the item the corresponding requirements should be considered still under risk. A specification including a risk ideally would not be considered for approval but, if there are good reasons for approval of a specification under risk, the principal engineer should check the history of any active TPM parameters and ensure that the goals expressed are feasible and a resolution on track.

15.4.5   Specification format check

The requirements must be fitted into a prescribed format defined by the customer for deliverable specifications and an internally defined format for others. As the requirements become available from the analysts, they should be assigned paragraph numbers from whatever format standard is used. Any paragraph titles from this standard format for which content is not available should have a good rationale for not including the requirement.

15.5   Engineering specialty integration overview

In Part five we discuss each specialty engineering discipline and how they contribute to the engineering effort. In this chapter we are interested in how the product of these analyses related to requirements definition can be blended into a coherent story. The fact that specialty engineers are tightly focused on their specialty and that frequently the effects of specialty engineering requirements are in conflict requires that the effects of their requirements inputs be integrated or coordinated to ensure that a condition of balance is realized in the final design and that an unnecessary cost burden is not placed on the program and customer. Failure to apply these disciplines to a program can easily produce an unfavorable life cycle cost result, unsatisfied operability needs, poor match between the system and its environment, and extreme support costs. Application of them with uniformly excessive zeal can result in added nonrecurring cost that does not contribute in fair measure to customer benefit. Application of the disciplines with irregular assertiveness can result in unbalanced characteristics of questionable utility. For example, you may end up with an extremely reliable system that cannot, in a reasonable period of time, be tested due to the multiplicity of redundant paths.

One way to integrate the specialty requirements is to hold periodic meetings with the specialty engineers during the requirements development effort and ask each engineer to defend his or her requirements under the critical review of the others. Figure 15.2 offers a more elaborate approach involving integrated checklists.

15.6   Interface requirements analysis integration

We need to check that we have identified all of the item interfaces and that we have adequately defined each interface. In addition, we should check the requirements sets for each item interfaced with and verify that their interface requirements are compatible with those for the item in question.

Each principal engineer or IPPT should cooperate in the formation of an internal interface working group run by the PIT as a means to evolve a joint agreement on each interface their item has with the rest of the system and the requirements for each interface. With the exception of interface requirements, all of the item requirements are commonly determined within the structure defined by the program's team structure. We purposely allocate the system's unfolding functionality to items with the understanding that the design of these items will be the responsibility of the IPPT assigned to

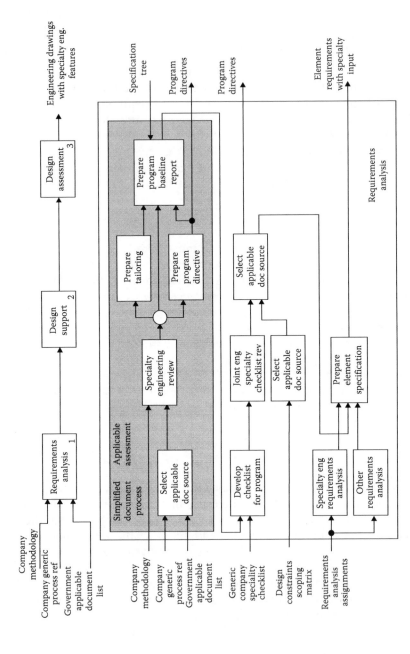

Figure 15.2 Specialty engineering integration process. (From Grady, J., *System Integration*, CRC Press, Boca Raton, FL, pp. 209–222, 1994. With permission.)

them. On large programs with multiple sites it may be necessary to federate the interface integration responsibility with some form of cooperation across the collections of site teams responsible for integration.

Interfaces commonly run at cross purposes with the cross-functional team organizational structure encouraged in this book, and the integration of the interface requirements can expose not only flaws in the product system but in the contractor's program and functional organizations as well. The program team and functional management should be monitoring this process carefully and react to any problems observed to improve the environment for successful teamwork. The symptom to look for is a tendency to withdraw from product interfaces that coincide with organizational interfaces. We need to encourage people to plunge across these gaps from both sides, not withdraw from them. Success here may increase inter-team tension on the program, but failure will increase product interface incompatibility problem seriousness. Program organizations focused on the product entity structure rather than company functional structure also help to abate this problem.

15.7 Environmental requirements analysis integration

Environmental requirements provide opportunities for much foolishness. Every environmental requirement should be checked to ensure that the language of the statement focuses on product characteristics and not on an attempt to control the natural forces of the world. Also, we should check carefully for conflicts between requirements that are out of synchronization with reasonable expectations (the forces applied by a snow load on the hot tin roof, for example).

In the process of integrating the environmental requirements, one has to determine exactly how they will integrate the union of the different environmental ranges influencing an item in different process steps. A common approach is to take the worst case in all cases. This solution can have a cumulative effect where each parameter is selected this way resulting in over specification and, consequently, overdesign or a null solution space. We should ask ourselves if some other method can be applied such as the root mean square or a weighted mean. Cumulative worst-case effects is a common path by which components are overdesigned.

15.8 Programmatic requirements integration

The same techniques indicated for specialty engineering integration are generally effective for programmatic requirements analysis. The principal engineer must ensure that the whole team remains focused on the same requirements baseline. Frequent informal meetings (that can be brief) of all of the team members that encourage synergism between the members are helpful such that they expose their understandings about this baseline for the education of their team mates and ventilate their own ideas with teammate criticism. Obviously, the principal engineer must encourage a spirit of togetherness and emotional safety on the team or the members will not share their views eagerly and openly.

The team must also have an effective means of communicating the baseline among themselves and with other teams. Common computer databases networked throughout the project, good voice communication capability, and physical collocation are helpful to this end. In addition, a team might use a war room concept to post a lot of information for simultaneous view. It is not necessary to have an actual room for this purpose. If the facility has any large expanses of wall, in corridors for example, they can be used very effectively so long as standup meetings are not called during the time there is a peak traffic load in the hall as at lunch time.

The principal objective in integration of product and process requirements is to identify inconsistencies between the requirements for each. Some obvious case studies of conflict may be helpful in suggesting many other possibilities:

1. The current product requirements can only be satisfied by a new metallurgical technology used successfully to date only in a laboratory environment. The item manufacturing schedule does not include time for acquisition of the special tooling required to perform the related operation and the tooling requirements do not list an appropriate tool needed for the operation. The risk list does not include this potential problem and no technology work is currently planned.
2. The system will require 132 articles of a particular product item and 25 spares at the level of the item in question. The principal engineer finds that material requirements call only for 147 due to an error in addition or failure to update the material requirements from 15 to 25 spares.
3. The wing of the aircraft is 53 ft in span but the logistics people are planning for a hangar with a door opening of 48.3 ft because that was adequate for a previous aircraft concept version that was compatible with an existing hangar structure at 13 sites where the aircraft is to be operated. The logistics requirements have apparently not caught up with the product requirements or have not impacted the product design for compatibility with customer resources. It is also possible that the decision to move to a design with a 53 ft wing span was ill-conceived given the customer ownership of hangers that will be incompatible.

4. The manufacturing tooling requirements call for a master tooling fitting at missile fuselage station 23.5 as a principal support point during missile fuselage build up. The NASTRAN model does not include this tooling point nor has it considered the resultant stresses during manufacturing.

There is a fine line between requirements integration and design integration across the product and process valley. Some of the cases included above are very close to this line and perhaps even across it into design integration. There needs to be this concurrent action between the requirements definition for product and process while the requirements are in a state of flux, and this same attitude must carry over into the concurrent design of the product and the process components. In actual practice it may be difficult to separate this into requirements and design integration, but it is not all that important that we do so. It is much more important in practice to have this activity working well for your program or company than to worry ideologically over whether, at any particular point, we are into requirements or design integration activity.

This work requires a green eyeshade approach to life and can be done very well by people interested in fine details who incidentally also need to be able to grasp the grand programmatic view. And, yes, there is a shortage of these people in every company. The best source for these people is among your most experienced employees who have not progressed in their careers to or above their level of competence from a management perspective. Hopefully, you have some of these people on the payroll who have also not been damaged over the years through poor management.

15.9   Hardware–software integration

At some point in the future not too far off we will have access to a universal modeling environment probably composed of some combination of what is now known as UML and SysML possibly augmented to permit modeling of the environment. This modeling construct will also be supported by an integrated tool set that makes it possible for a normal human being to establish and maintain traceability across the HW–SW divide better than the work-around included in Chapter twelve. What is described is viewed today as a utopia that has not yet been reached. Today many programs and organizations have serious difficulty integrating across the HW–SW divide. The fundamental problem is that the people on the two sides of this boundary have very specialized knowledge domains that are not commonly understood on the other side making communication very difficult.

In the requirements work the problem appears as a traceability problem from the higher-tier hardware specification and the highest-tier software specification within that hardware entity. The reality is that software can be contained in hardware but it is not feasible for hardware to be contained in software because software is a logical construct with no real physical composition. Therefore we will never have to worry about hierarchical traceability upward from the requirements for a hardware entity to the requirements of a software entity. Section 12.4 offers a discussion on one way to deal with a downward traceability from a hardware entity to its immediately subordinate software requirements. In Part four we will revisit this problem from a design perspective.

15.10   Structured constraints deconfliction

15.10.1   Can there be too many requirements?

The normal idealized situation with an unprecedented system at the beginning is that the only requirement that exists is the ultimate requirement, the customer's need. We apply the structured decomposition technique to identify performance requirements for things in the system exposed through structured analysis. We identify appropriate constraints as well. The point is that we create the requirements from an almost null condition in the beginning as a part of the process of defining the needed system.

Is the opposite condition possible at the beginning. Is it possible at the beginning for there to be too many requirements identified such that the solution space is null. It is hard to imagine this condition occurring with an unprecedented system but it could occur with a heavily precedented one.

An example of such a system is the nuclear waste disposal system in operation at the time this book was published. During World War II when the nuclear weapons development and production process began, everyone was too interested in survival in the bear term and winning the war to be overly concerned about long-term problems resulting from weapons production. This mind-set carried on during the selective war with the USSR that lasted until the late 1980s. As a nuclear power industry caught on and created additional waste products, they were stored like those from weapons production.

The end of the Cold War removed the motive encouraging a status quo forcing the Department of Energy (DoE) to place a higher priority on solving the open-ended problem of what to do with nuclear waste. Wear out of existing storage facilities and the increasing volume of the ever-expanding material requiring storage also contributed to moving this problem higher on the list.

Here is a system that arrived at a condition of over constraint because of the difficulty and cost of solving the system problem. The constraints applied to the solution space precluded a solution so the problem grew

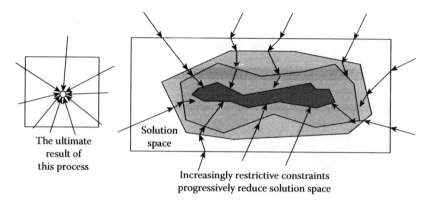

The ultimate
result of
this process

Solution
space

Increasingly restrictive constraints
progressively reduce solution space

Figure 15.3 Progressively restrictive constraint closure. (From Grady, J., *System Integration*, CRC Press, Boca Raton, FL, pp. 209–222, 1994. With permission.)

worse over time as shown in Figure 15.3 adding more constraints. An open loop situation like this can only get worse over time and in this example it has. The ultimate result of this kind of process is a null solution space. The nuclear power industry in the United States is caught in a similar trap.

15.10.2   Deconfliction

An over-constrained condition can be approached in a structured fashion. First, we identify all of the constraints that apply to the system and link those constraints to the key stakeholder who insists on that constraint but also has the power to change it. It is possible that this is two different people or agencies, but for the sake of simplicity assume they are one and the same. We also need to identify one or more alternative solutions to the ultimate requirement each in the form of a scenario, mission statement, or design concept. We can then map all of the constraints to the alternative solutions. We would be drawn to this mode of analysis because all of these currently defined solutions had already been declared unacceptable for none of them meet all of the identified constraints. The result is illustrated in Figure 15.4. Our goal is to expand the solution space for one or more solutions such that a solution is feasible.

At this point we must make some decisions about how much energy and money we wish to pour into the analysis. We can pursue all of the possibilities or exclude some alternatives based on a cursory review. For all surviving alternative solutions we now must evaluate the difficulty of changing each applicable constraint sufficiently to permit that alternative to be successful. We have a simple bookkeeping situation now with lists of things to do and these can be assigned to teams of people. Figure 15.5 shows one possible list arrangement for which a computer database would be a natural implementation. For each constraint–solution pair we

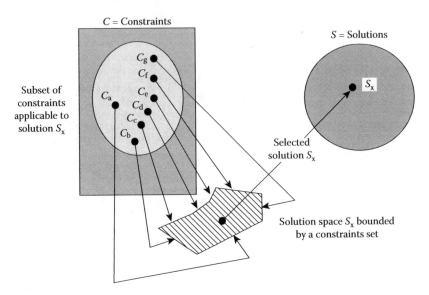

Figure 15.4 Solution–constraint mapping. (From Grady, J., *System Integration*, CRC Press, Boca Raton, FL, pp. 209–222, 1994. With permission.)

Constraint	Solution	Stakeholder	Reference	Difficulty	Effects
C_e	S_x	DoE	Federal Statute X	Easy	Reduced cost of interstate transport of waste.
C_r	S_x	State of Nevada	State Law 10.3.2	Difficult	LCC reduction of $7B dollars.

Figure 15.5 Constraints deconfliction tracking matrix. (From Grady, J., *System Integration*, CRC Press, Boca Raton, FL, pp. 209–222, 1994. With permission.)

note the stakeholder, source of constraint authority (reference), the degree of difficulty anticipated in relaxing the constraint, and the effects of the relaxation. We would probably wish to capture other information as well to support the study and a final decision.

At any point in our analysis there will always be at least one solution based on particular values of the constraints, where one of the constraints is how much money we do not wish to exceed, but we may not like any of the possible solutions exposed. At a particular point the remaining difficulties might be that the solution costs too much money, it will take too long to implement, or it exposes people to health risks. Whatever the remaining difficulties as we work to roll back the constraints, while the solution space may still be over constrained, it will be changing in the opening direction. From time to time we may establish baseline value combinations for the constraints and then continue work to deconflict the remaining constraints applied to one or more alternative solutions.

Some problems, it is true, could be so difficult that we eventually simply must walk away from them with the conclusion that the solution space is denied because our need is in conflict with the laws of physics or finance, the laws of man, or the aggregate sense of the public good. A problem like nuclear waste disposal, while it is very difficult, cannot simply be walked away from without placing ourselves in great peril. A problem like this requires an organized search along the lines discussed above where the constraints are consciously evaluated for rollback possibilities. Many, of course, feel that the Yucca Mountain facility is the result of having dealt successfully with many of these conflicts. A lot of people in Nevada, of course, do not feel this way.

part four

Product design synthesis

chapter sixteen

Program execution*

16.1   The central idea

This book talks all around a central idea that is very difficult to talk about because few of us fully understand what is going on relative to it. This central idea is the consummation of a design. The dictionary tells us that to design is to plan or fashion the form and structure of an object. Engineering is said to be the practical application of science and mathematics in the pursuit of a solution to a problem that can be solved by some form of machine. The question is just how exactly does this process occur within the mind of the design engineer. The author likes to believe that the design engineer comes to a clear understanding of the nature of the problem and then applies his or her command of the knowledge and techniques of an engineering discipline to gain insight into ways that the problem could be solved. Ideally, the engineer would conceive two or more alternative possible solutions and seek to determine which of these is the better.

Over the several years that the author has taught courses in system synthesis, he has greatly enjoyed one part of the work required of students. They are asked to interview a design engineer and report back to the class how they think they accomplish this process. Often the answers are that they do not know. Others have said that ideas come into the mind after intense thought about the problem that are explored further, leading to expressions of design features that might work. This is a creative process that involves what appears to be a chaotic disconnect in the system engineer's orderly world of deductive and inductive reasoning.

System engineers are vitally interested in establishing linear stings of traceability information, but it is very difficult to expose a clear case of traceability between the requirements and the design features. Eventually, as the tool companies consume one another into core businesses that cover both CAD and requirements work, it may come to pass that there will be a way to link the requirements with design features, but this will not change the existence of a discontinuity between the requirements and the creative process through which the design engineer conceives the features of the design solution.

---

* The material in this chapter is from Grady, J., *System Integration*, CRC Press, Boca Raton, FL, Chapter 8, 1994. Used with permission.

Given that this is an acceptable description of what is going on in the mind of the individual design engineer, still described here in a very incomplete fashion, we have to recognize that today the world engineers live in is so much more complicated than suggested by this simple description. The nature of the problem is commonly very complex to the degree that no one person can individually fully conceive the nature of the problem or a possible solution to it. Therefore, we have resorted to teams of people who collectively have the knowledge coverage needed to understand the problem space. But, because we have spread out the understanding of the problem space across several domains that different people have to deal with, it is necessary to patch together those thought processes in an integrated fashion. Whole design engineers are formed on programs through human communication that knits together the aggregate knowledge base of all of those cooperating on a particular task. Often this process of integration and optimization will proceed better if a system engineer, a generalist, is involved.

16.2   Controlling the well-planned program

16.2.1   Program execution controls

We will first focus on program integration when things are going according to plan. We arrive at a condition like this through good planning, faithful execution of the plan, and the good fortune to have developed precisely the conditions that we had planned for within an environment of imperfect knowledge of the future or the management agility to adjust when they do not. Throughout the period when things are going according to the plan, we need to have our sensors turned on to detect indicators of potential problems or risks, for risk changes over time as a function of what we do, and what happens in the world around us. During this initial discussion, we will assume that these sensors are reporting nothing but a Go condition.

Our program plan along with the integrated master plan and the program level schedule are the two principal program execution controlling documents at the top level. The author believes that the work breakdown structure (WBS) and statement of work (SOW) are subsets of the IMP properly structured so the IMP includes them. The combination of these documents tells what has to be done, when it must be done, and by whom to what end providing a structure within which program cost and schedule may be managed, collected, and reported to the customer. The contract data requirements list (CDRL) defines data products that must be delivered to the customer, and the schedule in the contract tells what product deliverables are required and when.

16.2.2   Alas, good planning is not everything

These documents do tell what should be happening but they do not, of course, tell what is actually happening during execution. So, in addition to good planning data, program management personnel need good program status tracking data. Good management skills and judgment are also needed to take advantage of accurate status information to steer the program along a sound course. Program status data must be compared with planning data and appropriate direction determined and given to encourage continuation on the plan. Most companies have cost/schedule control systems (C/SCS) or earned value systems (EVS) that relate to the WBS. These systems are designed to provide management with financial and schedule data that tells how the program is doing relative to the plan. Many of them at the time this book was published provided exactly that, data, in reams of computer print-outs. There is a lot of work for computer programmers to do in refining the presentation of C/SCS or EVS information to managers and workers in an easy-to-understand way that will lead to effective decision making. There are some good examples of a sound direction for this work in the form of the technical performance measurement (TPM) approach and functional management metrics. Some C/SCSC systems are still operating in the batch processing era while the world has moved to online access.

Program management must also have access to the third leg of program management information dealing with product technical performance progress. One very effective tool for managing product performance risk is the TPM process. In the TPM process, customer and company management agree on a relatively small number of quantifiable product performance parameters that will be controlled under the TPM program. These should be key requirements that collectively span the development effort, ones that will be adversely effected when trouble is encountered. The parameter values are tracked in time on TPM charts that are annotated to note critical events related to changes in values. When a reported parameter value exceeds an allowable deviation from planned value, it is clearly visible on a graphical chart. Specific actions are taken to drive the value back within limits in a way similar to cost and schedule variance reporting in EVS. The value continues to be monitored to see if the corrective action has the desired effect.

Weight, maximum speed, guidance accuracy, predicted turnaround time, throughput, and memory margin are examples of possible TPM parameters from several different fields. In picking these parameters it helps to know what kind of product requirements your company has had trouble in satisfying in your history. Some company people may be concerned that if the customer has access to this data it might make the company look bad to use these parameters for formal performance measurement. Some company program managers would prefer to have a contract set of parameters that are not so effective at spotting company problems supplemented by an additional list of parameters that are evaluated only internally that may recognize traditional problems. The customer that knows its supplier will not stand for this, of course.

The TPM technique is a closed-loop process and that is what program managers need for cost and schedule control as well. In these systems, an unacceptable deviation is visually obvious from the management information, specific action is taken to correct the indicator of unsatisfactory program performance, and the indicator then provides, with some time constant, feedback of the effect of the action taken. TPM and the application of this same approach to C/SCS are examples of what are called metrics. They all work this same way. Key numerical parameters are charted in time and used as a basis for management feedback to the process. The process reacts and the results are displayed by the charted parameter. We will return to TPM later in this chapter.

16.2.3   Implementing the IMP/IMS

In Part two we encouraged the use of the U.S. Air Force program planning initiative called the integrated management system using an integrated master plan (IMP) and integrated master schedule (IMS) or a variation developed by JOG System Engineering as the principal management tools. In so doing, we need the information inputs provided by the systems discussed above focused on the events identified in the IMP. The IMP provides not only an organized set of events to which we have hooked the tasks defined in the SOW, but for each event a list of accomplishments we should expect to see materialize, and specific criteria by which we may objectively judge whether or not those accomplishments can be claimed. The IMS fixes these events and accomplishments in sequence and time.

The IMP/IMS should have been organized such that it is perfectly clear which product team is responsible for the work covered. There should be an overall IMP sheet covering each team that may have to be expanded in detailed team planning data, but from which program management personnel may determine what should be happening on the isolated team and relative to work on other teams.

16.2.4   Controlling the advancing wave

One of the most confusing and distracting things about the development of a complex system is that at any one time all of the pieces of the development work are not at the same stage of maturity. They cannot possibly be if we are serious about the notion of top-down development since a sequence is implied. As we develop the product entity structure for our system it expands into the lower reaches as noted in Figure 16.1 originating

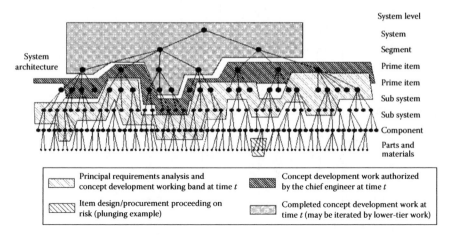

System level

System

Segment

Prime item

Prime item

Sub system

Sub system

Component

Parts and materials

System architecture

| | Principal requirements analysis and concept development working band at time *t* | | Concept development work authorized by the chief engineer at time *t* |
| | Item design/procurement proceeding on risk (plunging example) | | Completed concept development work at time *t* (may be iterated by lower-tier work) |

Figure 16.1 The advancing wave. (From Grady, J., *System Integration*, CRC Press, Boca Raton, FL, pp. 137–150, 1994. With permission.)

with the simple allocation of the customer need to the block called system. Subsequent allocations of identified functionality result in identification of lower-tier elements and requirements corresponding to those items. During the decomposition process there is an advancing wave of development downward through the product entity structure to identify items and requirements for those items.

When this process hits bottom, the wave rebounds upward and we move to the design and integration process. Hitting bottom is characterized by the following: (1) requirements are identified for everything in the product entity structure and (2) everything on the lower fringe will surrender to detailed design by an identified specialized design team or associate contractor or it will be purchased at that level. These two strokes or wave motions must be imbedded in our IMP/IMS, but we also need an effective means to track how well we are doing during execution. That means should also encourage a structured approach defined by requirements approved before design.

Figure 16.2 illustrates one version of a development control matrix designed for that purpose. The matrix tracks the development of each selected product entity structure item in three primary development modes: (1) requirements documentation, (2) concept development (preliminary design), and (3) design development (detailed design). In each case, due dates are identified from the IMS or lower-tier supporting schedules. The dates should be included but were dropped here for lack of space. Each development mode portion of the matrix also calls for an engineering review board (ERB) number entry where the item was approved to progress to the next mode and a status column where you could use numbers, letters, or colors.

| Product entity identification | | Principal | | | Requirements documentation | | | | | | Concept development | | | |
Model Id	Item name	Contractor	Team	Engineer	Type	Number	Rev	Due date	ERB nbr	Stat	Due date	ERB nbr	Stat	Remarks
A14	Cosgrove upper stage	GSTS	PIT	Jones	CI	72-1934	—	02-25-90	051	G	03-10-90	063	G	
A141	Equipment model	GSTS	IPPT2	Burns	DRD	72-2430	—	03-10-90	062	G	04-20-90	068	G	
A1411	EM structure	GSTS	IPPT2	Adams	DRD	72-1122	A	04-05-90	066	G	05-10-90	080	O	
A1412	Avionics subsystem	GSTS	IPPT2	Perkings	DRD	72-3323	—	04-10-90	067	O	05-15-90	081	R	Platform mounting scheme
A14121	Guidance and control subsystem	GFG	IPPT2	Smith	DRD	72-2579	B	05-20-90	083	R	06-25-90		O	
A141211	On-board computer	TELEBOND	IPPT2	Brown	PS	72-2582	—	06-10-90	091	O	07-10-90		O	Chip gidep alert
A141212	Inertial navigation set	REYNOLDS	IPPT2	Windham	PS	72-2585	A	06-15-90	091	O	08-10-90		O	
A141213	Star tracker	GFE	IPPT2	Fletcher	—	S-1267-3	A	02-10-90		G	02-10-90		O	
A14122	Electrical subsystem	GSTS	IPPT2	Blackmer	DRD	72-9333	—	05-10-90	082	G	06-10-90	093	G	
A141221	Battery	RPS	IPPT2	Gomez	PS	72-9354	C	06-10-90	090	R	06-15-90	095	O	Strike scheduled
A141222	Wire harness	GSTS	IPPT2	Chin	NA	72-9365	—	—			06-20-90			
A141223	Power control unit	ESI	IPPT2	Johnson	PS	72-9202	B	06-10-90	089	G	07-25-90			

Figure 16.2 Development control table. (From Grady, J., *System Integration*, CRC Press, Boca Raton, FL, pp. 137–150, 1994. With permission.)

There are many other things that a particular development program might choose to place on such a matrix. But matrix design is inevitably a compromise. You need to select things that help you to focus on status visually. You want to be able to tell quickly what the status is on every level and branch of the product structure shown in Figure 16.2 down to the lowest level at which you choose to manage. Naturally, the PIT might have a system-level matrix and each IPPT leader could have a matrix for their item, thus reducing the vertical dimension of any particular matrix.

Another technique would be to eliminate the status columns on the matrix (freeing up space for other text data) and place the status information on a graphic of the product entity structure block diagram by coloring the blocks in accordance with an agreed-upon code. This would require three different matrices for the three levels of status tracking shown in Figure 16.2 or some scheme to multicolor a block. Either of these techniques can be accomplished on the computer with network access to the graphics.

Whichever status coding scheme we select, we should always see the requirements for an item go green before detailed design work is allowed to move to an in-work status. The author would break up the requirements status tracking reporting into two steps: (1) a requirements list that captures the functional requirements for the item (what does it have to do and how well does it have to do it) and (2) the item specification. Concept development should be allowed to proceed based on an approved requirements list. Detailed design development should not be allowed to proceed without an approved specification. We could have a lively debate about whether preliminary design should be allowed to proceed in the absence of an approved specification. The author believes that it is during preliminary design that many requirements are finally validated suggesting that there could be some requirements volatility during preliminary design. Certainly the specification should be formally reviewed and released before the beginning of detailed design, but the author has no firm suggestion about the timing relative to preliminary design. If your habit is to approve the specification before preliminary design you simply have to be open to the likelihood of maintaining it actively during preliminary design.

16.2.5   Summing up

In summary, that worn-out maxim of good management comes to mind, "Plan the work and work the plan." To which the author would add, "... while keeping one eye on risk." Risks realized can cause a discontinuity in program execution relative to the plan. Therefore, the program must have some way of foreseeing discontinuity triggers and working to reduce the probability of them occurring and/or the degree of difficulty should they occur. Finally, we will discuss how to get back on track after having fallen off the plan.

16.3   Discontinuity management

Perfection is seldom attained. If a program was perfectly planned, executed with good fidelity to that good plan, and everything unfolded during plan implementation as conceived by the planners then everything would go according to plan. Has there ever been such a program experience—probably not. We have to plan programs with imperfect knowledge of the future and this reality practically guarantees that everything will not go precisely according to plan. Even if your plans were proceeding swimmingly your competitors may remain actively tuned to your progress and work to counter your best efforts with new products and advertising. Therefore, we must have a way to detect that we have fallen, or are in danger of falling, off the plan and have the machinery for getting back on. We must have a capability for replanning of any remaining work whenever serious problems arise. Perfection can be approached through good planning as discussed in Parts two and three, and good management leading to good execution can go a long way toward encouraging success, but it is just very hard to know everything that will happen during program execution at the time you are making your plan. We must plan a program with imperfect knowledge of the future.

16.3.1   Discontinuity defined

A program discontinuity is an interruption in planned activities. It is a condition that precludes continued efficient execution of work in accordance with the predetermined plan. There are three principal factors involved in determining whether conditions of discontinuity will develop during program execution:

1. Planning quality
2. The degree of correlation between the assumed conditions that will be in place during execution while planning and the reality during execution
3. The fidelity of plan execution

Figure 16.3 is a three-dimensional Venn diagram illustrating the possibilities from these three factors. In the simple case of binary possibilities for each variable, there are two cubed, or eight, outcomes. The preferred outcome during execution is, of course, that we had planned well, executed the good plan faithfully, and conditions in effect during execution were those for which we had planned. All of the other combinations will yield less-desirable results. One possible exception is that a failure to execute a bad plan faithfully amidst radically different circumstances than planned could conceivably turn out very well. This is a condition of almost total discontinuity, however, involving instinctive management without a plan that requires

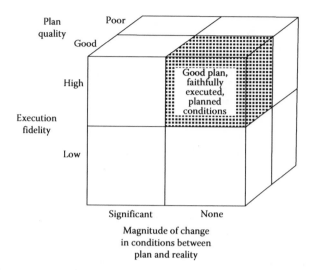

Figure 16.3 Discontinuity cause cube. (From Grady, J., *System Integration*, CRC Press, Boca Raton, FL, pp. 137–150, 1994. With permission.)

a lot of good luck and unusually fine intuitive management skills. The reader will have to look elsewhere for guidance for this possibility. If the management team is this good, think what they could do with a good plan.

We generally think of discontinuities as being bad, but it is possible that conditions can change for the better and we may wish to change our plan to gain full benefit from these changed conditions. For example, our plan could involve using a very expensive technology because another is thought to be too far out in development. During our program execution, that other technology might become available in practical terms making it possible for our customer to save hundreds of thousands of dollars. Certainly, we would want to embrace this new technology no matter what our current plan calls for.

16.3.2   Discontinuity detection

A program discontinuity can arise from several sources suggested from Figure 16.4 and we need a detector tuned to each of these sources. The three principal program planning focuses are technical or product performance, cost, and schedule. Most everything that can go wrong in terms of plan quality, execution faithfulness, and changed conditions can be expressed in these terms. Some people, including the author, would add technology list to this for many programs even though technology unavailability can be expressed in terms of cost, schedule, and performance. A new development program that is pushing the technology will benefit from this added trigger. A program that can rely on existing technology need not include

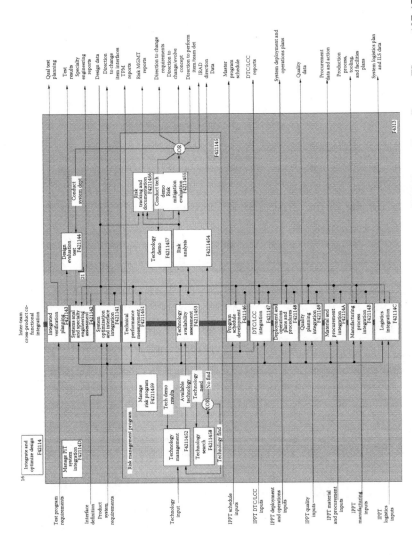

Figure 16.4 System integration process. (From Grady, J., *System Integration*, CRC Press, Boca Raton, FL, pp. 137–150, 1994. With permission.)

it. This gives us four triggers for which we must arrange detectors. For particular programs, or just because of historical precedent, a particular company may choose to add other specific triggers to this list.

The detectors we need are really tuned to what commonly are called risks. Risks are potential program discontinuity triggers that should be avoided, mitigated, or corrected for. Wherever possible, we should avoid them by detecting their potential appearance in the future and taking action in the present to ensure that the conditions needed to bring them into existence never occur. When a risk has been recognized, but is not yet disrupting our plans, we need to be actively mitigating it to prevent it from becoming an open discontinuity. A risk that is fulfilled is a program discontinuity that must be removed through replanning.

16.3.2.1 Cost and schedule triggers

The Department of Defense (DoD) requires its contractors to use a certified C/SCS or an EVS for tracking and reporting program cost and schedule information. The information these systems produce is adequate to detect possible or realized discontinuities from these two triggers.

Figure 16.4 shows one function of the PIT to be Program Cost/Schedule Control. This responsibility could, of course, be assigned to a program business team alternatively. This activity must constantly monitor the performance of all of the IPPTs and the PIT against planned expenditure and achievements using the EVS. Deviations beyond predetermined boundary conditions trigger a potential risk that has to be studied by someone responsible for program risk analysis. The corrective action generally applied is that the offending party must complete a variance repot that explains why the variance occurred and what that person is doing to cause future reports to be in compliance with the plan.

16.3.2.2 Product performance trigger

DoD customers also often encourage the use of a technique called TPM to detect problems in satisfying the key technical product requirements. TPM involves customer and contractor agreement on a small list of key quantified parameters, parameters the condition of which signals the general health or illness of the complete program and product system. Weight is a common parameter chosen where this is an important characteristic, such as in aircraft and missile systems. The contractor will track the value of these parameters in time against the required value and make predictions, based on planned work, how the value will change in the future relative to the required value.

The history of past values and future predictions of these parameters is reported periodically to the customer in graphical form. If a TPM parameter is outside predetermined boundary conditions, the customer will expect the contractor to explain the reasons and what they are going

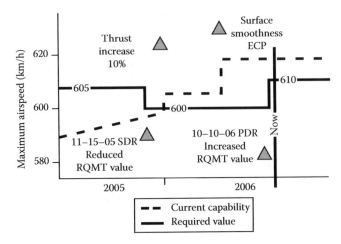

Figure 16.5 Typical TPM parameter chart. (From Grady, J., *System Integration,* CRC Press, Boca Raton, FL, pp. 137–150, 1994. With permission.)

to do about getting the parameter back on track with its planned value. As noted in Figure 16.4, the PIT TPM function feeds its concerns to the person responsible for risk analysis.

Figure 16.5 illustrates a typical TPM chart showing the requirement and current value historical traces. When the current value or prediction of future value exceed the agreed boundary conditions, the contractor will be required to brief their planned actions to bring the parameter on track. Boundary conditions may be defined to permit the contractor some slack within which they may manage the program without undue customer micromanagement.

16.3.2.3   Technology trigger

There is no comparable highly organized mechanism in common use for technology-triggered discontinuities. It is not difficult, however, to invent one here and now. First we need to know what technologies our design will have to appeal to. Each development team or principal engineer should be called upon to survey needed technology for the planned design concept. Concurrently, the manufacturing, quality, logistics, and operations people should be doing the same thing for their own fields based on the current product and process design. We then must compare the content of these lists against a list of technologies available to us. These technologies may be available to us because they are freely available for anyone's use, we know how to gain access to them on the open market, or we hold them as proprietary properties.

There is another class of technologies that drive program managers to insanity that are said to be just over the horizon. If we are depending on one or more of these coming in during our program, we are probably setting ourselves up for a grand discontinuity by plan. All immature technologies should be recognized in our risk management concerns and accompanied by risk mitigation plans corresponding to the case where they do not come online at the needed time.

As the first gear in our mechanism for a technology trigger, we should require design groups (manufacturing and logistics as well as product) to identify the technologies they intend to appeal to for each design concept during preliminary design. The system engineering community (in the form of the PIT) should review this mix against available technologies, as indicated in the Technology Availability Assessment block of Figure 16.4. The program must first have a way of knowing what technologies are available that might have a program relationship, and this may require a specific technology search or simply the application of an edited list from a comparable prior program.

In all cases where there is a match between required and available technologies, there is no triggering condition present. Where it can be demonstrated that a technology needed by an IPPT does not now exist on the program list or there is a likely probability that it will not be available in a practical sense at the time it will be needed on the program, there exists a technology-discontinuity trigger event and corresponding risk. The appropriate response is to conduct a technology search followed either by adding a found technology to the available list or signaling a technology risk to the risk analyst.

The outcome of the risk analysis process may be direction to the designer to change the design to avoid the unavailable technology or a technology demonstration to develop and report the new technology needed for the existing design concept. The later path should carry with it a periodic monitoring of the status of this new technology as the time approaches when it must be mature. If there is a risk that it may not be available in a timely way, it may be necessary to run parallel development paths to mitigate the risk of the technology not coming in on time. In this case, it may be possible to convince the customer that the alternate technology will be adequate for initial capability, even though it does not meet the system requirements, and that it can be replaced at a particular cut-in point in production. The customer may agree to a deviation from specification requirements for a particular number of articles that will be corrected when the needed technology becomes available.

The U.S. DoD has shown a great deal of maturity in the life cycle model being applied when this was written calling for employment of mature technology feeding into each cycle of the spiral development life cycle process. The author can recall programs where as many as three critical technologies had to be developed concurrently inside the program that depended on magic happening on schedule.

16.3.3   Risk assessment and abatement

When any of these triggers are activated for a specific risk, they should stimulate the PIT and/or program management to a particular course of action. First the risk should be studied to determine the nature and scope of exposure. What specifically will happen if the risk becomes a reality? Then we must try to ascertain the probability of the risk materializing. A risk with a combination of very serious consequences and a high probability of occurrence should be dealt with as a serious matter. A risk with minor consequences and low probability of occurrence may be set aside. In between these extremes we need a policy for selecting risks for mitigation. The problem is that we have limited resources and cannot afford to squander them on every possible problem. We wish to focus on real problems that we have to deal with now.

It is helpful to have a special form or worksheet for evaluating risks that encourages the person most knowledgeable about the risk to provide the information needed by those who will decide whether to spend scarce resources on mitigating it. Blanchard (2003) and the Office of Naval Acquisition Support pamphlet ONAS P 4855-X offer examples of worksheets for this purpose.

The program should also have a means to communicate in a summary way what risks are being managed and what their status is. Figure 16.6 illustrates a fragment of one graphical way of doing this developed or adapted by Rikha Patel, a system engineer at General Dynamics Space Systems Division for the Advanced Launch System (ALS) program following Willoby's risk template approach. Each major program area is identified in a hierarchical structure (this could be the WBS) with subordinate areas in each case listed. For each block on the diagram, three blocks corresponding to the three major risk types (technical, cost, and schedule), offer the analyst a place to enter the corresponding risk probability. The risk level is given by letters for the three risk types: low risk (L), medium risk (M), and high risk (H).

The PIT and program management use the status summary to determine on a program scale where the most serious risks are and how to allocate available budget in the most intelligent way for the best overall effect. Each active risk must have a principal engineer assigned to work issues associated with that risk and take appropriate mitigation actions. Periodically the PIT, program management, or a special risk management board should meet to assess the current status on all active risks and provide direction for future work on those risks. Where a risk has been fully mitigated, energy should be focused in other directions. As new risks become identified and determined to be serious, they should be formally accepted into the set being actively managed.

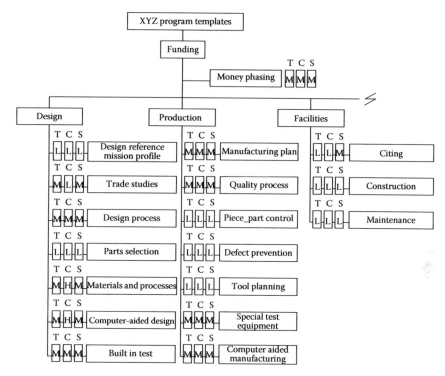

Figure 16.6 Program risk assessment status summary. (From Grady, J., *System Integration*, CRC Press, Boca Raton, FL, pp. 137–150, 1994. With permission.)

16.3.4   Formal risk identification

At any one time, a program may be managing 25 risks using the assessment summary illustrated above. If the risks are well mitigated and conditions permit, none of them may ever become program discontinuities requiring program replanning. At the same time, if we have done a good job of monitoring the discontinuity triggers and converting them into risks to be mitigated before reaching a discontinuity status, no discontinuities should befall us that do not flow from our on-going risk management program. An exception would be an act of God like an earthquake tearing our factory in half or a flood carrying away downstream the factory of a prime supplier.

These are cases where a discontinuity materializes instantly without passing through our best efforts to prevent them from happening. If we include every possibility, however remote, in our risk mitigation work, there may be insufficient resources to do the actual program work, so some nonobservable risks will remain despite our best efforts.

Our periodic review of the risks that are being managed should ask if any of these risks have reached a condition where we can no longer manage related progress in accordance with our current plan. This could be another indicator that is tracked on our chart. When we must answer this question with a yes, we must accept the need to replan part or all of the remaining program depending on the scope and seriousness of the discontinuity.

Risk is often measured using a dual axis criteria dealing with the probability that the concern will be realized and the degree of difficulty it will present if it does come to pass. This makes it a little difficult to track a single risk parameter over time and the way many people apply the dual axis system makes it difficult to accumulate a program metric that can be tracked over time. A variation on the safety hazard index described in MIL-STD-882 offers a way to measure risk with a single parameter that responds properly to characterize instantaneous values and a historical record for the program.

Figure 16.7 shows the risk matrix. Tables 16.1 and 16.2 provide the dictionaries explaining the values entered on the matrix axes that are intended to help people determine an appropriate value for a particu-

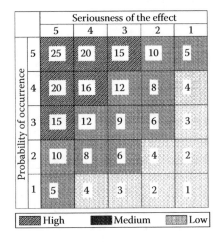

Figure 16.7 Risk matrix.

Table 16.1 Risk Probability of Occurrence Criteria

Cat	Title	P(O)	Description
5	Nearly certain	0.95–1.00	Will occur at least once during program
4	Probable	0.75–0.95	Will probably occur once during program
3	Possible	0.50–0.75	May occur during program
2	Unlikely	0.25–0.49	Will probably not occur during program
1	Nearly impossible	0.00–0.24	Will not occur during program

Table 16.2 Risk Effects Criteria

Cat	Title	Description
5	Catastrophic	Program in jeopardy of termination
4	Serious	Serious damage to program
3	Moderate	Problems cause program focus difficulties
2	Minor	Problems that can be easily overcome
1	Null	No problem

lar risk. The intersections of Figure 16.7 contain an index number that is simply the product of the axis numbers.

If you were to compare this information with the MIL-STD-882 safety hazard matrix, you would find that the safety hazard matrix offered in the military standard uses letters for one of the two axes and that the highest hazards (risks) have the lowest indices. Our matrix in Figure 16.1 uses numbers on both axes and the index values are higher for more serious risks. Therefore, it is possible to apply the index values in a mathematical sense as a program metric. Given the six program risks listed in the program risk list shown in Table 16.3 with the indicated risk index values, the instantaneous program risk index is 97. The author prefers the Table 16.3 display of risk to the Figure 16.6 display, but they are actually only two of many possible ways of exposing the current risk.

Thus, we have arrived at a program risk index or metric. If we maintain a chart of this metric over time we see that it characteristically will

Table 16.3 Program Risk List

Risk NBR	Risk title	Prob.	EFF	Index	TM	Principal	SUSP
2	Life cycle cost	4	4	16	1	Burns	02-10-07
5	Payload capacity	4	5	20	1	Adams	03-08-07
7	Stoddard supplier risk	5	4	20	3	Thornton	04-20-07
12	Program funding	3	3	12	0	Connolly	03-10-07
15	Computer software schedule	5	5	25	4	Sampson	05-23-07
21	Maintainability	2	2	4	1	Not assigned	—
	CURRENT PROGRAM INDEX			97			

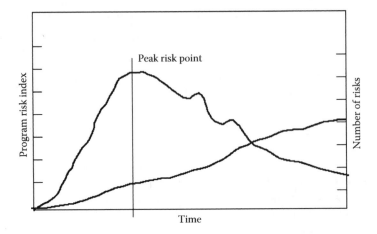

Figure 16.8 Program risk index tracking chart.

rise early in a program reflecting the identification of risks but a delay in mitigating them causing an early rising metric as shown in Figure 16.8. As a program progresses, this value will rise to a peak at some point in the program and subsequently start a long decline. Risks continue to be identified so we see the accumulated total number of risks continue to increase but the program is being successful in mitigating earlier risks and later risks are commonly lower in index value than those identified earlier in the program.

16.4   Program replanning

We can apply the same techniques in replanning that we applied in the original planning activity. Clearly, we must identify the scope of the tasks impacted by the discontinuity. These tasks, or an alternate set of tasks, must be defined in the IMP format as they relate to major program events. The interfaces between these tasks and other program tasks must be studied to determine if the critical path is influenced possibly triggering other schedule risks not previously identified. If possible, the original major events schedule should be respected unless the customer indicates otherwise. This may require parallel rather than serial work performance or overtime to avoid overall program schedule impact. Unless the schedule was very optimistic in the first place or prior replanning has absorbed all of the slack, there may be some margin in the schedule that precludes major program rescheduling.

chapter seventeen

Design modeling and simulation

17.1 What is a model?

A model is a representation of another entity to show the construction or appearance of that other object as a precursor to actually building the real object. In Chapter twelve, we discussed several models useful in gaining insight into the requirements appropriate in defining the problem that program design teams will be held to account to synthesize. Often, these same models can transition into useful design models. This is certainly true of UML and SysML models perhaps less so of traditional structured analysis models used in system and hardware requirements analysis by some. Models that have this characteristic are sometimes referred to as implementable models meaning that they can be operated reflecting how the intended reality that is being modeled will function. The software world has been fairly successful in evolving these kinds of models and it may come to pass that SysML will provide that capability for hardware developers. The closest to this capability in the system and hardware world are provided by behavioral diagramming used in Ascent Logic's RDD-100 and enhanced functional flow block diagramming employed in Vitech's CORE. These two modeling techniques are essentially the same just rotated 90° from one another. Refer to *System Requirements Analysis* (Grady, 1993, 2005) for details of these two models.

A prolific author in the modeling field Bran Selic, at the time this was written an employee of IBM Rational, has offered a list of characteristics for a useful model paraphrased as follows:

1. The use of **abstraction** to emphasize important aspects while removing irrelevant ones
2. Expressed in a form that is really **understandable** by observers
3. Fully and **accurately** represents the modeled system
4. **Predictive** such that it can be used to derive correct conclusions about the modeled system
5. **Inexpensive** meaning it is much cheaper to construct and study than simply building and observing the modeled system

17.2   How are models used in the synthesis work?

As we move from problem space definition during the system development process into solution space design we need a set of design models
to support our efforts because it is cheaper and easier to manipulate a
good model to observe the effects of different solution possibilities than
it is to build a prototype. While doing some work for Caterpillar several
years ago the author was struck by a comment from several engineers
and managers that they were tying to move from the use of as many
as four sequential prototype versions of a new product, referred to as
iron verification, to the use of simulation to evaluate different design
possibilities.

During the second proposal cycle for the Advanced Cruise Missile
at General Dynamics Convair in the mid-1980s, the estimating people on
the proposal team asked for help from system engineering to improve
their material cost estimate. The author applied a standard Convair solution to this problem called a system equipment list that was intended to
capture all of the entities that would have to be purchased and built during the development and production of the missile. During that cycle we
discovered a previously undisclosed list of design support resources of
considerable cost that included a complete software development lab with
special and costly features needed to support the development of classified software. But in addition, we discovered we needed several models of
various subsystems and design features. The U.S. Air Force insisted on a
set of full-size physical models of the missiles that had to have the external shape and smoothness, center of gravity, and mass of the missile that
could be uploaded to all of the attach points on each of the three bombers
that would carry it on its mission and actually launched in various points
of the bomber airspeed-altitude envelope singularly as well as jettisoning
the whole load to ensure that they would all part the bomber safely.

At the time the author was reminded of two stories from the past
that encouraged an attitude that these might be necessary. During the
development of a Ryan Aeronautical Model 147J low altitude photo reconnaissance aircraft on its first air launch from a wing pylon of a DC-130
aircraft, a flaw in the launch logic circuit resulted in it rising up to take
off the number 3 engine of the launch aircraft. The author also recalled a
story told by a former Marine fighter pilot in a bar in South Vietnam of
his experience on a close air support mission in Korea. He flew into the
bomb-drop position in his Corsair and released his weapon noticing that
it was rolling down the underside of the wing. It rolled to the end of the
wing and it started rolling down the top of the wing. He paused for a big
drink of beer while a younger pilot asked excitedly, "Well what did you
do Colonel?" He responded that he didn't know what happened but he
thought he had screamed so loud that it fell off.

Perhaps the proper release of Advanced Cruise Missiles could have been determined using a wind tunnel or unaided aerodynamic analysis, but there is something that is very reassuring about going aloft and dropping things. Realistic testing tends to produce credible results. The program could not, however, wait around until the flight test program because it needed to know early in the development work whether or not the shape defined by both aerodynamics and radar reflectivity would support clean separation.

The program also discovered a need for several models needed to understand and verify the design of the electrical umbilical interface between the three bombers and the missile. A fuel development bench was identified to model plumbing system use of orifices and redundant plumbing to encourage mission completion even with damage. Several other physical models were also identified as well as several math models and simulations.

Engineers need models to validate their ability to satisfy difficult requirements as early in a program as possible and to permit examination of alternative design concepts. Man's imagination is powerful, but when the complexity of the object of human thought rises to a certain level it cannot be depended upon to derive flaw-free conclusions.

Figure 17.1 identifies the activity involved in the development of a good design. Models should be used as a basis for deriving the requirements that define the problem to be solved because the simple graphics in these requirements models focus human thought on the important aspects of the problem space. Ideally, the use of these models could be extended into the design phase to examine potential solution space possibilities. As noted above, this is possible with some models particularly in the software arena. But, whether this feed through of requirements modeling is possible or not, the design community needs an effective

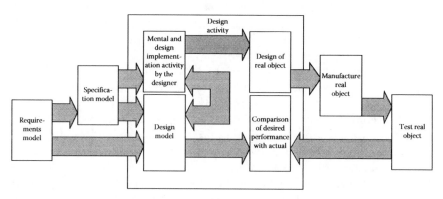

Figure 17.1 Modeling feed through.

design modeling capability to support their thinking and design efforts. When the real object becomes available and its testing reveals that it works exactly the way it was modeled and predicted, one cannot help but conclude that the organization that accomplished the development work was characterized by good engineering integrity. When the real results differ significantly from the modeling results it is a symptom of serious problems in the design organization. This is a very disturbing but vitally important lesson learned from a program signaling a flaw in the resources provided to programs that the appropriate functional departments should take action to prevent reoccurring on future programs.

17.3   Use of models and simulations

17.3.1   Math models

Many of the specialty engineering disciplines employ math models. The reliability and maintainability, the mass properties, and the DTC/LCC math models are representative. All of these disciplines should be using precisely the same product entity structure and collect numerical values in their discipline relative to these items. These allocations should fall into the corresponding specifications. A good model should permit capture of an allocated value, a margin, and the current achieved value (predicted value based on an evaluation/analysis of the actual design or a measured value on the actual product). With these quantities captured, it should be possible to use the model base for finding management space when it has been shown that the current requirements cannot be satisfied.

Ideally, these models would be interactive with the requirements database system such that the requirements values in the models may be used to refresh the requirements database at particular baseline points in time respecting a fundamental characteristic of a good information systems—a single entry and location for any one data element under the responsibility of a single authority.

Queuing models may be used to model many situations where a service is to be performed as in maintenance work on the product system. It is often possible to predict with fairly good accuracy how long it will take to accomplish some maintenance action under ideal conditions. But the dynamic situation with arrival of product items for service randomly in time is much more difficult to deal with without queuing models. Both the arrival rate and service rate in these situations in the real world are random variables, but it is possible to predict the performance of a particular service case and determine the best condition of balance between service and cost.

17.3.2   Physical models

Often it is necessary to build full-size or scaled physical models of the product though the need to do so has been reduced significantly by the use of three-dimensional computer-aided design (CAD) tools that permit engineers to grasp the physical relationships between entities and to design packaging solutions with clear knowledge of available space and the configuration of the available mounting surfaces. There are many other cases where physical models remain in demand. A scale model reflecting very accurately the external shape and surface characteristics is needed for wind tunnel tests. A scale or full-size physical model reflecting the external shape and radar reflectivity is needed to support signature testing. A full-scale aircraft fuselage accurate in its construction is needed to support structural and life testing. The reader can no doubt add to this list from his or her own field of experience.

17.3.3   Prototypes and pilot models

Over the years, since computers have been available in support of the development of product systems, a revolution has been in progress in system development. We have been able to create product representations early in the development period that are easily manipulated permitting evaluation of alternative design solutions before a large investment has been made in the design and manufacturing processes. Some companies are still developing two or three cycles of prototypes and pilot models in a progressive move toward the final production product while in other industries the heavy use of models and simulations has significantly reduced or eliminated the need for costly construction of physical prototypes.

Vehicles of all types seem to be one product sector where prototypes are still heavily depended upon because the use of models and simulations has not been sufficiently advanced. Auto producers still build prototypes and run them on test tracks and road routes to see how humans respond to the feel of the vehicle. Part of this reality perhaps is that a human is still the best judge of such an intangible characteristic as ride quality. But, part of the problem also no doubt is that adequate models and simulations have not been perfected to permit direct observation of the driver effects of particular combinations of steering ratio, power assist boost, spring rate, shock absorber tension or orifice size, tire pressure and side wall structure, front-to-back weight balance, and numerous other factors.

The auto industry in particular has not yet found the right product interfaces for procurement purposes. They began by procuring master brake cylinders, hoses, brake pedals, and brake calipers from various vendors and assembling them into brake systems on each vehicle produced. They have begun procuring brake systems where all of these parts and Automatic Breaking System (ABS) controllers are purchased as a system. However, the system is cut at the wrong interface planes making it difficult to apply models and simulations effectively. They would be better served if they elevated the procurement boundary higher still to purchase vehicle control systems that included the four corners complete with tires, wheels, brake assemblies, suspension components, steering components, and all other parts that couple these elements to the frame of the vehicle and the driver of the vehicle. At this level it would be possible to create a simulation that modeled the whole vehicle control system. If they wished to produce all wheel drive systems, then the plane might have to be moved further still to embrace the axles, differentials, and possibly even the transmission.

With the procurement boundary aligned with the true subsystem boundary, it will be possible for the supplier to develop a subsystem simulation accounting for all of the engineering factors related to that subsystem reducing the dependence on physical prototypes and pilot models and encouraging reduced cost and schedule impacts in the development of better systems.

One problem with this solution is, of course, that the auto final assembler will want to retain system-level mastery over key elements of their product line. If the interface is cut as high as suggested above, they will lose that mastery of perhaps the whole control system. Therefore, a cooperative relationship between Original Equipment Manufacturer (OEM) and supplier may be the best position where the OEM shares models with the supplier.

17.3.4   Descriptive models

System and software engineers build descriptive models of a system to be as a means of gaining insight into what the system must do and how well, what it should consist of in terms of physical entities, and how the system should behave. They include problem space models that help to investigate the problem to be solved and solution space models that explore the product entities required, the relationships between them in terms of interfaces, and the environment within which they will function. The problem space models include traditional structured analysis (TSA) employing functional flow diagrams of various kinds, state diagrams, modern structured analysis (MSA), process for system architecture and requirements engineering (PSARE), object-oriented analysis (OOA), unified modeling language (UML), system modeling language (SysML), and Department of Defense Architecture Framework (DoDAF). Each of the models noted include their own space models, but the author's preferred set include product entity diagrams, schematic block (or n-square) diagrams, the specialty engineering scoping matrix, and a three-layer environmental model. All of these models are explored in *System Requirements Analysis* (Grady, 2005).

Generally, these models are not executable in that they cannot be caused to operate in a dynamic fashion. They are useful as a means to come to an understanding of the static features of the system being created and are generally satisfactory as a basis for deriving the requirements that will populate the specifications that synthesis work described in this book will attempt to translate into an appropriate design solution.

17.3.5   Executable models

Using the definition of the word model meaning a simplified representation of a system or phenomenon, all models, most often including prototypes, are not fully representative of the final intended product which, after all, is not fully understood at the time models are contemplated, and this is one of the reasons why models are contemplated. Physical models appeal to multiple human senses and therefore have a tremendous influence on the mind of the designer. Increasingly, computer simulation is capable of stimulating multiple senses as well and doing so with tremendous flexibility and realism.

An executable model can be manipulated while operating them simulating the operation of the real system such that the observer can deduce useful features of the system before a lot of money has been spent designing the system.

Computer simulations may be as simple as a few lines of code or very substantial and complex constructs. The latter permits extensive examination of alternative design solutions as well as a search for optimum values of key characteristics. It is necessary to describe the problem to be modeled, define the input data required, the mathematical and logical relationships between the data entities so as to produce the desired output data or condition. There will be a set of independent variables that are defined for a particular condition of interest that results in model determination of desired dependent variables. When the inputs are changed the operation of the model can involve a simple response to a new static equilibrium or some form of continuous operation at a different set of values of the dependent variables.

For many of the problems found in systems that involve very complex problems the simulation will involve sets of differential equations dealing with accelerations, velocities, directions, and positions. The model may involve human inputs or human in the loop operating in response to stimulus from the model.

A simulation should grow with the program. Its initial manifestation may be relatively simplistic but still yield useful results during a time when product knowledge is fairly immature. It may also consist entirely of software. As the system design matures, the prototype can evolve into a combination of software and hardware reflecting a progressively more perfect model of the intended system. In a system involving a control system implemented in electronics and software, for example, the simulation can be upgraded from pure software to replacement of some of the software with breadboard units. As qualification units complete testing, and to the extent that they have residual life remaining, they can be substituted for the breadboards. The result will be increasingly sophisticated answers in step with the increasing knowledge about the problem to be solved.

17.4   Representation configuration control

During the early design phase we will use many models and other representations of the product system. It is extremely important that those responsible for those representations maintain configuration control. At the time we begin to use these representations unfortunately we commonly fail to fund configuration management on the program, so we have no one with those skills available. A simple way to maintain representations configuration control is to first list all of those used and then assign a responsible engineer for each. Make each of these principal engineers responsible for knowing the configuration at all times and how they compare with the current product baseline. Earlier the notion of engineering integrity was mentioned and the way it is often violated is that someone responsible for a representation applies it in the wrong configuration producing results that are erroneously believed to be appropriate to the current configuration and those results ripple through other disciplines causing a distribution of errors. If this happens again in another area it will eventually come to the surface, but with such a clouded pathway that it will be very difficult to determine how the program got itself into this horrible situation.

Throughout the development of the product, the models and simulations will provide a basis for decisions leading to the final design configuration. Once the product has been produced and subjected to testing, the results of the testing of the real product should be compared with comparable results from the models and simulations. If they are not the same, it is necessary to find out why and make such changes as necessary to cause them to produce the same results. The result will be validated models that can be subsequently used to study future modifications, evaluate reports related to system use, and develop additional operational scenarios.

chapter eighteen

Product design decision making

18.1   Concept development

Concept development is the first of three stages through which designers and the teams on which they serve pass between completion of the requirements analysis for the product and availability of the final design ready for verification that it satisfies those requirements. The other two design stages are preliminary design and detailed design, both of which will be covered after the concept stage.

18.1.1   The requirements

Specifications define problems. Ideally, they would be prepared before any money is spent on detailed design. Some would say that the specification should be released before any budget is spent on design, period! But, that is probably not realistic. While performance requirements can be defined prior to allocation of functionality to a physical entity, it is nearly impossible to define design constraints (interface, environmental, and specialty engineering requirements) without some knowledge of the design concept for the item covered by the specification. The interface requirements must be defined in context with the technology associated with these interfaces (electrical, mechanical, hydraulic, etc.), for example.

Now, engineers are not naturally gifted in problem defining. Engineers endure a 4–5 year education moving from textbook to textbook each containing problems to be solved at the end of each chapter. An engineer is a problem solver applying today's science and mathematics to the development of useful artifacts. This poses a problem for the development organization. The design engineer could be made responsible for leading the development of a specification for the item in question prior to shifting to the design process. The strength of this approach is that the engineer will understand the requirements very well. The weakness is often that the designer slips into the design process before the specification is complete. In some organizations, the specification may even be released after the engineering drawings. The other alternative is that someone else be placed in charge of the specification development. In this case, the designer may not as fully understand the requirements, but the specification may be developed with full content that is design independent as a prerequisite to detailed design.

There is a particular way that design engineers responsible for specification development become convinced that it is okay to release the specification subsequent to the engineering drawings. The Department of Defense often insists on the preparation of two part specifications. The first part variously called a development, part I, or performance specification, should contain design-independent content only and be prepared prior to the beginning of detailed design. This part is the basis for the design and qualification of the item. The second part variously called, a product, part II, or detail specification, contains design-dependent content and is the basis for acceptance. This specification is commonly released after the engineering drawings since its content is dependent upon the design solution.

Many organizations prepare one-part specifications for procurement and in-house purposes and an engineer may fall into a habit of rationalizing that it is okay to release the whole specification subsequent to the design because part of the content is design dependent and it will cost less to go through the release process only once. Where one-part specifications are used, they should be released in two steps with the design-independent component released prior to detailed design and then reissued with design-dependent content concurrently with or subsequent to drawing release.

18.1.2 The bridge between problem and solution

Most system engineers would hold that requirements must be defined prior to design and this is a good general rule. The realities of the design process encourage a compromise on this principal, however. Design of things to satisfy complex needs is seldom accomplished in one continuous flow of thinking. Many engineers and development organizations have concluded that a two- or three-step process is preferable. Three design steps commonly applied are concept development, preliminary design, and detail design. Some would merge the first two into one. In either case, there is a proper interaction between the early design process for an item and the requirements for subordinate items. When this process is applied well, it flows like a kind of zigzag movement down through the requirements and designs. We define the requirements for an item, hopefully through some form of structured analysis, but the design concept for the parent item will commonly drive identification of some derived requirements for the item and will definitely do it for the subordinate items. We therefore alternate between requirements analysis and design concept development, which uncovers some necessary characteristics of the next lower-tier requirements.

Figure 18.1 illustrates this process. Ideally, the program will apply a structured front end if it entails solving an unprecedented problem. Functional analysis of some form will give us insight into product entity decisions, possibly arrived at through a trade study, and appropriate performance requirements. Once the product entity structure is known, it is

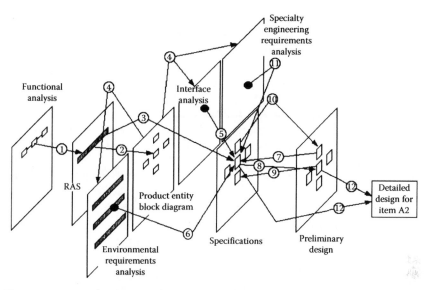

Figure 18.1 Design and requirements interaction.

possible to define the design constraints in the form of interface, environmental, and specialty engineering requirements. Ideally, the design concept data would be captured in a design concept record of some kind. The software world recognizes this kind of document but sadly the hardware world commonly does not. The content is exactly what is needed to support the preliminary design review so it is money well spent. In fact, the author conceived Figure 18.1 while observing the excellent concept record documentation approach applied by what is now Electromotive Diesel, previously the EMD division of General Motors, that develops locomotives.

The product entity structure is determined through functional analysis and allocation (1 and 2) in the unprecedented development case. The performance requirements captured in the item specification come from allocation of the functionally derived requirements (3). Knowledge of the needed product entities triggers identification of specialty engineering requirements (4) and environmental requirements (4) for the item. Interface analysis is based on the analysis of function–product entity pairs interface requirements (5). A three-tiered (system environmental standards, service use profiling, and zoning) environmental requirements analysis will give insight into item environmental requirements that flow into the item specification (6). The specialty engineering requirements developed by the many specialists flow into the item specification (11). With the parent item installed in the system product entity structure, the design people should be enabled by budget to accomplish the concept development work for the parent item (10). The features of the parent item design will drive some specialty engineering requirements for the item at the next lower tier (7) and may encourage other design-derived requirements for the item. The requirements for the item hopefully will drive the preliminary design (8) and detailed design (12) of the item. That design will also influence the design of the subordinate items (9).

For a precedented problem space, it may be possible to de-emphasize the functional analysis and concentrate on performance requirements differences driven by the functional differential between former and new system functionality. Higher-tier design decisions based on these changes will drive lower-tier design constraints. Where the precedented system structured analysis was not preserved prior to the precedented development, it should be back filled for the new system if possible and released in the form of a system definition document (SDD) fully described in *System Requirements Analysis* (Grady, 1993, 2005). The design concept record should expand upon the name of each of the product entities identified in the product entity structure describing briefly the chief characteristics of the solution to the problem defined in the item specification, an insight into the technologies to be appealed to, and some ideas about its physical form.

Once the responsible team has an item specification and their inputs to the design concept record ready for release, they should be reviewed at what we might call an item requirements and concept review by the appropriate authority and if acceptable the team should be empowered to proceed in accordance with their part of the integrated master plan and schedule for the program to accomplish preliminary design where they will remove all risk from the detailed design phase to come. A preliminary design review should review and approve or redirect the team moving on to detailed design resulting in formally released engineering drawings or lines of code, analysis reports, qualification verification plans and procedures, and detail specifications or the increment to the previously released performance specification. These requirements will all be design dependent and should be selected based on some minimal set that must be proven to be satisfied as a precedent to delivery of the particular manufactured article to the customer. Acceptance test plans and procedures will be based on the content of the latter.

Many system engineers develop the false notion that they are at the center of the design process. This simply is not so. Everyone in an organization that develops products to solve complex problems, including the system engineer, is in the service of the creative design engineer. The design engineer must synthesize a design that complies with all of the requirements for that item. True, today, designs are influenced by all of the members of a team, but the design engineer is at the center of this process putting together the complete story.

The design comes from the creative mind of a design engineer by whatever route. The system engineer's role should be to provide the design engineer and team with the largest possible solution space with boundary conditions clearly defined. This is counter intuitive to many system engineers who have come to believe that the best specification is the most complete one, the heaviest one. Actually, a specification should clearly define the necessary characteristics only and the smaller the specification the better. This kind of specification will provide the designer with a large solution space while protecting him/her from failure. The smaller the solution space the fewer the possible alternative solutions. The limit of this condition in the extreme is a null solution space where a solution is incompatible with the laws of science and economics. The common result of over constraint is unnecessary cost escalation, schedule problems, and most often failure in meeting all of the performance requirements stated in the specification, some of which might not have been essential anyway, of course.

18.1.3   Preferred solution selection

The designer or team should avoid a leap to a point design solution based on past experience because these solutions commonly do not take advantage of up-to-date technology. An engineer or a team should develop alternative solutions and trade these alternatives against a set of selection criteria. If the design is developed by a cross-functional team, the members of that team can provide a wider range of alternatives than any single person because all of these people will have taken different paths through life leading to different experiences. An approach to selecting the preferred concept that has become very popular in engineering organizations is referred to as performing a trade study or simply a trade. We will consider four different models for a trade matrix running from the very simple to a fairly sophisticated structure.

18.1.3.1   Trade fundamentals

A trade study is an organized process for arriving at a decision between alternatives based on best evidence within a context of incomplete knowledge. Several approaches to satisfy this goal have been popularized, and any of the four offered below are better than an ad hoc decision-making process. All of these trade approaches involve a matrix as a way of concisely summarizing the decision drivers. One of the matrix axes is for identifying all of the selection criteria through which we will determine the best solution. The other axis of the matrix lists the candidates. Therefore, the cells in the matrix provide places in which to record the value of each criterion for each candidate. We then need a means of combining the values for each candidate in such a way that it is clear which is the best overall.

18.1.3.2 *Trade requirements*

There are two sets of requirements of interest in a trade study. First, there should be some requirements developed that apply to all candidates equally, absolutes that must be satisfied by all candidates or they are not accepted as candidates. These are key characteristics that are used as a basis for candidate entry into the study. The second kind of requirement is permitted to vary over a range defined in the study and used as a selection criteria, a part of our value system.

18.1.3.3 *Our value system*

We will decide which candidate is the preferred one based on some value system, hopefully the right one. So, we might first inquire, "Whose value system should we use?" It could be the program manager's, the customer's (if we knew what it was), our understanding of the customer's value system, or that of the person accomplishing the trade study. Generally, it should be based on some mix of customer needs and sound business rules. When numbers are used to indicate value, the value system is captured in how we identify relative criteria value (weighting) and how we normalize the criteria. In addition, what we choose to use as the selection criteria may influence what candidate will in the end be selected. It is critical to clearly identify before embarking on the trade study work details, whose value system will be used and what it is composed of.

It is not an easy thing to define. A good first step is to inquire of the decision maker what the value system should be. In the author's experience that question is usually followed by the answer, "I don't know. Come up with one and lets talk about it." The system engineering functional department of an enterprise would do well to prepare a trade study manual or include practices in a system engineering manual. Over time these practices should be improved based on lessons learned on programs. This may include guidance on weighting factors and utility curve shapes.

18.1.3.3.1 Selection criteria We must score each candidate in each criterion in some fashion. The best way to score the candidates in each selection criteria is to use numbers such that each candidate column may simply be added and the candidate with the largest score is the better one. The obvious problem in doing this is that the criteria do not all have the same units. It does not make any sense to add 35 miles to $102,000, for example. In addition, for some selection criteria, a larger value is better than a smaller while for others smaller is better. These problems can be overcome by normalizing the criteria such that they are all represented by unitless numbers with increasing numerical values of higher selection value.

The simpler trade matrices avoid these problems by not using numbers. One could use colors to indicate preferences, simple words such as bad, good, better, best, or pluses and minuses as some of these models do. These tend to be weaker approaches than those that use numbers, but they have the advantage of being quick and therefore cheap. Quick and good seldom go together and they seldom do in this case either. There is a natural programmatic phenomenon that encourages us to use a trade study approach in making tough decisions. Early in a program we know very little about the problem and its solution. This knowledge increases over time but is seldom sufficient at the time we have to make tough choices between product development alternatives.

18.1.3.3.2 Weighting Generally, the criteria are not all equally important, and to score the candidates without taking this into consideration gives the less important one an upward bias that can lead to selection of an inferior candidate. In order to overcome this problem where we have used numbers to indicate goodness, we can introduce a weighting column in our matrix. Each criterion is identified with a weighting number that is used as a multiplier for the candidate scores in that row.

One can arrive at a weighting by first ordering the criteria in decreasing order of importance. Then pick some number like 10 and partition it into values for each criterion. Alternatively, one could group the criteria into n subsets, where all of the criteria in a subset have essentially equal importance, and then assign n to the highest value criteria, $n-1$ to the next, and so forth. Some engineers prefer to use the factors 1, 2, and 3 with 3 awarded to the most important criteria. Do not apply too large a spread to these numbers. For example, if you conclude that criterion 1 is 100 times more important than the other criteria make the selection based on criterion 1 only.

18.1.3.4 Identifying candidates

The process of identifying and developing candidate solutions is a creative one. Therefore, our trade study team had best have some creative minds aboard. We could simply allow each member to ponder the possibilities individually or within a group and develop a list of preferences. This process could be organized as a brainstorming or senetics session. But, there exist several structured approaches from which two are discussed below.

18.1.3.4.1 Trade trees It is important to visually identify alternatives and one simple way to do this is a trade tree illustrated in Figure 18.2. The solution one is looking for is written down with alternative solutions branching out from it. This diagram can be multilevel in nature where one alternative solution becomes the parent of other possibilities.

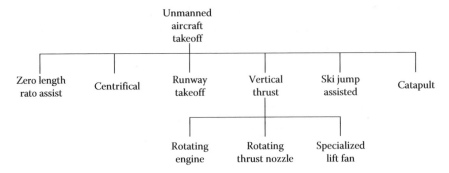

Figure 18.2 Trade tree example.

This technique is only useful for displaying alternatives and we need some other machinery for encouraging a sound selection between alternatives.

It is important to note that we should consider not only single alternatives shown on the trade tree but alternatives composed of combinations of those listed. There are no obvious combinations in aircraft takeoff trade tree offered in Figure 18.2, but consider the AIM-9X Sidewinder missile development problem of directional control. The missile had to be very maneuverable in order to attack hostile aircraft far off to the side of the firing aircraft's direction of flight (off boresight). All of the competing contractors found that none of the single alternatives such as gimbaled engine, engine exhaust paddles, or aerodynamic control fins would solve the problem and they had to pick the combination of exhaust paddles and aerodynamic fins.

18.1.3.4.2 Morphological chart   Nigel Cross in his book *Engineering Design Methods* (Cross, 1994) offers a simple yet powerful way to expose a large number of possible design solutions in what he calls a morphological chart illustrated in Figure 18.3. It provides a visual way to couple combinations of functions and means into a fairly comprehensive design concept while encouraging examination of several alternative combinations.

In the example shown, we are trying to develop a new machine and must decide on the machine's motive power, steering, drive train, and other factors. We simply list all of the functions (Cross called them features) the machine must satisfy (only three listed but the reader may wish to extend this list from their own experience) and then we list all of the possible means for each function. There is no need to list these means in any particular order; just as we think of them is fine. When we have exhausted our mind of possibilities, we can then start connecting means combinations in stings as illustrated for the combination of articulated chassis, wheels, and turbo diesel. This is only one possible combination

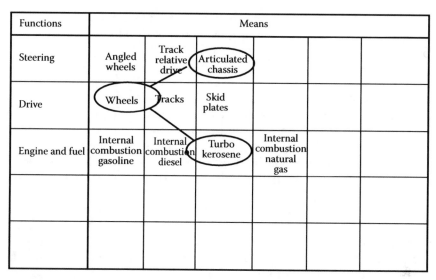

Functions	Means					
Steering	Angled wheels	Track relative drive	Articulated chassis			
Drive	Wheels	Tracks	Skid plates			
Engine and fuel	Internal combustion gasoline	Internal combustion diesel	Turbo kerosene	Internal combustion natural gas		

Figure 18.3 Morphological chart. (From Cross, N., *Engineering Design Methods,* John Wiley, New York, 1994.)

and we can form others in a similar fashion. We may then course screen out weaker alternatives thus formed and trade among a few selected candidates.

18.1.3.5   Candidate development

The design engineer(s) on the team or the engineer doing the trade independently must synthesize two or more candidate solutions. One of them might be the most recent design solution to the problem while others may appeal to more advanced technology. As they are developed, the responsible person must make sure that their candidates satisfy all of the trade requirements.

Some organizations make this a competitive process choosing different persons to develop each candidate and asking them to defend their candidate through the selection process. On very large programs, each candidate may have their own team. This can result in a strange selection process if the trade takes some time to complete. Intermediate in the process the responsible engineers may become aware of candidate preferences on the part of the decision maker and move their candidate features closer to those of the preferred candidate. By the time the final selection is made, all of the candidates have essentially the same score. The selection decision has been made through the movement of all of the candidate engineers toward what they perceive to be the winner. This is, of course, a case of a kind of distributed groupthink and it may simply focus on the decision maker's desires rather than the best alternative.

Those responsible for developing each candidate should use all of the power of their profession just as they would in developing any real preliminary design solution except that there will be a very short time in which to do it. If tools exist such as pertinent simulations and models, they should be taken full advantage of. The candidate should be represented by a series of engineering sketches and corresponding analyses. The responsible engineer should be able to state the requirements as he or she understands them, the features of his or her design, and how the design satisfies them. Also, the engineer must determine honest candidate values for the selection criteria based on an analysis they can defend against questions from people who understand the engineering and science behind the problem.

Trade studies must be accomplished in an environment of truth, honesty, and integrity rather than win at all cost. The decision maker should be honestly interested in the best candidate and not their predetermined preferences. The engineers working on the trade study should be briefed on the need for integrity in the results of the trade unless this is a normal or routine demand on all of the employees of the enterprise, as it should be.

18.1.3.6   Candidate values

Given a selection criteria and a set of requirements that all candidates must satisfy, the responsible engineers or team must develop proper values for each criteria for each candidate. The candidates need not necessarily satisfy the content of the complete specification as the complete specification may not apply to the aspect of the item under trade. If there is no specification, there should be some requirements developed for trade study purposes. Some of the trade study matrix methods discussed below do not require numerical values, rather relative values expressed in some nonnumerical way. These matrix models tend to be less precise, of course, but they also require less time to develop.

18.1.4   Trade study matrix models

Several different trade matrix models have been developed over the years. Some involve numerical values for candidates while others apply relative methods. An engineer who must accomplish a trade study should determine which approach is most appropriate for their situation as a function of how much time and funding are available and how much risk the program can tolerate in a bad decision.

18.1.4.1   Simple entries

Some decision makers frankly do not trust numbers or those who come up with them. We should try to discover, in the process of developing the trade, whether the decision maker is an intuitive or analytical thinker. Some decision makers would be very content with colors entered into the cells of the matrix. Some colors and related alternatives often used are shown in Table 18.1.

18.1.4.2 Pugh concept selection approach

Stuart Pugh has offered a simple trade decision process involving relative values between a set of alternatives and a datum concept that one is trying to improve upon. A plus sign means the candidate in question is better than the datum and a minus sign means the candidate is not preferred compared to the datum. The use of an "S" means the two concepts are essentially the same. Some analysts prefer to also use the symbols ++ and −− to mean a lot better and a lot worse respectively than the datum candidate. Once the matrix is complete, one adds the number of symbols of each type in each column and tries to make a decision on preference. Figure 18.4 illustrates this model.

An alternative matrix could use +1, −1, and 0 for the entries along with a weighting column as discussed earlier. In such a matrix the totals row could be a single row with the entries being the simple sum of the weighted values. In such a matrix we should probably include a raw and

Table 18.1 Alternative Nonnumerical Value Systems

Quality	Colors	Symbols	Words
Outstanding	Blue	++	Very good
Good	Green	+	Acceptable
Marginal	Orange or yellow	−	Marginal
Unacceptable	Red	−−	Bad/unacceptable

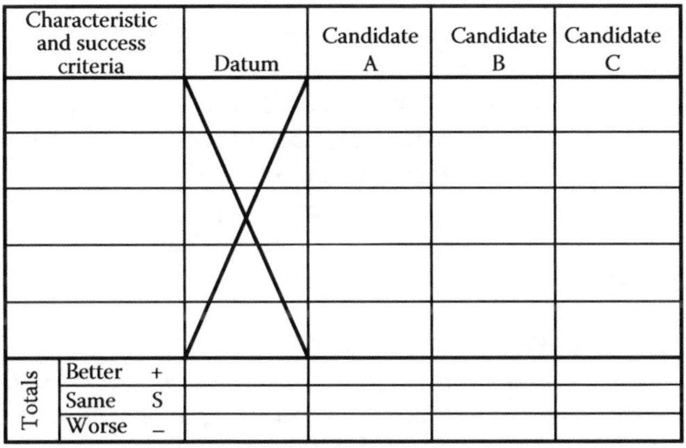

Figure 18.4 Pugh concept diagram.

weighted value for each candidate where the weighted value is the raw value times the weight in that column.

The advantage of the Pugh concept over the other approaches discussed below is that it is very simple and requires little preparation. It is also easy to perform in a team meeting environment very quickly. The disadvantage is that it tends to be less technically precise and may entail significant sensitivity problems that remain undiscovered by the team. There is, therefore, a potential for more risk in the use of the Pugh concept than the other two discussed below. However, we have to recognize that time and money are scare resources at the time we must make this decision and we wish to derive the best decision we can afford.

Where the Pugh concept can be used with relatively low risk is in industries where the principal development approach involves prototyping and model year improvements rather than development of new concepts from scratch in a straight through development effort. The more uncertainty in the requirements and the more demand on advanced technology, the less advantageous the simpler trade approaches are.

18.1.4.3   Technometric trade study approach

A. B. Arieh, H. Grupp, and S. Maitial describe a trade approach that adjusts candidate values to a range of values between 0 and 1, thus removing units and adjusting all values for an increasing sense such that they may be added or averaged, that is, normalized. Figure 18.5 illustrates an example using four characteristics two of which have a positive slope (greater range and payload is desirable) and two of which have a negative one (less cost and radar signature are desirable).

First, raw values (RV) are determined for each characteristic then technometric values (TV) determined from these. The rules for calculating the TV parameter for a characteristic with a positive slope is $TV = (RV_I - RV_{MIN})/(RV_{MAX} - RV_{MIN})$. For a characteristic with a negative slope the formula is $TV = (RV_I - RV_{MAX})/(RV_{MIN} - RV_{MAX})$. For example, range is a positive sloped characteristic, so the range TV value for candidate A would be determined by $(1200 - 900)/(1500 - 900) = 300/600 = 0.50$. In the example included in Figure 18.5, the better candidate appears to be A in that it has the highest average. The preferred candidate might change if we assigned non-unity weighting factors to the four characteristics. Clearly, this is a game of ratios and appears somewhere intermediate on the scale of precision between the Pugh and objective trade approaches.

18.1.4.4   Objective trade study approach

This approach is often applied in the aerospace industry under any number of different names including multivariate trade approach. The author is partial to this approach because it is effective but also because of a personal experience while employed at Teledyne Ryan Aeronautical.

Characteristic	Candidate A		Candidate B		Candidate C		Candidate D	
	RV	TV	RV	TV	RV	TV	RV	TV
Range	1200 km	0.50	1100 km	0.33	1500 km	1.00	900 km	0.00
Payload	150 lb	0.45	200 lb	0.91	100 lb	0.00	210 lb	1.00
Cost	$25 K	0.62	$28 K	0.25	$22 K	1.00	$30 K	0.00
Signature	$0.5M^2$	0.50	$0.6M^2$	0.33	$0.8M^2$	0.00	$0.2M^2$	1.00
Totals		2.70		1.82		2.00		2.00
Averages		0.67		0.45		0.50		0.50

Figure 18.5 Technometric trade study diagram.

A system engineer with what appeared to be a terminal illness called the author from the hospital and asked him to get a copy of a book on this subject from his desk and send it to him in the hospital. The author first thought it a little cruel that their employer would ask a sick employee to do a trade study in his hospital bed but the reality was that the engineer applied it to three alternative medical treatments and told the doctors to proceed with the one he picked as a result. The fact that the system engineer passed away was in no way a failure of the trade study process in this case in that it was essentially hopeless no matter the alternative. The booklet in question was *Multivariate Evaluation* by Ward Edwards and J. Robert Newman that was in a Quantitative Applications in the Social Sciences Series from Sage Publications printed in 1983, a copy of which the author keeps in his library to this day.

The first step in this approach is to decide on a set of requirements that all candidates must satisfy. If a candidate does not satisfy these requirements, it should be course screened from the list of candidates. Secondly, we must identify a selection criterion. We may choose to weight this criterion for relative importance or allow them all to have the same weight. We then must come up with a list of candidates through brainstorming, a trade tree, morphological chart, or other means. The candidates are then subjected to analysis to determine values for each of the selection criteria in the units of those criteria.

We would prefer to simply add up the numerical values for each selection criteria for each candidate and select the winning alternative. Unfortunately, the selection criteria will seldom encourage this simple alternative. Some of them may be positive increasing while others are positive decreasing. Commonly, a simple sum of the criteria values also will not make any sense. The sum of weight, dollars, and speed figures, for example, has no real meaning. Utility curves can be used to translate all values into a unitless positive increasing function, often called normalizing the criteria.

Figure 18.6 illustrates several possible utility curves and Figure 18.7 illustrates a trade matrix using all of these concepts. Several curves are superimposed in Figure 18.6 to illustrate different possible shapes. For

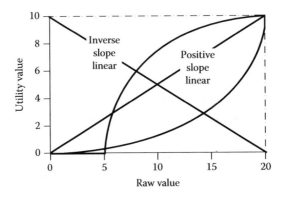

Figure 18.6 Utility curve examples.

Selection criteria	Weight	Candidates											
		A			B			C			D		
		RV	UV	TV	RV	UV	TV	RV	UV	TV	RV	UV	TV
1 LCC	2.0	6.7	8.0	16.0	7.8	6.0	12.0	8.5	4.0	8.0	7.8	6.0	12.0
2 Weight	1.8	4.0	5.0	9.0	3.3	5.3	9.5	3.5	6.0	10.8	5.5	3.0	5.4
3 Speed	1.0	14.0	10.0	14.0	13.0	7.3	7.3	13.5	9.0	9.0	10.5	3.8	3.8
4 Range	3.0	10.0	8.0	24.0	11.0	9.0	27.0	10.0	8.0	24.0	10.0	8.0	24.0
5 Payload	2.2	3.6	8.0	17.6	3.4	7.0	15.4	3.2	6.0	13.2	3.8	9.0	19.8
Score				70.6			71.2			65.0			64.0

Figure 18.7 Typical trade matrix.

a particular trade study it would be better to include a separate curve for each selection criteria. You enter the raw value for a particular candidate on the horizontal scale and trace up to the utility curve and pick off the utility value (UV) on the vertical scale. This number goes in the trade matrix UV column for the candidate. In this example, the curve that produces a utility value of zero between raw values of 0–5 and then increases in a curvilinear fashion to a utility value of 10 is for range. In this case, our value system appears to accept that we must have at least 5 miles of range and that range has an increasing value beyond that up to about 15 miles where there is no significant increase in value for marginal increases in range.

The figure also illustrates three other possible curve shapes that might be applied to other criteria. One is an upward opening parabola and two are linear. Note that the negative slope would allow us to transform a variable characterized by larger values being less advantageous into better trade study values for smaller values. This is illustrative of how we normalize the value system through utility curves such that we can add the values in the candidate columns.

Utility curve building is encouraged by a good knowledge of analytic geometry that helps one to relate mathematical equations and their graphical portrayal. A good policy to adopt is to bracket UVs between 0 and 10. The process of building a utility curve entails the following steps:

1. Decide on a raw value range
2. Build a graph with the raw value range on the horizontal axis and a utility range of 0–10 on the vertical axis
3. Decide on a curve shape
4. Translate the curve into an equation (if you choose to use a spread sheet) and plug the equation into the spread sheet UV column cells

Some system engineers refuse to use any shapes other than linear ones recognizing that they do not have enough information to know that another shape would be better. These same engineers would properly shy away from worrying about values accurate to several decimal places. We simply do not know the problem space with that kind of accuracy at the time we have to accomplish this work. Given that we select only linear curves, we have then to select a slope either positive or negative, fairly easy to determine based on the parameter, and the slope magnitude. One simple solution is to simply limit the curve by the minimum and maximum range of raw values selected and the vertical range of 0–10.

Over time an engineer who does this work on a particular product line will begin to observe that some selection parameters can be more accurately portrayed in a trade study by curvilinear shapes. In the case of range in Figure 18.6, the engineer might have observed that the customer had indicated in their request for proposal (RFP) that the radar set that was going to be used by the launching aircraft would have a range of 15 miles so a missile range approaching 15 miles had a decreasing improvement in value. Also, the customer may have said that the missile range had to have at least a range of 5 miles for some particular reason. Thus, any range less than 5 miles would have zero utility. Experience with some of these selection criteria frequently used in program trade studies in an enterprise should be generically described in a system engineering manual to provide guidance on future programs.

It is extremely important to begin the trade study with a clear idea of whose value system we are dealing with. Is it the decision maker's value system, the value system of the engineer doing the trade study, the customer's, or the decision maker's understanding of the customer's value system? Where do we get the weighting factors and the utility curve shapes? These reflect our value system and should be hotly or at least thoughtfully debated in the team doing the work. They must reflect the value system defined or allowed by the decision maker. The weighting factors and the curve shapes do have a note of subjectivity in them and therefore we must subject the results to a sensitivity analysis as explained shortly.

There are at least two kinds of decision makers and it is important to know which kind one is dealing with in preparing the trade study matrix. Many engineers are analytical decision makers wanting to know all of the numbers and related technical rationale. The numerical trade matrix is very useful to such a person. Program managers may begin life as an engineer but over time many become very intuitive in their decision-making process if they did not start that way because they have to deal with people problems and often have to make decisions based on incomplete information. Some intuitive decision makers do not like to deal with numbers at all and would prefer colors or simple words in the matrix.

The engineer doing the technical trade study work should talk this over with the decision maker as a prelude to doing the work. Where the decision maker is reluctant to use numbers the engineer should show him/her that they will be exposed to how those numbers were derived and nothing will be withheld. Often, the intuitive decision maker is reluctant to use numbers because they know the technical people had to have made a series of decisions to derive those numbers and they will not have access to the logic behind all of those decisions. Most often if the engineer explains how the weighting, utility curve, and sensitivity analysis approaches work and gains the decision maker's support in determining the curve shapes, the decision maker will insist on using the numerical approach.

Figure 18.7 offers a matrix for the objective trade study approach. Note that three columns are provided for each candidate. The first one is the raw value (RV) column that is determined by a study of the features of the candidate relative to the selection criteria. Second is the UV column that is determined by entering the raw value into the horizontal axis of the corresponding utility curve and noting the vertical axis value (UV). The third column is the result of multiplying the UV times the row or criteria weighting factor. A given cell value, therefore, is determined by the formula $T_{ij} = W_i f_i(R_{ij})$ where T_{ij} is the trade value for the ith criteria and jth candidate, W_i is the ith criteria weighting factor, f_i is the ith criteria utility function, and R_{ij} is the raw value for the ith criteria and jth candidate.

There are trade study computer applications available, but many people build their own spreadsheet applications. Even if you built one for each trade study, it would probably save time and improve study results over the use of pencil and paper combined with a hand calculator because with one it is so easy to play what-if games compared to the use of a calculator or pencil and paper mathematics. The formulas are fairly straight forward. The UV column is simply $U_{ij} = f_i(R_{ij})$. The trade value (TV) column is $T_{ij} = W_i U_{ij}$. The only hard job here is the utility function formula, and one can restore their Calculus 101 skills in analytic geometry fairly easy. This restoration is supported by the understanding that perfection is the enemy of good enough in this case, because you simply do not understand the problem with enough detail to argue over small matters. Simple shapes are fine. You can probably get by with straight lines and the conics.

Figure 18.8 offers a set of utility curves for the trade study illustrated in Figure 18.7. Each curve includes the equation that would be entered in the TV intersections of the related row such that when the analyst enters 6.7 in the candidate A LCC cell, the equation for LCC in that cell will generate a value of 8.0 as suggested by the markings on the LCC curve in Figure 18.8.

In Figure 18.7, it would appear that candidate B is the winner in that its point total is higher. However, at the time a trade study is accomplished we seldom have all of the information we would wish so we should be very suspicious of the results. One approach to gaining confidence in the results is called sensitivity analysis. We vary the weighting, utility curve shape, and candidate raw values over some range (5%, for example) and to observe if the preferred selection changes as a result. If we can vary these figures over some range and the preferred selection does not change, we say that the matrix is insensitive to variation of the value system suggesting a good solution. The wider the value range that retains the same selection, the less sensitive the matrix is to variation and the better the confidence in the result. If the selection changes with small variations, it is very sensitive to variation and the preferred selection should be treated as suspect. The use of a spreadsheet for the matrix will permit easy variation of these factors and observation of the sensitivity of the preferred selection

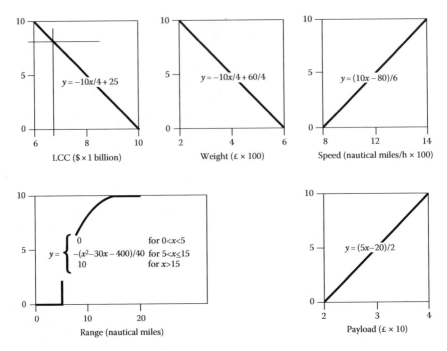

Figure 18.8 Trade study utility curve set.

to variation. It is unlikely an analyst will run the sensitivity study if they must use a pencil and paper or even a hand calculator.

18.1.5   Trade study or design concept review

No matter which trade model has been applied (including none), the design concept should be subjected to some form of review and approval. In the case where the concept was determined through trade study, the results of that study should be reviewed by the decision maker and additional people selected by the decision maker. Someone from the trade team should brief the study requirements, selection criteria, weighting factors, utility curves or other means of normalizing the candidate scores (if numerical scores were used), the candidates, candidate capabilities relative to the study requirements and selection criteria, and a recommendation for selection. Finally, the team should expose any residual risks and how they plan to mitigate them. This review can satisfy the need also for a design concept review unless the preferred concept from the trade study needs additional work to evaluate potential risks. If the risks exposed are substantial in the review, after the selection has been identified, the risks of alternatives rejected should be exposed and possibly the trade decision reopened. A less-than-perfect candidate with low risk may be preferable to a candidate with great performance potential buried in considerable risk.

18.1.6 Post-concept action

The work that goes into a trade study or concept development activity produces valuable information often hard won. There should be surviving documentation about this work. In the case of a trade study, there should be a trade study report. In the case of a non-traded determination, the concept development work could be captured in a memo to the appropriate team leader or program manager depending on the scope of the work and internal policies. The preferred concept must also be protected from purposeful or unintentional alteration after the selection process through configuration control.

18.1.6.1 Concept documentation

Many organizations that perform excellent trade studies can produce no evidence of having done so. One client that the author visited wanted to improve their trade study process through a workshop. The author asked to see a copy of a good study report that had been done in the past in order to understand something about the process that was being applied. Someone in the room left and came back in 15 min saying that he could not find a report from a past study. Now, this was a great company doing a lot of quality system engineering work. Too often we do not place adequate value on documentation at the time we have the knowledge, time, and money to accomplish that documentation. Later we cannot backfill the documentation when we find we need it. Documentation often falls under the utility that we only recognize through delayed gratification. At the time that you can do it there is no immediate need for it. It is only months or even years later that you may need that documentation. Therefore, it is easy at the time to believe that it is not adding value, which may be the case in the near term.

Therefore, we should establish a trade study documentation scheme that coordinates with a reporting model such that in the process of developing the trade study artifacts they fall right into the reporting model. Thus, we can have a good report with relatively little pain. Figure 18.9 illustrates a suggested outline for a report. An alternative to this structure that could be applied, and is probably a better choice where there is a separate team or principal engineer for each candidate, is to assign an appendix for each candidate that would be the responsibility of the team or engineer for that candidate. Each appendix then would contain all of the information about the corresponding candidate.

1	Introduction
1.1	Title
1.2	Purpose
2	Programmatic considerations
2.1	Study resources
2.2	Study schedule
2.3	Study risks and uncertainties
3	Study parameters
3.1	Study requirements
3.1.1	Requirement 1
3.2.2	Requirement 2
3.1.i	Requirement i
3.2	Selection criteria
3.2.1	Criteria 1
3.2.2	Criteria 2
3.2.j	Criteria j
3.3	Criteria weighting factors
3.4	Utility relationships
4	Candidate descriptions
4.1	Candidate A
4.3	Candidate B
4.J	Candidate J
5	Candidate raw and trade value development
6	Sensitivity evaluation
6.1	Weighting factor sensitivity
6.2	Utility curve sensitivity
6.3	Raw value sensitivity
7	Conclusions and recommendations
A	Trade matrix
B	Study utility curves
C	Supporting rational

Figure 18.9 Sample trade study outline.

18.1.6.2 Configuration management of trade results

At the time most trade studies are performed, there is seldom a full configuration management process funded on a program. This commonly is not done until the drawings start to roll out of the preliminary and detailed design process. By some means, however, we must maintain configuration control of the trade study documentation. A simple way to do this is to fractionally fund configuration/data management to support an online program library that the program should have anyway. The trade study documentation simply becomes a part of that library. The released report is entered into the library and is available to all for read-only access.

At some point after the initial release of the trade report, it may be necessary to revisit the study and possibly change it even to the extreme of selecting a different candidate. The revised report should carry a revision letter or version number and be released into the library as well.

chapter nineteen

Product design integration in an IPPT environment*

19.1 What is the principal problem?

If we have defined the requirements well and planned the development process well and execute the plan faithfully within an environment identical to the one for which the plan was intended, the integration process during design development will be very simple. The reason for this is that there will be no discontinuities during the design process because we will have perfectly understood the problem prior to beginning the design work. This possibility of future perfection is music to the ears of those who have experienced several imperfect program efforts. But, the realities of our past experiences are evidence of the great difficulty in ever attaining perfection. While the development of requirements before design, sound planning, and effective concurrent engineering practices will take us far in our quest for perfection, we should not totally ignore the potential for errors and changed circumstances, and fail to provide for possible course correction.

Given that we have organized our program by IPPTs, coordinated our IPPT structure with the system product-entity structure, and provided for supplier and customer involvement in these teams, we will have produced the perfect environment for product integration. We now have to implement a good design solution, which, unfortunately, is even harder than organizing for integrated development. As we said in earlier chapters, the principal mechanism of integration is human communications. In Chapter seven, we discussed some specific communications aids that we assume have been taken advantage of.

It is a valid premise, and perhaps an overly used cliché, that the biggest problems in design development will appear at the interfaces where different parties are responsible for the development of the design. In Chapter fourteen, we used the term cross-organizational interface to name this kind of critical interface. We want very much to know where these interfaces are in the system with respect to the IPPT responsibility boundaries. In fact, in Chapter thirteen, we especially defined the evolving system product-entity structure, and aligned our IPPT with these entities, such

---

* The material in this chapter is from Grady, J., *System Integration*, CRC Press, Boca Raton, FL, Chapter 14, 1994. Used with permission.

that we minimized cross-organizational interface density within the system. You will recall that we cannot eliminate this kind of interface because the richness of the system capability depends on these interfaces as does the existence of the system itself.

The reason that we worked toward minimized cross-organization interface was to reduce to the maximum practical extent, consistent with maximum system capability with respect to its requirements, the communications problems during design development. Our theory is to minimize the need for the teams to communicate while we maximize their capability to communicate. We do this to fight against the potential for the humans to withdraw from the critical interfaces rather than to dash across them to understand their fellow's problems.

This is a valid concern. We can overcome it by the techniques discussed above and by clearly defining the interface responsibilities for each IPPT. Each team must be held fully accountable for their innerface (both terminals touch items within their responsibility). We must allow them to avoid any responsibility for their outerface (neither terminal touches an item under their responsibility), and hold them jointly responsible with another IPPT for their crossface. It is the aggregate crossface at the IPPT level that constitutes the cross-organizational interface for the system and the principal area of employment for system engineers performing integration work on the PIT and IPPTs during product design development.

Figure 19.1 illustrates the power of organization by product-entity-oriented cross-functional teams on programs. It is perfectly clear that an easily understood and contiguous portion of the program WBS, SOW, and IMP/IMS is the responsibility of a particular team manager because there is perfect alignment between the WBS and the teams, both oriented as overlays of the functionally derived product-entity structure. The team managers can be held accountable for cost, schedule, and performance, that is, everything related to the product entity their team is focused on. But, also it is perfectly clear that in order to develop the cross-organizational interface between segments 1 and 3 marked on the lower (product) n-square diagram, teams 1 and 3 must engage in integrating technical conversations reaching compatible conclusions for their interface terminals indicated on the top-level n-square diagram. In that this is very clear, interface development accountability is encouraged relative to the most difficult development effort on the program.

There should be a specification crafted at the level of each IPPT defining, among other things, the interfaces with items under the responsibility of other teams, associates, the customer, and major suppliers. These IPPT-level specifications will define the interfaces in question. Each team must now clearly understand who is responsible for the other terminal of each of these interfaces defined in their specification and accept the responsibility to jointly work with that party for the integrated development of the design

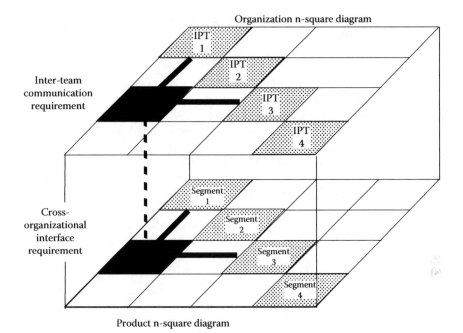

Figure 19.1 Dual n-square perspective.

solution on each terminal. Methods were discussed in Chapter fourteen for bringing this condition about. If we are successful in applying this method to every cross-organizational interface, our past product integration problems will be overshadowed by other, less severe problems. Alternatively, these interfaces may be defined in interface control documents with specification reference to them for the interface requirements.

19.2 How do we accomplish crossface integration?

Chapter fourteen includes some suggestions and tools for accomplishing integration across the cross-organizational interfaces. We absolutely must organize the interface responsibilities and make them perfectly clear. The fundamental key once again, however, is human communication. Every cross-organizational interface element must be the subject of one or more conversations or meetings between the opposing parties with an effort by each party to understand the interface from the perspective of the other's position. The team leaders must also be attentive for signs that team members are subconsciously withdrawing from these interfaces.

Cross-organizational interface is an area where an online development information grid (DIG) described in Chapter seven can be very helpful. Each IPPT should have a set of interface design charts on the DIG during concept development and preliminary design. Each pair of teams with a joint responsibility for cross-organizational interfaces, as defined by the PIT, must be held accountable for being familiar with their opposing party's interface concept/design and the degree of compatibility of that design with their own. The two teams must also meet from time to time to discuss and resolve interface issues that may be exposed through this joint cross review of the other's interface concept.

A program will do well to encourage distributed team meeting times such that teams may cross meet for the purpose of interface integration. If democracy rules when each team will meet, they could conceivably pick the same time to meet making it impossible for persons from one team to meet with another team for this purpose. Often a shortage of meeting rooms will force a dispersal of meeting times, but this is so important that we should not rely on chance resulting in opportunities for cross-team meeting.

Each pair of teams responsible for a cross-organizational interface must reach agreement on the existence of a particular set of interfaces between them under the encouragement of the PIT, mutually agree on the requirements for those interfaces, have a means to detect when and if their design efforts come to cross purposes with the agreed-upon requirements, and a method by which they may come to a mutual agreement on any actions necessary to resolve any issues across their shared interface.

These are all actions that can be done by a pair of teams. A moderately complex system may require several of these relationships. Let us say that our program has five IPPT and that interfaces exist between every pair of items corresponding to our teams. This makes a total of 10 interface relationships that must exist between the teams. Each team will have to maintain four interface relationships. Figure 19.2 shows the relationships for two through eight IPPT. As you can see, the maximum number of interface relationships that any one team must maintain is $N - 1$, where N is the number of teams. The maximum number of program team cross-organizational interface relationships is $(N)(N - 1)/2$ where directionality is of no interest.

As the number of teams increases, obviously the difficulty of performing product design integration increases. But, the problem is not as bad as the mathematics of the possible predicts. The reality is that many of the cross-organizational interfaces will be voids, that is, no interface requirement, as indicated by a void in the corresponding n-square diagram cell (or the absence of any lines between two items on a schematic block diagram). By the way, you should note that a 5-square matrix has 20 non-diagonal cells that is twice the number predicted by the formula above and in Figure 19.2. You will recall, of course, that the n-square matrix includes two cells for each pair to account for directionality. The divisor of two in the formula above is included for this reason. So, all of these techniques for identifying the number of possible interfaces are consistent.

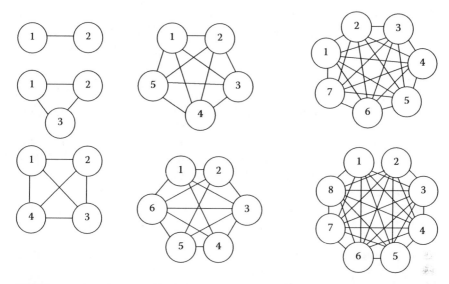

Figure 19.2 Interface possibilities explosion. (From Grady, J., *System Integration*, CRC Press, Boca Raton, FL, p. 225, 1994. With permission.)

As the number of teams expands beyond some number, perhaps two or three, there must be an outside agency, the PIT in our story, to manage the interface development activities, to audit team compliance with their interface responsibilities, and to track and resolve difficult interface problems. These cross-organizational interfaces must be singled out by the PIT for auditing with the IPPTs responsible. The PIT must assure that the integrating discussions and meetings are taking place between the teams and the IPPTs responsible have reached, or will reach, a condition of compatibility. This responsibility should flow down to the higher-level teams when they establish sub-teams.

19.3   There are more interfaces

The PIT must take full responsibility for interfaces external to the system for which the overall program is responsible. These interfaces may be with associate items, hostile systems, noncooperative systems, or the natural environment. Where associates are involved, the interface development process may require an interface control working group (ICWG) to formally resolve the interface design. These are all cross-organizational interfaces at a higher level than those discussed in the previous paragraphs. The PIT should be responsible but can delegate technical discussion and agreement efforts to the team where the interface attaches to its system.

Each IPPT should replicate the cross-organizational interface solution discussed above at their own level. Where one IPPT has two or more subordinate IPPT reporting, they should hold each subordinate team responsible for their component of the major teams responsibilities. This pattern could be broken down further, but at some point we must come to the major system components like computers and engines that may have a principal engineer named. This principal engineer, likewise must be held accountable for all of his or her innerface, protected from any responsibility for the item's outerface, and held jointly responsible for developing the crossface at that level. The IPPT within which that component fits must act as the PIT for everything within the team's area of responsibility. Commonly these situations will coordinate with procurement of these entities in accordance with a procurement specification.

There are some special cases where a team or principal engineer may have to show interest in an outerface. Let us say that a team is responsible for the flight control software in an aircraft system or the towed implement control software in a farm tractor. This team is going to have to show interest in control loops that extend beyond the computer and its I/O ports. The software design must take into account the dynamics of the actuator in either of the cases noted while the control surface passes through the air at 500 knots or the plow passes through the earth at 10 miles per hour. This is an extended interface. The principal must extend their interest beyond the immediate physical limit of their terminal. These situations can result in friction as well as omission.

If the software engineer shows intense interest in the dynamic characteristics of an actuator, the engineer or team responsible for that actuator may become irritated not understanding the software engineer's need for information. On the other hand, if the software engineer mistakenly believes his or her responsibility in this situation stops at the I/O port and the actuator engineer believes his or her responsibility stops at the electrical connector, the end item control system will surely be incorrectly developed resulting in significant performance risks exposed late in the program that will lead to cost and schedule problems as the fix is concluded.

The Russians believed they could build and successfully fly a Moon rocket that used many liquid engines fix mounted to the structure and control the direction of flight by throttling the engines on the periphery. Meanwhile the Americans were using engines that gimbaled, a technique perfected on the Atlas ICBM design. The Russians finally had to accept after two crashes that the control system could not handle the delay in throttling the engines.

19.4　System optimization

The handmaiden of system decomposition and interface integration is system integration and optimization. We carve up the system functionality into a product entity structure assured that the aggregate of the elements will satisfy the system need in an optimum fashion entailing maximum capability for minimum cost. We may not have allocated the needed functionality so as to yield a well-optimized system in the process, however. It is also possible, during the design evolution, that one or more IPPT may develop a design solution, within the boundaries established by their requirements, that places the overall system in a condition of non-optimization. The design solution in one team may force a solution on another team that is unnecessarily costly or force a severe weight penalty on another team. Each team should be forced to carry a reasonable share of the design difficulty if possible.

It is relatively easy to detect the symptoms of non-optimization during the design process. One or more teams will be struggling to meet their requirements while one or more others are having no trouble satisfying their milestones, possibly under-running team budget and schedule allowance in the process. Mathematical models developed to solve requirements and design problems are also excellent tools to help identify suboptimization as well as to look for optimizing alternatives within the bounds for which the models are appropriate. The PIT must be alert to detect these symptoms and react to them, to understand the driving forces and bring about a balanced condition between the teams if possible.

It may be possible to rebalance the requirements for reliability, mass properties, design to cost, or other parameters such that the teams are equally sharing the burden and a more balanced system will result. If this cannot be done, it may be necessary for one or more teams to alter their design concept or even the technologies to which they are appealing. These actions can ripple into the risk assessment and reshuffle those priorities.

On the other hand, it may not be possible to rebalance the pain across the teams, but it may be possible to identify areas of growth potential, corresponding to the over-achieving team areas, that the customer may choose to exploit while providing additional funding. It is also possible to reduce the budget and schedule allowance for one or more over-achieving teams leading to increased margin (or profit) and schedule slack that may have to be spent at a later date to cure a discontinuity. The point is that we should be actively seeking out these conditions rather than letting the corresponding opportunities slip through our fingers from indifference. The alternative to an active program of optimization is that suboptimized conditions will materialize in a time of panic and have to be solved by throwing money at the corresponding problem.

19.5   Other PIT actions during design

In addition to the top-level interface integration and optimization work for which the PIT must accept full responsibility, the PIT must also provide many other integration services. These include test planning integration, system level analyses, the risk management program, and the several nonengineering integration functions as well.

The PIT must be constantly alert for opportunities to integrate the design work between product and process. The production process member(s) of the PIT must remain aware of the unfolding production solutions from the IPPT and integrate them into a system production solution. At the same time, they must remain alert to possible opportunities to better balance the design solution between the product, operational employment, and production. There are any number of opportunities for production and maintenance engineers to cooperate over producibility and maintainability issues. Safety and maintainability share many potential coincidences of interest.

This same attitude must be present in the material procurement members, quality members, integrated logistics members, and others. Each PIT member must remain attuned to what their counterparts are doing on IPPTs and integrating those solutions into the system solution while working as a team at the system level to constantly question whether they have achieved the best condition of balance between the evolving designs for the product, the production process, and operational and logistics deployment considerations.

19.6   Special hardware–software integration needs

All of the previous content of this chapter applies to computer software as well as hardware, but software does impose some special concerns and opens up some special opportunities. Commonly, a software entity can be properly assigned to a specialized IPPT focused on that one item. This team and the PIT must adhere to the same prescription discussed earlier where the cross-organization interfaces are clearly defined as are all of the other requirements that the software entity must satisfy. A software person might call these requirements collectively their customer requirements.

A program or customer that mindlessly requires a single prescribed functional decomposition methodology, like functional flow diagramming, will immediately run into difficulty at this juncture. The computer software engineering community has developed several effective specialized decomposition and development methods that the community would prefer over functional flow diagramming and they should be allowed to use these techniques. One of the big problems at this interface is the difficulty of requirements traceability across the abyss between the parent hardware entity and the software element. Some companies have integrated their tools for system/hardware requirements and software requirements to the extent that they can capture the flowdown relationships between them in one unique location. Chapter twelve offers some ideas about how to encourage success in this matter, but it is not easy at the time this book was being written.

Software interfaces are a little more difficult to characterize than hardware. The software community may need information about an interface that extends beyond the computer I/O port. It may be necessary to characterize the delays throughout a one-line circuit diagram all the way to the ultimate load (an actuator, for example) to properly define the software needs. The author refers to these as extended interfaces. A simple reading of the interface responsibilities rules may exclude software people from the discussion between the actuator team/principal engineer and the computer team/principal engineer who may jointly feel that their interfaces are in the software outerface set. This is true as far as the n-square diagram or schematic block diagram might show, but we have to be attuned to special needs for extended interface information on the part of the software teams.

There is a feeling among system and hardware people that software offers a safety valve for mistakes in other parts of the system development process and this sometimes results in late changes to software that are very difficult to make. This attitude is driven by knowledge that the software manufacturing process is embedded within the software development process and software changes are focused less broadly across an organization than in the case of hardware changes late in the development process. Hardware changes commonly drive manufacturing changes (possibly extending to tooling and test equipment), procurement changes (with significant cost and schedule impacts), and test plans and procedures changes, and quality inspection changes. Computer software changes, on the other hand, offer an irresistible alternative that can be completely addressed with the software IPPT. Development teams should consciously recognize the attraction for late software changes and carefully weigh the real relative benefits with accurate inputs on the software impact of the needed changes.

Effective software development work requires adequate testing and simulation capability as do some forms of hardware development. There may be a tendency in the software and hardware communities to isolate on their immediate needs in this area resulting in specialized resources that may duplicate some functionality and expend available money unnecessarily. For example, systems involving guidance and control functions, and computer software implementation of those functions, are a good candidate for simulation and test resources shared by hardware and software communities.

The hardware and software communities also respect vocabulary differences that can lead to communication problems across their interfaces. It is extremely important that engineers involved in discussions across these interfaces challenge each other to ensure understanding. The phrase, "I understand you to mean....," should be often heard in these discussions. The V words (validation and verification) offer one of these vocabulary differences that can lead to misunderstandings, but there are many.

The greatest problem in hardware/software integration, however, is a separation between hardware and software people that sadly is becoming a tradition. We said earlier that it is not uncommon for engineers at the cross-organizational interfaces to withdraw rather than plunge across these interfaces. This is nowhere more pronounced and fatal than when the interface is between hardware and software. The PIT on a program and the leadership of each IPPT must be sensitive throughout the development activity to any symptoms of separation between these parties and encourage in every way possible a close working relationship between them. The functional system engineering department must be especially vigilant that its staff does not become full of hardware-dominated engineers who have no interest in software.

Software developments commonly include in-process reviews by several different names. The IPPT with software responsibilities should invite related hardware people to these reviews to educate them and to get useful feedback from them. Likewise, hardware-dominated teams should encourage participation in their meetings by software people even though there may be no software under their responsibilities. Communications is the machinery of integration.

Hardware-oriented system engineers are often not well educated in software methods and models despite the fact that the software community has gone to great lengths to evolve models that are not only useful to software people in understanding very complex problem spaces but also are very effective in communicating their understanding of that problem space to other persons of normal intelligence who may or may not be gifted software engineers. System engineers who advanced to that job from a hardware background have an obligation to learn something about these software modeling approaches to encourage good communications across the hardware–software gap. With more and more of the more difficult functionality being allocated to software, it would also be in the hardware-oriented system engineer's career interest to expand their knowledge base in this area. The good news is that the software people are right that their models can be understood by normal people.

The isolation of all of the software people on one team could be questioned. The software running in a single on-board computer is related to several of the systems that are by themselves hardware dominated perhaps. But it would be possible to partition the software development responsibility and assign software engineers to a propulsion team to developing propulsion-related software. This will probably result in better integration within the propulsion system, but it will probably result in poorer integration within the software aggregate running in the on-board computer. We have a choice in how we assign functionality to entities that determines the interfaces that must exist between the entities, and we have a choice in how we assign responsibility for development of the entities that determines where the cross-organizational interfaces ill occur.

Increasingly, the system development work can involve more than one processor forcing those responsible for each to have to cooperate in some fashion. The teams responsible for the different processors may even be from different companies that have to negotiate mutual agreements through interface control working groups that join the company-oriented programs to negotiate mutual agreement on interfaces.

chapter twenty

Preliminary design

20.1 The purpose of preliminary design

The purpose of the preliminary design activity is to develop a credible design sufficiently advanced to permit identification of risks that may remain and to encourage specialty engineers to reach initial conclusions about the likelihood of the design satisfying their requirements. In order for this to occur, the designers must capture their ideas in a tangible fashion and share that information with their team members. This purpose carries with it some potential for pain on the part of the design engineers. To the extent that they fail to capture the meaning of some of the requirements, their best effort will, and should, draw critical review, and few of us take criticism with gladness. Team leaders have to prepare their members to accept the reality that criticism is the least expensive commodity available to a program and if the team can handle it well it will move the team toward success.

20.2 Requirements validation, risk identification, and risk mitigation

Throughout the period when requirements are being identified, those responsible for requirements writing must also take responsibility for identifying the degree of risk associated with complying with them. This falls into five specific areas as follows:

1. Is this requirement essential to the item design and its value achievable?
2. Is the requirement consistent with the cost and schedule resources available for the design?
3. Is the technology available to support a design satisfying this requirement?
4. Does the responsible design engineer have any concerns about being able to synthesize this requirement?
5. Can it be shown that this requirement is traceable to a parent item requirement?

The principal element in determining a requirement validation state is the designer's attitude toward the requirement. If the principal designer is confident it can be satisfied, one may consider it validated. If the designer is concerned about being able to design so as to satisfy the requirement, we should look into it more deeply in terms of the complete criteria listed above.

Requirements that will be difficult to satisfy could be accepted into a formal risk program where progress is tracked in accordance with a specific plan to mitigate the perceived risk. The requirement could be tracked as a technical performance measurement (TPM), as discussed in Chapter sixteen and *System Requirements Analysis* (Grady, 2005). We could also perform a technology demonstration to gain confidence in being able to satisfy the requirement. Special analyses might also be performed.

20.3   Design ideas capture and baseline control

Initial design ideas will often be captured in informal sketches rather than formal engineering drawings during preliminary design. Formal release of drawings and configuration control of them is generally not necessary during this work. Packaging concepts are explored and solutions developed for the most challenging requirements as a way of removing risk concerns. However informal the documentation process applied, it is very important that the program have some organized way of defining the current baseline such that all of the design support work remains coordinated with that baseline. A baseline is defined by a list of documents in a particular release status where the master is under some form of configuration control. If the program has not included adequate funding for configuration management in this phase there is obviously a conflict.

The preliminary design period should permit some flexibility for the design team while not permitting the baseline to be corrupted, so the teams should apply a team document release concept, at least, where they maintain a list of documents, listed by some unique identification (could be a block of engineering document numbers set aside for preliminary design), as suggested in Table 20.1. This could also be used to capture the definition of the models and simulations under the team's responsibility.

Table 20.1 Team Document Release Record

Document number	Rev	Title	TM	Concerns
85E002	C	Electrical Power Subsystem Sketch	3	Battery capacity
85F001	A	Fuselage Section 25.0–55.0 Sketch	2	Stiffener weight
85W002	B	Wing Stores Support Structure Sketch	2	Material trade— steel or titanium

These documents could be CAD drawings or PowerPoint charts. Ideally, they would be maintained online so that all team members could call them up in read-only mode and consider their features relative to the requirements in which they are interested.

20.4   Design communication

Each IPPT manager should have full authority to manage their team in accordance with the IMP/IMS as it relates to their team. The team will have been formed of members from different functional departments as a function of the technologies that will have to be applied to solve the problem defined in the item specification. It is understood that all of these people are specialists and will have some degree of difficulty describing their design and analysis ideas to others in other specialties. However, it is assumed that all of the team members speak the same language as a first or second language and this along with the documents developed by them is the principal means of communication between the team members through which the aggregate knowledge of the team members will be poured into the design solution. There are three modes in which the team members can cooperate toward a solution to the problem defined in the specification.

20.4.1   Independent work

The specialists on the team do require some time to work alone within the confines of their specialty. One option is that they be allowed to spend all of their time at work on their view of the problem and solution with only incidental interaction with other team members. Clearly, there is not a lot of cooperative effort going on here.

20.4.2   Continuous meeting

The team members must communicate between themselves so as to arrive at a consensus about the design and could conceivably be in a continuous meeting discussing the different parts of the problem and solution. Clearly, there is not a lot of individual effort going on here. The author actually worked in an environment like this on one small project. It took place in open area with desks in the old style engineering office arrangement. As engineers arrived in the morning, conversations would start up about the current problems. Initially, the conversation might be about the location of the mounting locations for the engine gimbal hydraulic actuator ends. Four engineers might walk over to the wall where full-size drawings were taped up. As that conversation tapered off another might start up with two of the same people but two others not in the earlier conversation. They might find it useful to call up something from the network and project it onto another wall during the conversation changing a sketch while they talked. This pattern could continue with the center of the conversation shifting around the room. The author found that he could focus on his work while these discussions were going on but if a topic of interest came up he could join it for the period he was interested and then return to his own work.

It happened that this team was composed of fairly senior people all of whom had full confidence in their knowledge and skill in the product line. The author found this to be a stimulating and useful environment but concluded that people right out of college might have some difficulty being productive in it.

20.4.3   Cyclical work and meetings

The two modes offered in the previous two sections have some serious problems in that they are at the two extremes, one all individual work and the other all joint meeting. Most people would agree that we need both individual work time where we can concentrate on our own area of specialization and from time to time meet together to gain the advantage of synergism of the team. The question is the duty cycle of the activities. We could meet daily for 1h and work for the other 7h, for example. The reader can imagine many other combinations. The optimum duty cycle probably has something to do with how much time has passed since the team was formed. A new team may require a more frequent meeting schedule than one that has been together for a period of time. Another factor is the current degree of difficulty the team is having with a really serious problem, meeting more often and longer when being stressed with a serious problem.

In the earliest meetings, the team leader would do well to ask each specialty engineering discipline to offer a 5 or 10min pitch on what their requirements mean and how they have been responded to on some other programs. This will help the design engineers to understand these requirements that can be a challenge in some cases and also communicates to the design engineers that the team leader values their opinion. These meetings should also spend some time on risks or concerns with which the team is currently dealing. There should be a brief discussion of the current design concept/baseline led by a design engineer followed by specialty engineering questions or a general discussion.

20.5   Preliminary design review

The preliminary design review (PDR) offers a formal opportunity to examine risk and the maturity of the design concept as a prerequisite to fully funding the design effort. At the PDR, we should assure that the corresponding specification is complete, all TBDs (to be determined) removed, and the document formally released. MIL-STD-1521B (recently resurrected from handbook status) offered an excellent outline for things to review at the PDR for hardware items, and these elements have been selectively included and edited below:

1. Trade studies accomplished.
2. Analytical basis for the requirements in the specification in the form of functional analysis models, item architecture, and interface diagrams.
3. Layout and preliminary drawings and sketches.
4. Environmental control and thermal design concepts.
5. Preliminary mechanical and packaging concepts.
6. Safety engineering considerations.
7. Security engineering considerations.
8. Survivability and vulnerability considerations.
9. Preliminary lists of parts, materials, and processes, their inclusion in a program or company PMP standard, their availability, and long lead status. Planned use of COTS and nondevelopment items.
10. Preliminary plans for acquisition of major items in the item and planned sources.
11. Reliability, availability, and maintainability estimates and concerns.
12. Preliminary weight, center of gravity, and form factor considerations.
13. Item interface relationships with other items.
14. Item development schedule and budget status and projections. DTC/ LCC considerations.
15. Models, simulations, prototypes, breadboards, mockups developed or planned and their application.
16. Producibility and manufacturing considerations both from the product and process perspective.
17. Transportability, packaging, and handling considerations.
18. Human engineering and biomedical considerations.
19. Preliminary verification considerations identifying any testing that will require special facilities or ranges that require long lead scheduling.
20. Logistics considerations related to spares, technical data, facilities, support equipment, and personnel selection and training.
21. Software and firmware relationship considerations.
22. Quality considerations.

The first thing to consider when preparing for a PDR is the exit strategy. The contractor should find out from the customer what their minimum exit criteria is. If the customer does not have one written down, the contractor should offer them one such as

1. All action items assigned with response dates defined.
2. All items on the approved agenda discussed.
3. Meeting minutes released and approved.
4. All customer direction found to be in scope integrated into program planning.

The next task is to develop an agenda and gain customer approval. Once approved, the contractor can start filling in names of presenters for the agenda items, call the presenters together in a kickoff meeting offering them guidance on PDR materials preparation standards and schedule. The materials should be reviewed either in a mock PDR or a storyboard format on the wall.

Once the materials are approved, they must be collected in the form of a master and submitted to a reproduction service for copies needed. Depending on the contractual arrangement it may be necessary to have to deliver the materials to the customer 30 days prior to the review as well as providing copies at the review. If the review results in significant changes, the materials may have to be run again. Theses materials could be in paper format, but a CD might make a more useful format as well as cost less to reproduce and distribute.

Prior to customer arrival for the review, the contractor should plan on a meeting room and meeting-support resources. If it is intended to record the meeting on audio, video, or both, arrangements will have to be made for equipment installation and operation. This kind of equipment has a way of changing the review by amplifying normal human characteristics. Quiet people will probably be more quiet and extroverts may be more so as well possibly calling for the cochairmen to be more forceful in drawing people out or quieting them down. The key to reducing this effect is to locate the recording equipment with the least possible visibility. Company audio–video people generally will not voluntarily conform to a low profile and will have to be instructed to do so. This is how very sensitive group meetings in weight, stress, and other kinds of clinics are often recorded, and the same techniques will work in design reviews.

The meeting may have to be captured in video in order to be transmitted in real time to other sites, and if this is the case, it can be taped at the same time. Those attending should be informed of any intent to record and/or send meeting audio/visual (A/V) data anywhere else. Recording or camera capture for transmission must be coordinated with presentation media. The best approach is probably computer projection coordinated with video and audio capture where the presenter need only advance the charts with a button. A computer should be part of the presentation equipment in this case networked into project servers for access to backup data if needed. Fully equipped presentation rooms include a camera stand from which paper charts can be digitized and projected. Other arrangements might simply pick up the screen in the video scene where the presenter is using plastic charts on an overhead projector. There are many possibilities and they should be selected to be both effective in presenting the review and easy for the presenters to use. When in doubt, bring in a few presenters and see how easily they adapt to the equipment and its use. If there is any indication of difficulty, bring them all in and hold a school.

Even if recording is used, it is advisable to have someone with technical and program credentials available to record minutes of the meeting. This is not a job for a secretary, a recent college hire, or someone who can be spared that day from other work. This meeting secretary should also remain tuned in to capture action items insuring they are properly documented. The meeting secretary should serve both cochairmen by asking for clarification when the meaning of any meeting comments are not absolutely clear. Finally, the meeting secretary must capture any and all decisions and directions. The person selected has to be able to respectfully but confidently interact with high-ranking company and customer people.

If the review is to be attended by many people and there is any chance of contentious situations, it will be a good idea to schedule not only the main meeting room but one or more caucus rooms besides. The customer may wish to meet periodically in one of these and they offer an opportunity to defuse potential time-consuming discussions in the main meeting by shuffling the people involved in those discussions into a caucus room for a period of time with instructions to return to the main meeting room at some point to brief their results.

Each presentation should cover three fundamentals. The principal design presenter should first convince the reviewers that he or she understands the item requirements by summarizing them. Next, the presenter should show features of the design and tell how this solution satisfies the previously stated requirements. Finally, if there are any residual risks, they should be stated along with a plan for mitigating them.

It is very important that the meeting run as scheduled so that presenters can come to the meeting a little before they are scheduled to present rather than be present throughout the meeting. If the actual schedule becomes disconnected from the planned one, it will be necessary for meeting coordinators to keep the presenter staff informed of the situation via phone or e-mail leading to a lot of extra work. Therefore, the cochairmen should be briefed on escape strategies such as (a) move the discussion to a caucus room with a subsequent report, (b) make out an action item, (c) declare the discussion out of scope, and (d) delay discussions until the topic comes up on the agenda.

Immediately at the end of the meeting, or each evening if it is a multi-day affair, the cochairmen should meet to go over all action items offered for the day to assign responsibility and rule on their scope. The final meeting of this kind should reach agreement on the extent to which meeting closure criteria negotiated prior to the beginning of the meeting has been satisfied and what actions remain to reach closure.

Subsequent to the end of the meeting, a series of administrative actions are necessary to conclude the review. The minutes should be offered to the customer as soon as possible, certainly within 5 working days. If changes were made in the presentations, those changes should be provided to the customer. The customer and contractor must both work their action items energetically to close them as scheduled or before. In the end, the contractor should receive some kind of formal letter stating the review has been completed since this will often trigger progress payments and other positive program actions.

chapter twenty-one

Detailed design

21.1 Information and communication importance

The detailed design work will, like the earlier work, be accomplished by many people working together for the same reason, no one knows everything. Therefore, the need for continuing tremendously effective human communication is critically important. This communication includes, of course, the spoken word but also the written word in the form of documents whose configuration is maintained by configuration management. All or most of the documents developed for the program should be in computer media wherever possible and the masters of those documents maintained under configuration control. The author would encourage that data management take over the global program data responsibility maintaining a formal program library for all released or approved documentation on the program. Commonly, data management only feels responsible for documentation that crosses contractual boundaries, but they should be funded to extend that license to include all program data.

The design teams should have the facilities needed to work and meet together. They need wall space for the big picture and computer projection in their meeting space to permit any team member to project any computer image onto the screen (a white wall may suffice) for discussion. Above all, the members of the design team should be located in the same physical area to encourage effective human communications.

21.2 Integrated team activity

The design will commonly be developed in the mind of one or two designer engineers assigned to a team based, hopefully, on the requirements in the item specification. Also, ideally, the several specialty engineers on the team will have had an opportunity to influence the mind of the designer(s). The best way to do this is to encourage the specialty engineers to offer comment on design briefings offered at each meeting.

The team members will require independent work time but must also come together in meetings periodically for the purpose of synergism. In those team meetings, they should discuss the current baseline and suggested changes to it from the several perspectives represented. Everyone should have an opinion and the team leader or facilitator should specifically ask everyone who does not volunteer a remark what their position is on important topics.

The teams at the top level must interact across their boundaries for interface purposes and may have to have representatives cross meet with other teams with whom they share interfaces. Top-level teams may have sub-teams and so have to act as an integration agent for their top-level product entity. The PIT in the author's mind is the top-level system integration agent that must work across the top-level IPPTs to ensure they are on track to satisfy system requirements. The team structure could be three or four layers deep but in each case, a team should act as the integration agent for their team and all subordinate teams one layer down.

Throughout the detailed design, the teams must maintain a current risk list and, as it changes, brief their immediately higher-level team of those changes. Any new risks that are identified must be assigned to a responsible engineer who develops an action plan to mitigate the risk and then follows through in accordance with that plan.

Depending on how the verification work responsibility has been cut up, the teams should participate in the development of the qualification plans, procedures, and management information driven by the content of their specification. Each team at the top level, and possibly at lower-tier levels as well, should hold a critical design review (CDR) upon completion of their design work. This review should be chaired by the next higher team all the way to the top-level teams presenting to program management and the PIT.

21.3   Specialty engineering relations

The author has the greatest respect for design engineers and their creative genius, but there is a side of many design engineers that is a bit dark. Design engineers tend to be very positive about their creations as well they should be but this can become a problem in IPPT work. In many companies, design engineers are assigned to lead cross-functional teams. The team will commonly have several specialty engineers on it or supporting it. Specialty engineers tend to be very negative about a design not because they do not like designers or their work, rather because it is the nature of their work to consider how things can go wrong. The reliability engineer considers all of the ways the design can fail and the consequences of those failures. The maintainability engineer deals with how to restore a broken item to useful service. The safety engineer considers ways that the design could cause bodily harm or loss of valuable resources. This deluge of negativism is more than some design engineers can handle while trying to maintain their positive attitude toward their work.

If the design engineer reacts protectively with energy and in a hostile fashion, the specialty engineers may be suppressed and quieted down causing the design engineer less grief in the near term, but the risks associated with the specialty engineering concerns will not go away by themselves or by shouting loudly at them. If they are not dealt with in a timely way during the early design development, they will come back later in the program when it will be very expensive and time consuming to solve the related problems.

This problem gets so bad in some programs and companies that the specialty engineers cease going to team meetings in order to avoid being treated badly if they open their mouth and utter negative thoughts, which are the language of their trade. Team leaders and system engineers must recognize that there is a natural conflict between designers and specialty engineers. They need to try to help these people understand the nature of this conflict, its necessary existence, and the value the program and team can gain from it. Criticism is a valuable part to the design process. It is a reality that a person who creates and utters a thought fraught with errors commonly cannot see the errors that are so easily observed by anyone else.

The team needs to come up with a way to foster cooperation between the designers and the supporting staff including specialty engineers. One good way to do this is to have each specialty engineer go over their requirements and what they mean as early as possible. This may take 10 min for each specialty engineer with one discipline briefing at each meeting until they have all given a pitch. These pitches can be preceded by a talk by the team leader encouraging the designers to consider their work in a positive light while noting the need for someone to find any flaws that might exist in the design. The team leader must fashion acceptance of the members being part of a team and that team success is more important than individual ego problems. The same motives that bind the members of a championship sports team together can work, but they must start with mutual respect of the team members for each other and what they bring to the team.

21.4   Configuration control

21.4.1   Representations control

The team will use many representations of the evolving design in their work. It is not uncommon practice for these different representations to be controlled in a very slipshod fashion despite their importance to the overall effort. The PIT should develop a list of all of these representations used by all of the teams including themselves and identify a responsible principal engineer for each one who must accept the responsibility to know the configuration of the representation at all times and its relationship to the current product design configuration. These representations may include physical models, computer models and simulations, math models, and breadboards.

21.4.2 Continuing specifications maintenance

The top-level teams should each have been given a specification by the PIT for the top item in the system product entity structure for which they are responsible. This specification will very likely require continuing maintenance through the detailed design phase. In all cases where it is changed, they should gain the approval of the changes from the PIT.

If the team is responsible for a very complex entity in the system product entity structure, they will likely have had to develop lower-tier specifications during preliminary design and possibly have assigned the development responsibility for some of those items to lower-tier teams. In that case, that team will have to act as the integration and approving agent for lower-tier specifications.

21.4.3 Product design control

The PIT is the principal system design agent and as such, top-level IPPTs must gain approval of their designs from the PIT. Once approved, those designs or changes pass under formal configuration management control, which should be an ancillary function of PIT. All original designs and changes to those designs should be reviewed by the system agent for the item. This is PIT for top-level teams and the superior IPPT for all lower-tier IPPT. The PIT must have the ability to overrule lower-tier configuration rulings in the interest of system optimization.

21.5 Technical control of hardware sources

21.5.1 Internal teams

A program that organizes all work through cross-functional teams would do well to look upon everyone acting as a supplier as being a team. This includes in-house teams, sister division sources, associate contractors, customer-furnished material, and suppliers and subcontractors of all kinds. The same set of controls is appropriate across all of these sources and includes a formal specification and some form of control over the supplier process to ensure that the supplier follows a disciplined approach to product development including development of a design that is consciously compliant with the requirements in the specification and eventually proved to be so through a formal verification process producing convincing evidence of that fact that is made available for audit.

21.5.2   *Design to specification procurement*

Where the enterprise developing a product at product entity level *N* understands the problem at level *N* but does not have the right knowledge, resources, and skills mix to design and produce a subordinate element at some lower level, the enterprise could choose to develop a specification to be provided to a supplier as a part of a contract to deliver an element satisfying the requirements contained in that procurement specification. In some very specialized fields, common industry practice is for the supplier to prepare the specification subject to buyer's approval and change control through a configuration control board (CCB). This often happens in guidance set, jet engine, and rocket engine fields, for example.

An engineer familiar with the technology of the item being procured and a member of the parent item team should be identified as the buyer principal technical agent and this engineer should coordinate with the assigned procurement and contracts people, as members of an extended sub-team including the supplier throughout the design development period. This team should review the design concept, preliminary design, detailed design (drawings essentially complete), qualification planning, qualification results, and production planning. In order to accomplish this oversight activity the buyer needs certain information, and these items should be identified on a supplier data requirements list (SDRL) included in the contract.

Periodic meetings should be held through travel, Internet meeting software, or videoconferencing to ensure that everything is remaining on track. The buyer team members should be especially intent on ensuring that item interfaces evolve in a thoroughly compatible way.

The buyer could organize the qualification process in one of two ways on this kind of program. First, they could require the supplier to deliver one or more qualification articles and accomplish the qualification testing work themselves reaching a conclusion about whether or not the design satisfies their requirements. Ideally, the supplier would be permitted to participate in this kind of qualification process. As an alternative, the buyer could require that the supplier accomplish all qualification work and report the results at a buyer audit when all of the evidence is available. In this case, the buyer should be permitted to participate in the qualification process if they choose to do so.

Some large organizations have gone too far in off-loading the design responsibility for their products while working toward becoming a cost-competitive assembly agent. Every development organization probably should retain a competent design capability for some range of key technologies that sets them apart from their competition and through which certain key product qualities are preserved and enhanced.

21.5.3   Manufacture to print procurement

Where the enterprise possesses the necessary command of the technology and chooses to do so, they may develop an in-house specification and accomplish the design resulting in a drawing package. For whatever reason, the enterprise may then choose to off-load the manufacturing responsibility to a supplier requiring them to manufacture to the drawing package supplied under contract.

The design should be qualified prior to the product manufacturing process beginning. This can become a difficult problem. If the buyer was qualified to manufacture one article for qualification purposes, then perhaps they should have accepted the manufacturing responsibility in the first place. The qualification article or articles could be built in an engineering shop but they should be manufactured in essentially the same way the final product will be manufactured and to the extent that these differ, the qualification evidence can be properly called into question. If the supplier manufactured item is qualified, then either the supplier or the buyer could accomplish the qualification testing.

Clearly, there are potentially significant problems in any of these arrangements. The buyer is responsible for the design, and qualification is a process of verifying that the design satisfies item development requirements. The problems with this kind of procurement are driven by partitioning the development responsibilities at a point where there is a cross-over interest no matter where you place the contractual boundary. This form of procurement is best applied only for relatively simple items with little development risk.

21.5.4   Commercial off the shelf

Commercial off the shelf (COTS) is a Department of Defense (DoD) term but it can apply to any system and customer base. As we accept COTS into the design solution, we must assure ourselves that it will be adequate to satisfy requirements. The requirements area where COTS is the most likely to fall short is environmental when the application involves high stress, as in military applications. Commonly COTS is designed for fairly benign civilian situations, but a portable PC in the hands of a farmer in the fields is not much different environmentally from it being in the hands of a soldier on the battlefield. Many items of commercial equipment can take a significant beating. We may find in testing COTS items that they do in fact satisfy some fairly demanding environmental requirements. In other cases, it may be possible for the buyer to obtain a waiver from their customer for some requirements where the COTS item fails to comply. A third option entails altering the COTS item such that it can withstand the field environment.

Commercial suppliers may be reluctant to divulge the design margins and capabilities of their product considering these data to be proprietary and competition sensitive. One approach to this problem is to select COTS hardware and software for early use in the system where it makes sense. During early system testing and deployment, the COTS can be evaluated. If it performs acceptably, it is made part of the final system. If it exhibits failures and shortcomings that cannot be tolerated, an item may have to be changed from COTS to an altered item where the commercial item is altered in some way after purchase from the commercial source and before delivery to the customer. Alternatively, a new item may have to be designed that fully satisfies the rigorous military environmental stresses in the application with the COTS item filling the functional bill through a requirement waiver until the new item is available. This pattern fits into the DoD application of spiral development perfectly.

21.5.5   Parts, materials, and processes

At the bottom of the system product entity structure food chain, we have detailed parts and materials from which the system items are built. There are many advantages in standardizing the component parts, raw and finished materials, and the manufacturing processes used in product systems. This control is implemented by building a list of these parts, materials, and processes that have been approved for use on a particular product or within a particular company. If this is not done, individual designers will commonly draw on their own experiences with various suppliers and products. As a result, our design may use 10 different parts to satisfy the same function. When you multiply this effect on one component type by hundreds, the supply and maintenance chains can become heavily burdened unnecessarily. Also, different engineers calling for different manufacturing processes intended to achieve the same results causes unnecessary training and certification problems for manufacturing engineers and personnel as well as added difficulty for quality assurance personnel.

In many organizations, it is difficult to merge parts engineering and materials and processes engineering efforts because these organizations are widely separated with parts engineering being in an electrical engineering organization and materials and processes located in a structural design organization. There is some sound logic supporting this organizational structure clearly, but it can impede evolution and implementation of a sound parts, materials, and processes standardization program.

One alternative is to have separate parts standardization and materials and processes standardization programs, but where organizational differences persist, the enterprise would be better served to insist that those two departments cooperate across an organizational bridge in accordance with a common plan. It is possible to retain these functional organizational differences if necessary but cooperate in all program assignments.

21.5.6 Sister division sources

In the 1990s, many very complex enterprises from previously independent companies were formed. All of the great names in aviation pioneering are now hyphenated into division name strings. In many cases, the need for specific product entities can be satisfied by these sister divisions. The program manager must not, however, treat these sister divisions any differently than nonrelated suppliers. There is a natural tendency to treat sister division suppliers with kid gloves even though they may have performed badly on prior programs. We may not feel it necessary to provide a specification for something that is going to need some development work. We may manage the sister division with less energy than a nonrelated supplier providing less oversight than warranted. It pays to treat sister divisions exactly the same as any supplier, no better and no worse.

21.6 Software design

All of the comments in this chapter were frankly formed in the author's mind in context with hardware item development, but all of these design process controls and risk abatement techniques apply to all product design activities including computer software design. The software product is, clearly, very different from hardware in that it is an intellectual entity rather than a physical entity. The fact that hardware has a physical form that can be taken into the mind through vision or imagination, depending on the stage of development, encourages a much wider understanding of the hardware product than seems possible with computer software.

In early software development, engineers use models to understand the problem space, just as hardware people do. These tend to be different models and the software models are seldom understood by hardware people. Therefore, the software modeling driven requirements analysis process even when properly done is difficult for hardware people to fathom. If the system engineers, who should be able to bridge this gap, are former hardware people as is commonly the case, hardware–software integration during the requirements analysis process can be very difficult. Hardware dominated system engineers have to force themselves to come to an understanding of the software development process because the systems they will be working on in their career will accomplish all of the hard functionality in the software. One approach to closure on this endeavor is to study software models commonly applied by the enterprise such as UML.

Following a sound software requirements analysis process resulting in a formally released specification, software people must begin the design of the software by defining the software architecture, software–software architecture element interfaces, hardware–software interface planes and details, languages to be used, and preferred algorithms. Ideally, the computer within which the software will run will be jointly selected by hardware and software people rather than the computer determined independently by the avionics people during this design process.

The author maintains that the writing of software code is akin to the hardware manufacturing process while others accept that this is part of the design process. It is unfortunate that all too many lines of code are written by analysts as their first development act without having previously defined appropriate requirements or accomplished conscious design work as prerequisites. The software development process should follow essentially the same four-step process used for hardware: define the problem (a specification), solve the problem (synthesis of the requirements including design, possible procurement of some of the software code, and manufacturing), and prove that the design satisfies the requirements (verify) all within an infrastructure of sound technical management.

21.7   Critical design review

All of the discussion under PDR applies to CDR as well except that the state of the design should be complete rather than preliminary. Upon completion of CDR, approval should be forthcoming to manufacture such equipment and software that is needed to qualify the item for the intended application. The CDR must, therefore, approve the design but also the verification, manufacturing, quality, and logistics planning for that design.

During the CDR, the design features should be presented, of course, but the emphasis should be on those elements that have been a cause for concern during the detailed design period. A large portion of the review should be devoted to assessment of the design by the analysis and specialty engineering community. As in the case of PDR, the first planning action should be to identify the exit criteria followed by building the outline and assigning people to be responsible for the items on that outline. Those responsible for preparation of the presentation should be given a period of time to develop their components followed by a poster session or dry run, critical feedback, a repair period, and a final review and approval.

The review could be held at the contractor's facility with the physical attendance of customer personnel or accomplished through the use of teleconferencing possibly using meeting software with participation from several sites connected via the Internet. The review documentation could be provided by a paper document or CD. The meeting could be recorded with the recording made available as an attachment to the minutes. A video recoding will provide visual evidence of who was talking on the audio track but be careful if using only audio recording. In an effort to reduce the time for preparation of minutes from 2 weeks to some lesser period of time, the system engineers on the Advanced Cruise Missile program at GD Convair decided to record the audio. GD Convair security would not allow any recording device to cross the facility fence line, so a wire recorder from the security office of World War II vintage had to be used. No record was kept of who was talking and it took 3 weeks to complete the minutes.

When using recording it can have an amplifying effect on those attending. The outgoing personalities tend to be more so as do those more withdrawn. While attending and coordinating several of these large meetings over a 30-year tour as a captive employee of three aerospace firms, the author discovered how to deal with meeting recording and it was an odd experience that offered the method. He attended a stress management session at his Health Maintenance Organization (HMO) with his wife because he was clearly his wife's stressor. When entering the room, he noticed there was a lady quietly sitting in the corner with a video camera. He was very worried about this, of course, but by the time the meeting started he had forgotten she was there. He recalled major reviews where the audio–video people were making a spectacle of themselves making it impossible for anyone in the meeting to forget they were there.

During the CDR meeting, the cochairs from customer and contractor need help managing the meeting by a contractor person, ideally a system engineer, who can run a small staff to include a meeting secretary who must acquire the information needed to prepare minutes, one or more runners to make contact with presenters in the event the schedule changes and accomplish many unscheduled tasks, and an action item data entry person who can also staff incoming phone and e-mail lines for possible messages for the busy people in attendance.

The review must remain on schedule in order to control the flow of presenters into and out of the meeting room. The contractor person supporting the cochairs is in the best place to regulate this but the cochairs should be briefed that they have the key role in making this happen. They have to take action to avoid schedule discontinuities. Long discussions about a presentation can be tabled by a statement inviting an action item on the part of the meeting attendee who will not be satisfied by a speaker's position. Where there is a healthy debate in the room, those participating can be asked to form a subgroup to meet in a caucus room (that should have been provided for in meeting planning) and return to the main meeting for a report out at a specific time. Other discussions may be completely out of scope relative to the contract and it is not a bad idea to have a contracts person sitting close to the contractor cochairman for this reason.

The enterprise should be served by an action item system residing on a networked computer infrastructure such that action items can be entered at a major review at 10 AM sent to the person responsible for responding and possibly getting a response that afternoon in time for a post-session executive session attended by the cochairs at the end of the meeting for that day. At this executive session, the cochairs should review any action items created during the day and reach an agreement of which ones are out of scope and which side should respond to those that are in scope.

The GD Convair program manager for the Advanced Cruise Missile program applied an effective technique that avoided meeting scheduling problems with the speakers. He required all of the speakers to be present throughout the several-day meeting. The common reason why program managers would not want to do this is, of course, budget. These engineers in the room are not accomplishing program work but they are charging their time to the program. The program manager's motive was that he realized that this PDR was the only opportunity he would have to ensure that his staff, recently expanded, would all go into the detailed design period with the same program perspective. The PDR acted as a training opportunity.

chapter twenty-two

Integration of test and analysis results*

22.1   Two "V"s for victory

We may have to wait until the customer finds the delivered system fully satisfactory to declare final victory, but an initial victory in the quest for integration excellence can be claimed when (1) it has been proven that the product articles perfectly satisfy the requirements for which they were designed, (2) that they perfectly reflect the controlling documentation, and (3) that we have accomplished these feats within budget and schedule limitations. Something pretty wonderful has taken place in order for these conditions to be realized. It means that hundreds, possibly thousands, of people have found effective ways to cooperate to reach a collective goal that none of them individually could have attained. In order for this to occur, the team must have sure knowledge throughout the project of every representation of the product articles and the correlation of these representations to a sequence of specific product baselines.

It is a very difficult task to maintain configuration control of the product design. During the period prior to preliminary design review (PDR), the design concept may be very volatile. Once the product is deployed into the customer's environment and placed in service, modifications may expand the configuration control difficulty for both the engineering and manufacturing data and the actual product articles in use. When we think of configuration management we most often think of maintaining the relationship between the actual product articles and the engineering drawings representing that product article. This is a proper role of a configuration management organization: to assure that the product and engineering documentation agree for specific articles. But, there are many other representations of the product articles that should be configuration managed and configuration management organizations do not always accept responsibility for these other representations. Someone must accept the responsibility.

In addition to the engineering drawings, specifications, manufacturing planning, and inspection data representations, we must maintain the configuration of our test and analysis data, models, simulations, mockups,

---

* The material in this chapter is from Grady, J., *System Integration*, CRC Press, Boca Raton, FL, Chapter 15, 1994. Used with permission.

and test articles that are used to validate and verify the design, and the computer programs associated with these items.

We will use these "V" words here in the MIL-STD-1521B and EIA 632 sense rather than with the DOD-STD-2167A meaning. The first two and the latter document use these words in almost the exact opposite fashion. Since 2167A has been discontinued, it is not an authoritative source today. MIL-STD-961E, replacing MIL-STD-490A, offers some support for the author's preferred meaning of these words by naming Section 4 of a specification verification. We said earlier that integration depends on the specialists having a common language through which they may communicate with each other at the edges of their specialties. Integration work is much more difficult when the specialists use different meanings for words in their common vocabulary. You should be very careful to understand another person's intended meaning when they use the popular words: validation, verification, function, traceability, and integration.

According to the dictionary, the word validate means to compel assent because of the soundness of the convincing reasoning offered. Verification means to substantiate or prove a premise to be true. Clearly, right thinking people could decide to couple either word with exactly the same process intended to produce evidence of compliance with some standard. Both of these words do require some objective standard of truth against which the premise is compared. ANSI/EIA 632 is offered as that standard in this book.

To validate a set of requirements according to 632 means to prove that it is possible to satisfy them with a particular design through analysis of that design or test of an article produced through a traceable process from its requirements and engineering. This standard poses another form of validation where the user evaluates the product relative to their user requirements. The author includes this operational test and evaluation (OT&E) process with development test and evaluation (DT&E) that together are part of verification along with item qualification and item acceptance. Verification, in the author's view, is a process of proving that a product article, produced in accordance with the engineering, manufacturing, and inspection documentation is truly representative of this documentation.

These words also find their way into technical manual proofing. Government customers will require the contractor to validate the technical data (maintenance and operations instructions, for example) prior to exposing government personnel to it. This step might be accomplished at the contractor's plant using contractor technicians and technical publications personnel. The data is then taken to a customer facility for verification where customer personnel attempt to apply the data to practical situations involving the product. When the data has passed both of these steps and corrections completed derived from these experiences, it is both validated and verified and ready for use by customer personnel.

These words also are used in the context of an independent team that evaluates some activity or product to assure that those intimately involved in the item being evaluated have satisfied a particular condition. These teams are commonly called independent verification and validation (IV&V) teams. A customer might contract with such a firm to help them evaluate the contractor's process and product. A commercial company might hire such a firm to cross check their findings or work on a new product before placing it on the market.

Set aside the technical data and IV&V applications for these words now and let us return to the definition and context of these two words included above. In this context, validation should take place prior to detail design work as a means to prove that a particular design concept has a very good chance of success. Verification follows the design process as a means to prove that the final design satisfies its original requirements. Both of these processes are accomplished through test and analysis of product articles or representations of those articles. Throughout this sweep of events we must know what we are talking about. We must understand what the product configuration is in all of its representations.

22.2   *Configuration control*

We must have an absolutely foolproof way to unambiguously identify every representation of every product article and to correlate these with each other at specific product baselines. These representations could include

1. Actual product articles themselves with possible differences between articles
2. Engineering drawings
3. Specifications
4. Manufacturing planning
5. Inspection checklists
6. Procurement statements of work and supplier reports
7. Test plans, procedures, and data sheets
8. Specialty engineering analyses and reports of those analyses
9. Models (physical, mathematical, and software)
10. Test articles
11. Mockups, and breadboards

For a given category on this list, we need to be able to identify and define each item in that category. For example, we need to be able to identify every model used in association with the product development process and define each one unambiguously such that there is no doubt about which model we are concerned with. One way to do this is to apply the same configuration management techniques that almost everyone has mastered for engineering drawings to our models. Each model may be assigned a drawing number and the actual model and the corresponding number of the drawing representation kept in synchronism using dash number changes or version numbers. In addition to the controlling number, each model should have a unique name that we humans can easily relate to its function or purpose.

This same pattern can be extended to all of the other product representation categories listed above. For most of these categories it is obvious how this pattern can be applied. But even analyses can be numbered through the analysis report numbers assigned whether they are memos or formal reports. Though many will say that it is simply a matter of good engineering discipline that each person responsible for one of these items would want to maintain clear knowledge of its configuration, it is very hard to realize in practice. Everyone will simply not maintain effective configuration control of their tools voluntarily. Even if they would, there is a larger issue of configuration control that no individual responsible for one of these entities can satisfy. There is a need for external integration of these V&V articles. Generally, this is co-product, cross-functional, cross-process integration, but the reader will be able to imagine other possibilities from his or her own experience.

22.3   *V&V article control matrix*

In the context of our plan expressed earlier to apply integrated product development, the PIT should be held accountable for maintaining a list of all of the representations of the product articles. It may very well be the configuration management representative on this team that actually does the work or it could be a system engineer. The work consists of listing all representations coordinated with specific articles, versions, and persons responsible. In addition, the work should also call for the correlation of the configuration of combinations of these V&V articles with particular product development milestones or baselines.

There are very few problems that a system engineer cannot ventilate with a flow diagram, Venn diagram, or a matrix. True to this premise, Table 22.1 offers a partial solution to this problem with a matrix. The left-hand column identifies particular product development baseline milestones that should be chosen to coincide with major customer reviews or marketing milestones and possibly some intermediate events (such as an arbitrarily chosen date of July 10, 2007). These are moments in time when we chose to coordinate all system representations.

Each V&V article has a column and in the intersections we place the dash number or version number of the article that corresponds to that baseline. Note that each column is identified with a drawing number and a brief item name. In an actual situation, you will need many analyses

Table 22.1 V&V Article Configuration Control Matrix for Item XYZ

Master configuration baseline name	Validation and verification articles			
	52-10001 model	52-10210 analysis	52-20235 test art	52-9012 mock up
PROPOSAL BASELINE	NA	−1	NA	NA
PDR BASELINE	−2	−2	−1	NA
JULY 07 BASELINE	−2	−3	−1	−1
CDR BASELINE	−3	−3	−2	−2
FCA/PCA BASELINE	−3	−4	−3	−2

Source: Grady, J., *System Integration*, CRC Press, Boca Raton, FL, p. 235, 1994. With permission.)

the names of which will have to be differentiated. The matrix should also have a column for responsible person, team, or department. This matrix should be maintained in electronic media using a database or spreadsheet and made available to everyone in read-only mode via network access. A particular baseline is composed of each of the indicated V&V articles in the dash number configuration noted on the matrix in the row corresponding to the baseline. If you combine with this information the corresponding product configuration definition data (drawings, specifications, etc.), you have a totally integrated view.

It is also useful to know the current configuration of each of these V&V articles. It may be possible to maintain several versions of a mathematical model, but it will be very difficult to maintain several configurations of a mockup or test article. These high dollar items, for which there is only one copy, generally will have to sequence through a series of versions representing the latest product item configuration. From time to time, it will be necessary to determine what it will take in time and money to change a particular V&V article from its present configuration to a new (or even older) desired configuration. Prior to changing any of these V&V articles, the PIT should review and approve the change and update the matrix. We could include a row at the top or bottom of the matrix for current V&V article configuration.

These resources also need to be coordinated with the formal verification planning work as discussed in *System Verification* (Grady, 2007). The planning data can be captured in a series of relational databases that also provide traceability from the product requirements in specifications Section 3, to the verification requirements in specifications Section 4, assignment to a verification task in a verification compliance matrix with linkage to a verification task matrix that coordinates plans, procedures, and reports to the tasks that will employ the resources noted in Table 22.1 and through which we acquire the evidence of compliance.

22.4   Test integration

Testing cost is generally higher than analysis cost for similar objectives, but the results are commonly accepted with greater credibility. Government customers commonly have the attitude that contractors will avoid a test whenever possible substituting an analysis if they can. Good contractors will want to apply the test approach when no other less expensive method of gaining information about something will work. The contractor will want to group all testing work possible into the smallest number of test events in the interest of cost. This is a typical conflict on programs where extreme positions on the part of the customer and contractor are possible and the right path is characterized by a condition of balance.

There are three generally accepted kinds of tests, at least in the aerospace industry. They are design evaluation testing (DET), qualification testing, and acceptance testing. In addition, large systems, or major components of those systems, may be subjected to operational testing by the builder in DT&E or customer in OT&E. Where the item is an aircraft, this is called flight testing.

DET is accomplished to validate design concepts before committing completely to a particular detailed design. These tests applied to particular design concepts also effectively validate that the requirements will yield to effective design solutions. It is always possible that some of our requirements defy the laws of science. It is entirely possible to over constrain the solution space driving it to zero. It may require testing to discover this unfortunate condition.

Design evaluation or development testing commonly requires special test articles that can be very complex and costly in their own right. These articles are designed for test purposes and need not include all of the characteristics of the final design. They must reflect the specialized requirements associated with the test article function. A hydraulic test bed, for example, need not abide by mass properties requirements for the final article but should faithfully reflect the volume, flow, and pressures corresponding to the final product.

There are great opportunities for effective integration of test articles. First, we should work to minimize the need for these articles by studying ways to cause one item to satisfy multiple needs while not compromising the needs of either function. We should challenge the risk basis for particular test articles. If it can be shown that the risk that a particular product design feature or subsystem is characterized by little risk, out of balance with the cost and schedule impacts of the associated test article, then a good case can be made for challenging that article. Finally, we should evaluate the availability and applicability of related test articles that may be available from prior programs and customer resources before committing to build an entirely new test article.

Qualification testing is accomplished to prove that an item is adequate for its intended operating environment and application. Flight or field testing of a complete system is an extension of qualification testing to prove that the system composed of the many configuration items satisfies the customer's operational needs and to prove out operating and maintenance procedures. Qualification testing is one powerful method for verifying that a particular product solution does satisfy the requirements driving the design. It is at the tail end of the overall system engineering process entailing decomposition of the need into a set of smaller problems defined by sets of requirements, design to those requirements, and testing to verify that the design satisfies the requirements.

Acceptance testing is applied in the manufacturing process to each production article, or in some sampling fashion to selected items, to prove that the production item is acceptable for delivery to a customer in accordance with a set of requirements expressed in acceptance test procedure steps. These steps should be chosen in response to a systematic identification of what is important in determining product quality. Commonly, this process involves identification of product requirements in a product specification referred to by DoD at the time this book was written as a detail specification. The content of this specification is translated into specific test events and procedures. A subset of these activities should be selected as a basis for formal acceptance testing upon which customer acceptance is based for each article offered up for shipment to the customer.

22.5 Non-test integration

Many system engineers have broken their pick trying to apply the same degree of planning and organization to analysis, examination, and demonstration events as is commonly applied to test events. The test world has the advantage in many enterprises of a single cohesive functional department that is responsible for all test work. The non-test world is fragmented into system analysis, specialty engineering, and other departments, each of which accomplish some analytical work. It is very hard to create a comprehensive plan for all of these analysis activities. The difficulty of the task should not dissuade us from trying.

On a given program, there will be a finite number of analysis events. Some will be planned ahead of time while others just happen driven by evolving circumstances on the program. Each analysis will be accomplished or controlled by one analytical discipline under the management of one of the IPPTs or the PIT. It should be possible to make a list of these analyses coordinated with the name of the principal engineer, analysis name, responsible team, objective or purpose, analysis report identification, date completed, current status, and any other information needed by the program.

Some of these analysis reports will have been stimulated by an interest in the design while others will be a part of the formal qualification stream of evidence. The latter reports produced from these analyses are for a subset of the tasks listed in the verification task matrix. Those not involved in the verification work should be captured in the program library outside of the verification documentation stream.

An analysis control matrix could be used to list all of these documents and help us to maintain them in some orderly fashion for easy retrieval. We should start this analysis control matrix during the proposal period based on what the customer has required us to do in the way of analyses plus any others we feel we must accomplish. This matrix should be maintained current throughout the period leading to CDR. This is an important element of the design decision or rationale traceability information that some customers might expect maintained. No matter what the customer requires in this area, it is very important to the contractor to know at all times the particular condition of all planned analyses.

part five

---

*Specialty engineering
methods and models*

chapter twenty-three

Introduction to specialty engineering and concurrent engineering

23.1   Specialty fundamentals

23.1.1   Systems and their development

A system is a collection of things that interact to achieve common purpose. System development is a process for creating man-made systems. This is a problem-solving process. Where the problem is complex and is to be solved by the application of engineering knowledge, the system development process is a candidate for the application of the systems approach to problem solving. The systems approach entails three fundamental steps wrapped in a framework of engineering and program management. These three steps are

1. Define the problem in a specification consisting of requirements derived from models ideally using a structured analysis approach.
2. Solve the problem through the application of creative thought by one or more design engineers possibly supported by others.
3. Verify (prove) that the design satisfies its requirements.

The same may be said for other kinds of systems besides those applying engineering knowledge, but this book is focused on systems that require the application of engineering knowledge. The techniques covered in this book are also most appropriate for systems where failure of the system may entail very serious consequences such as unplanned loss of life (some military systems have as their principal goal the intended loss of life on the part of the adversary); serious injury; and loss of economic, military, or business assets or position. It is appropriate when there are difficult cost and schedule constraints applied to the development of the solution and adverse consequences when these limits are exceeded.

Engineers solve difficult problems by translating today's knowledge of science and mathematics into practical artifacts that solve specific problems. Engineers are, therefore, problem solvers. The profession of engineering has been pursued for a very long time. One could argue persuasively that the Egyptians and Romans in ancient times practiced what one might today call civil engineering in the development of many awe-inspiring public works impressive to this day. But since those times, man's knowledge has expanded tremendously and we are able to bring to bear powerful techniques in search of solutions to very difficult problems posed by particular human and organizational needs.

23.1.2   The knowledge foundation

There was a time when the engineering profession was un-partitioned, when one engineer could master all of the engineering knowledge then accessible to mankind and apply it toward solving problems consistent with the available technology. There was a time fairly recently when one engineer could complete the design of a fairly complex artifact within one engineering domain such as mechanical, electrical, or civil engineering. The principal thing that separates us from our Cro-Magnon ancestors is our access to the cumulative knowledge of the ages that has monotonically increased over time because of continuing efforts to uncover new knowledge, often motivated by economic or military advantage, and some experience in using it to man's advantage.

Our accumulation of knowledge has many advantages in that better solutions are derived through the application of more and better knowledge to our problems. There is a downside, however. We have been so successful in the accumulation of knowledge that no longer can any one man or woman master all of man's engineering knowledge base and apply it in an economically advantageous fashion. If the problem is sufficiently complex, it may not be possible for one person to most advantageously independently solve a problem even within one domain.

Man's individual capacity for knowledge has changed very little in thousands of years but the available knowledge has outstripped our individual capacity. Man's universal solution to this problem is specialization. We partition all of knowledge up into subsets. Engineering was one of those sets at one time, but today it has been further partitioned into many kinds of engineering, some of which have subspecialties. The other part of this solution that is more difficult than partitioning is integration across these different domains to gain the effect of one great mind that has access to all of our pooled knowledge. This process takes place largely through human communication and the deeds of system engineers who specialize in boundary conditions, boundaries between knowledge partitions, between product partitions, and especially where these coincide.

23.1.3   Enter the specialty disciplines

During World War II, military experience with weapons systems revealed that some of these systems were difficult to operate, were prone to failure, and were difficult to maintain. The design of these systems was accomplished largely by mechanical and electrical engineers few of whom had any special training in ways to design systems such that they would be supportable once delivered. The emphasis was on satisfying operational functionality and speed of production and deployment. As these fields like reliability, maintainability, logistics support, and system safety came into being, they became populated by engineers and analysts from the existing engineering staff.

Over time, societies were formed as a means to promote good works, provide a publishing opportunity, and standardize the approaches for these new fields. The organizational structures in companies dealing with the Departments of the Army and Navy and their follow-on agency the U.S. Department of Defense and procurement organizations in them unintentionally encouraged that walls were built around each of these disciplines. A matrix organization matured in many defense contracting companies with emphasis on functional organizational structures. Insular attitudes often developed in these departments as their members were badly treated by the design community. Societies and their members populating both sides of the contractual border cooperated in developing extensive standards and called for compliance with them on contracts.

Many design engineers became hostile to people in these specialties because they all appeared to be so negative about their design, finding ways it could fail, kill someone, or be heavier or more costly than planned. Design engineers are very positive people, at least about their own design. This is one of several fundamental conflicts that arose between members of the design and specialty disciplines. Over time, this led specialty engineers to evolve one-way attitudes about design engineers to the effect that they must satisfy their specialty requirements regardless of the overall effect. Sometimes, the specialty requirements were themselves in conflict but no matter.

These attitudes and their organizational expression led over time to fort-like walls between the several specialty and design organizations in many companies to the detriment of their ability to collectively attack and solve difficult development problems within contractual cost and schedule constraints. The evolution of specialty engineering disciplines was the right thing for industry to do, but the implementation was often flawed.

It was not until the early 1990s that this began to be corrected through the recognition of a need for what came to be called concurrent engineering, which was what the system engineering process was initially intended to provide. Thus, one must conclude that system engineering implementation was flawed resulting in a flawed specialty engineering introduction. The same forces that spawned the specialty engineering disciplines also had brought into being system engineering. Perhaps, it is no wonder that in many organizations, the specialty disciplines are in the system engineering organization.

23.1.4   Several kinds of specialists

Since no one knows everything, everyone is a specialist, we could simply inventory all of the disciplines needed in a development organization and that would be the specialty engineering discipline list. While probably true, there is a particular subset of all engineering disciplines that are referred to by the term specialty engineering.

23.1.4.1   Every engineer

Does there exist an engineer who knows everything about all facets of engineering. Some you have met probably think they are such engineers, but the author has never found one and certainly not in himself. You may say, "How about a system engineer?" System engineers, unfortunately, do not know everything. Their knowledge tends to be T-shaped, as noted in the *System Engineering Handbook*. The system engineer's vertical knowledge member (depth of specialized knowledge) tends to be weaker and their horizontal member (breadth of knowledge) tends to be stronger than conventional specialists. Persons named specialty engineers tend to have the opposite relationship between their vertical and horizontal knowledge members.

Even mechanical engineers are specialists within the broader context of man's total knowledge base even though they are generally pretty good all-around engineers. While every engineer is a specialist, we do not commonly consider them specialty engineers. This is especially true of design engineers, of whatever stripe, who are responsible for an overall design solution. These people have to synthesize the requirements for an item into a successful design often making use of interpretations of those requirements by people that we do call specialty engineers.

23.1.4.2   The system analyst

There is a whole set of disciplines often called system analysis disciplines that are populated by very specialized engineers. This includes aerodynamics, thermodynamics, structural dynamics, thermal analysis, and mass properties. These disciplines tend to have been recognized engineering necessities over a longer period of time than those known by the term specialty engineering. They also tend to use mathematical techniques similar to those used by design engineers often involving the calculus and differential equations. These organizations tend to be installed within the design organizations rather than collected in some kind of specialty structure.

23.1.4.3   The specialty engineer

The term specialty engineering tends to be reserved for a specific set of specialists. These engineers often do not use differential equations in their work, rather the mathematics of probabilities or qualitative analysis. Each discipline focuses on a narrow aspect of the evolving product design. A list of these disciplines would commonly include the following: availability, human factors, maintainability, producibility, reliability, system safety, system security, supportability, and survivability. Chapters of this part provide an appreciation for what these specialty engineers do on a program.

In this part of the book, we will address the more general and broader notion of specialty engineering embracing both these specialty disciplines and the system analysis disciplines. They are all performing a similar function actually. They focus tightly on one narrow aspect of the problem space and the design solution identifying related requirements and ensuring that the evolving design embraces those requirements.

23.1.5   The past

The term specialty integration was popularized by MIL-STD-499 and 499A. One of the three major sections of a system engineering management plan (SEMP) described in that standard had this title. Specialty engineering disciplines arose as it became obvious that the design engineer could not master the creative process of design synthesis and the ever-expanding number of specialized design views. As these specialized views, like reliability and maintainability, became accepted, it soon became obvious that someone had to integrate the work of these many people and the system engineering process was born.

Specialty engineering disciplines developed effective methods to define their requirements, communicate them to designers, and assess the design for compliance. Each specialty engineering discipline became a separate house focused narrowly on its own agenda. Societies came into being encouraging this inward focus. Functional organizations were founded in development enterprises for each discipline encouraged by DoD customer organizations in these same fields. DoD prepared military standards and handbooks on how to perform these disciplines and required their application and compliance on contracts.

Something went wrong in the evolution of the systems approach that resulted in the specialty disciplines acquiring the habit of serial work performance with physical collocation by functional organizations rather than product teams. Designers would create designs and the engineers from the many specialty engineering disciplines would independently review the designs for compliance with their requirements. This pattern of work is called by many transom engineering. The designer throws the drawings over the transom to an isolated specialty engineer who notes problems that must be corrected and throws the drawings back rejected. A tremendous amount of time is wasted in this process that could be avoided by a simple conversation between the designer and the specialty engineer before the designer committed to a particular design feature expressed on engineering drawings. The transom engineering analogy is very apt in illustrating the antithesis of what the system engineering process was supposed to be.

Figure 23.1 illustrates the specialty engineering problem from the designer's perspective. The author was first exposed to this view by Steve Landry, at the time a system engineer from General Dynamics Convair Division, at a General Dynamics Systems Engineering Seminar that at the time brought together system engineers twice a year from all of the divisions of the company in a week-long training program. Many years ago, one designer could master everything needed to synthesize the requirements into an effective design solution. Times were simpler and so were systems. The designer could master what was known about reliability and safety. If the product embraced these characteristics, it was because the designer put them there. In the1980s, there were many old designers still around who had lived in this world and resented the specialty engineering crowd trying to influence what they saw as their prerogatives. In large

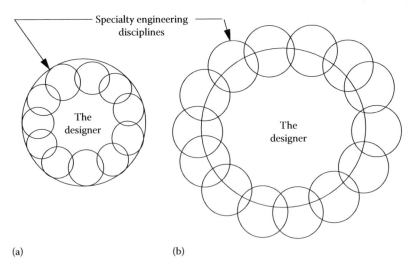

(a)   (b)

Figure 23.1 A changing world: (a) the generalized engineer and (b) the specialization effects.

companies that develop systems today, the number of design engineers who lived in this isolated world are not so many, but it is still not uncommon for design engineers to yearn for those days to return.

Today, the designer simply cannot master his or her engineering discipline and all of the specialty fields as well. This is the very reason for system engineering and structured decomposition of large problems into related smaller ones. The challenge is to provide an environment within the work place such that the designer and the crowd of specialists become one again. The stage is now nearly set for the entry of the concurrent or simultaneous engineering initiative and a return to what the founding fathers of the system engineering process had in mind.

Those programs that have been well executed from a specialty engineering perspective and produced the required product within cost and schedule constraints have very fortunate customers. Let us picture for a moment that the customer for program XYZ has prescribed through the program SOW and CDRL that there shall be 20 specialty engineering disciplines on the program. Since the reader will immediately say that this is an outrageous number, let us list the 20 the customer has predetermined in their contract language. Do any of these seem outrageous? No, most of them are fairly common on programs. It is not hard to arrive at 20 specialty engineering disciplines on a program.

Reliability	Human factors	Logistics support analysis (LSA)
Maintainability	Parts engineering	Survivability and vulnerability
Availability	Materials and processes	Transportability
Mass properties	System safety	Life cycle cost
Aerodynamics	System security	Quality engineering
Structural dynamics	Thermodynamics	Producibility
Guidance analysis	Contamination control	

In most engineering organizations, the older disciplines requiring use of mathematical modeling involving differential equations and calculus as well as very technical engineering knowledge and skills (aerodynamics, thermodynamics, etc.) would be grouped into a system analysis department. These are the disciplines that the design community has more or less accepted. The newer disciplines that do not normally require design skills, like reliability, maintainability, and safety, would commonly be grouped in one or more system engineering functional departments. These disciplines have yet to prove themselves to the design community, and practitioners of these disciplines in many development organizations have to fight every day for their self-respect.

The work of all specialty engineers also brings their attitudes into conflict with those of the designers. The designer commonly has a very positive attitude about their evolving design solution and thinks of it in terms of it operating normally. The specialty engineer, on the other hand, thinks in very negative terms about those designs. They think of all the ways they can fail, kill and maim the operators and maintainers, and otherwise create havoc. You can see the natural clash between these views that often results in open hostility without either party really understanding the cause. Therefore, there is a real need for program managers, team leaders, and system engineers to be watchful for the signs of conflict between the designers and specialty engineers and to take action to make the parties aware of the cause and to make clear to the workforce that unprofessional hostility will not be tolerated.

During team meetings, the meeting facilitator or team leader should make sure that the specialty engineer's views have been addressed calling upon those who have not voluntarily come forward to give theirs to do so. Specialty engineers long treated badly by designers can become reticent about speaking in public forums about the design and must be brought into the conversation and shown that it is both necessary and safe to offer their professional opinions. In one otherwise excellent company with a big government contract, the government demanded that the contractor improve their system engineering practices resulting finally in a more systematic work process and formation of integrated product teams. Everyone went to classes about cross-functional teams and things seemed to be moving in the right direction. After a period of time, symptoms of poor specialty engineering integration started appearing in the form of manufacturing and test anomalies. On looking into the causes of these problems, management discovered that on some of the teams, designers had achieved the role of team leader, not an unreasonable outcome, and were encouraging attitudes and actions in the meetings that lead to specialty engineers choosing not to attend the meetings. These actions probably had not been consciously taken by the team leaders, but design engineers naturally will feel in a positive way about their design and the specialty engineers will naturally approach any design with a negative attitude. In this case, management insisted that everyone be passed back through a team training exercise.

But, let us accept that it is a perfect world that program XYZ is functioning within. Each specialty has the following approach to their activity:

1. Define specialty requirements
2. Communicate specialty requirements to the design engineers
3. Ensure the design engineers understands the requirements
4. Interact with the design engineers as the design concept evolves
5. Offer examples of how these requirements have been satisfied on similar systems

6. Review the engineering drawings for compliance with specialty requirements
7. Identity for the design engineers specific failures to comply with the requirements
8. Reevaluate changes for compliance

Each of the process steps in bold requires communication with the designer. If there are 20 specialists assigned to program XYZ, each of which must communicate a minimum of four times for an average of 1 h each with each design engineer. This is 80 interactions and 160 program man-hours for each design engineer. Expand this number by the number of design engineers simultaneously working on the program, say 40. We now find the specialty engineering designer contact event count is up to 3200 (20 × 4 × 40) on the program through CDR requiring 6400 man-hours.

Let us say that each specialty engineering discipline is especially effective in ensuring its requirements are clearly understood by the design engineers. Each discipline has prepared a checklist and they will expect each design engineer to read and sign the checklist signifying his or her understanding. These checklists commonly represent a digestion of the content of a standard called out in the SOW or system specification. Let us say that the average checklist has 50 items. The designer will have to read and understand 1000 checklist items (50 items/checklist times 20 checklists) in items b and c in the list above in addition to the other requirements in the item specification. The total program checklist event count becomes 40,000 (40 principal engineers times 50 items/discipline times 20 disciplines). Heaven help the designers if all of the specialty engineering disciplines were to become uniformly effective in influencing the design process in this fashion.

In what became the traditional specialty engineering approach, each specialty engineer would talk to the designer independently yielding no opportunity for interaction and synergism between specialty engineers. The only one who was exposed to all of the conversations (given that all specialty engineers pursued their responsibility with equal vigor) was the designer. If only one person on a program is to have this experience, it should be the designer, but why must only one person have this experience? Concurrent engineering implemented with a DIG or Web site, war room, physical collocation, and good leadership will permit an explosive simultaneous conversation with everyone's information available to all.

Clearly, our history has shown that products that comply with specialty requirements provide better value than those that do not. The challenge is to provide a teamwork environment that frees the designer to apply his or her creative genius to synthesizing the complete set of requirements within the safety of boundaries protected by the specialty engineers. This environment must ensure that all needed specialty views are energetically pursued yet ensure that the designer has adequate time to think and to do the creative design work. We need an orderly environment as far as the specialty community is concerned. But, the designer needs a creative environment with an absolute minimum of constraints. Somewhere there must be a balance point between these conflicting needs. That is our challenge.

23.1.6   The future

Before we approach a practical solution that can be applied today, let us first consider what might be the ultimate solution to this dilemma. The author was convinced, after seeing some of the work going on at General Dynamics Space Systems Division (similar to work at other firms no doubt) in the area of RAMCAD (reliability, availability, and maintainability, computer-aided design) several years ago, that the ultimate solution to specialty engineering integration will take the form of design rules embedded within the CAD tools that protect the designer from violating the constraints now expressed in many organizations in checklist form.

The designer need not read and understand any of these checklists (though the designer might be more efficient the more he or she understood those requirements). If the designer of a circuit board places a high-wattage resistor too close to a temperature-sensitive junction device such that it is exposed to an excessive thermal condition, the CAD tools will not permit the error to be made. A signal will flash on the screen noting the error and what to do about it. The design would never have to be reviewed by the reliability or thermal experts, or the review could exclude certain tedious elements. Specialty engineers in this environment ply their trade by helping to program the CAD tools for error avoidance.

Some of these kinds of features are being built into CAD tools today. It is likely that this environment will be very effective for some specialty engineering disciplines and not so effective for others as a function of the ease or difficulty in translating the specialty discipline requirements into CAD rules. Many aspects of system safety and human factors engineering will continue to demand human judgment for a very long time, perhaps forever. At the same time, some human factors elements like the size of access openings and accessibility for adjustments can be very easily introduced into 3D CAD tools.

These demonstrations of future capabilities are very exciting to behold, but generally are not uniformly available now for deployment into effective tools for use on programs. In the near term, until the ultimate solution is available, we must find another solution to effective specialty engineering integration.

23.1.7   A generic specialty engineering process

A system engineer charged with the responsibility of integrating the work of several specialty engineering disciplines should be familiar with the work these specialists do. Summary descriptions are included in Chapters twenty-four through thirty-one for several disciplines. All or most of these disciplines follow the work pattern presented in Sections 23.2 through 23.4 applying their own special techniques in doing so.

In summary, this pattern focused on identifying requirements, encouraging understanding of the requirements, concurrent support of the designer while they synthesize requirements into a concept followed by expression of the concept on design drawings, and assessment of the design solution for compliance. There is a spectrum of implementation possibilities in this process from focus on compliance assessment to effective concurrent engineering. We will assume here that the organization is capable of concurrent engineering using cross-functional teams.

Specialty engineers responsible for logistics support, operational employment, and product manufacturing should be simultaneously designing the logistic support system, operational employment process, and manufacturing processes (tooling, facilities, manufacturing flow, etc.) while the product design is in progress. As a result, the design engineer functions as a specialty engineer with respect to their activities. Taken together, we see that all of these people must forget about who is a designer and who is a specialty engineer and form a cross-functional team producing a coordinated product design, manufacturing process, and employment process. The goal for the team should be to become the equivalent of one all-knowing engineer whose interest stretches across the complete system.

23.1.7.1   Concurrent development and IPPT overview

Throughout the period of work characterized by the application of specialty engineering integration work, the emphasis will be on the specialty disciplines imposing their requirements on the product designer. The logistics engineer is supposed to influence the product design to be responsive to support concepts. The manufacturing engineer has to ensure that the product design includes producibility features. Similarly, other disciplines attempt to influence the product design in a unidirectional fashion. The author recognizes the concurrent development approach as an expression of the originally intended system engineering process with one exception. That distinction is that concurrent development encourages a bidirectional development conversation between the product design engineer and those responsible for the design of the production process, the logistics support concept, and the operational deployment and employment plans.

The emphasis today is on all of these designs being developed together, each influencing the others to arrive at a true system-optimized solution, not just an optimized product design solution. Just as a manufacturing engineer may ask a product designer to replace a forged beam with a larger cast beam to simplify the manufacturing process, it is just as valid for the product designer to ask the manufacturing engineer to use a forged beam rather than a cast one to preclude a beam size that adversely influences the space available for needed instruments. This should not be a one-way conversation only addressing the design of the product. The integrated product and process team (IPPT) responsible for a product element should also be responsible for the corresponding factory, logistics, and operations designs wherever possible. The team must develop a unified solution to the overall problem, not just a product design that respects specialty requirements.

This significant difference of a bidirectional rather than unidirectional conversation between the product designers and specialty engineers has the effect of encouraging a better degree of timely teamwork between the team members, whereas the old view of the process encouraged shell-shocked product designers overwhelmed by checklists.

Figure 23.2 illustrates an overall view of the IPPT-implemented concurrent development process. It is very difficult to illustrate the linear relationship of tasks and the explosive vertical dimension of the decomposition and integration process comprising the structured development process. The spiral diagram included in *A Spiral Model of Software Development, Tutorial: Software Engineering Project Management* (Boehm, 1988) and the vee diagram encouraged in "The relationship of system engineering to the project cycle" (Forsberg and Mooz, 1991) attempts to solve this difficulty. Both of these diagrammatic treatments

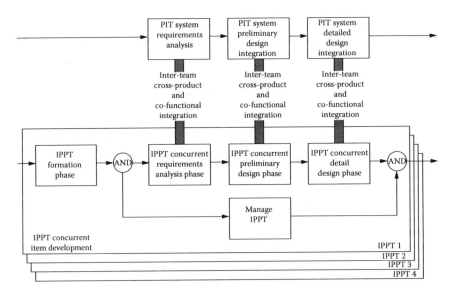

Figure 23.2 Overall concurrent development process.

are effective, but the author has chosen to use the simple relationship depicted in Figure 23.2 though the work can be explained in a spiral or V as well.

You will note in Figure 23.2 that there is one PIT process but many IPPT processes, one for each product team recognized by the program. Further, an IPPT may find it necessary to establish sub-teams and principal engineers below that level. The PIT is responsible for developing the system product entity structure and assigning the responsibility for development of those product entities to IPPTs. These teams come into being through this process. The PIT must concurrently develop the system requirements and monitor the lower-tier development of requirements within the IPPTs for traceability to the system requirements. As the IPPTs complete their requirements analysis process, the PIT must review and approve those requirements and authorize the IPPTs to initiate the design phase. During the design process, the PIT must monitor the IPPT concurrent design process. In so doing, the principal PIT targets are the detection of suboptimization and interface problems between items for which the different IPPTs are responsible.

23.1.7.2   The concurrent engineering bond

Product requirements analysis, concept development, preliminary design, and detailed design work must be accomplished in concert with a host of specialty engineering disciplines. Figure 23.2 illustrates this activity in the form of a concurrent engineering bond depicted as a dark band joining tasks that must be performed in a concurrent fashion involving intense cooperation between personnel responsible for the tasks thus joined. You will note that the integration work taking place between PIT and IPPTs on Figure 23.2 involves primarily inter-team cross-product and cofunctional integration. The concurrent engineering bond within an IPPT involves primarily intra-team co-product and cross-functional integration. It was the recognition of these different views of the integration process expressed in this diagram that eventually led to *System Integration* (Grady, 1994).

This bond signifies the shared experience and knowledge of the team members. Our intent is to create a work environment within which many individuals can cooperate toward common goals. This cooperation requires communication, the keystone of concurrent development and the medium through which the team members share and thus become one powerful development force.

The concurrent development process does not encourage concurrent definition of requirements and design development, a common mistake that results in a leap to point design solutions later supported by requirements that are satisfied by the design solution. Requirements definition and design solution must each be accomplished concurrently, but there should be a sequential relationship between requirements definition and design work with requirements work leading to recognizing that the concept development work must follow closely behind the requirements work because the lower-tier requirements work will depend on higher-tier design concepts to some extent.

23.1.7.3   Team formation

The IPPTs comes into being as a result of a PIT decision. Two team staffing alternatives are offered that the program and company management staffs must decide upon: (1) team leadership selection and (2) imposed leadership. Some integrated or concurrent product development authorities have encouraged that the team should pick its own leader. That may be the best approach, but program and company management are going to have to decide if that is within their range of acceptability. It is only suggested here that the program manager, to whom this person will report, should have something to say about the selection. Otherwise, the team may flounder for some time until it can deal with the leadership issue. A bad choice of an imposed leader, initial or permanent, may, of course, have an even worse influence on the team's prospects for success.

The solution offered in Figure 23.3 is to initially staff the team from available personnel (supplied by the functional departments) and appoint an initial leader who will have the responsibility to complete the team formation process. If this person gains not only program management's confidence but the team's as well, then he or she is probably the right one to lead the team to the completion of its work. The last block on the right of the first tier allows an opportunity to change the leadership based on the team's experience in forming. The author has not met any program managers that he can recall that would have permitted this action.

The team members must reach a common understanding of their goal expressed in the IMP and agree upon the process to be used to attain that goal. They must all understand the communication aids available (computer networks, bulletin board, telephones, specifications, the IMP/IMS, and meeting area tools) and how they will use them. Most of us humans are wonderfully effective goal-striving creatures. We can together satisfy very challenging goals if we understand and agree with them, are well led, and have the necessary resources. Good leadership will have the effect of infusing us with a common purpose so strongly felt that we must attain it. The concurrent bond reflects this intensity of feeling through which we are encouraged to share our specialized knowledge toward a common purpose.

The final step in team preparation is to train the team in concurrent development practices. The nature of this instruction is a function of the current position of the company on its continuous process improvement path to IPPT capability. If the company has only recently shifted to this

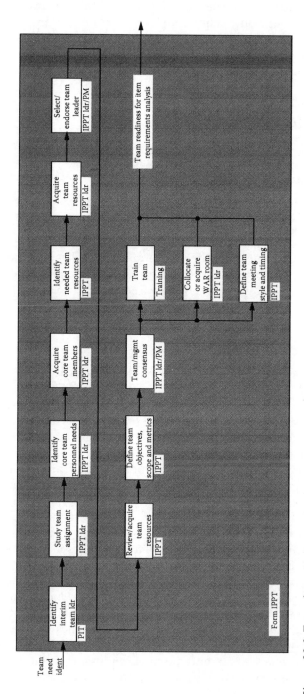

Figure 23.3 Team formation process.

approach, a course may be needed. If this is the third program employing IPPT, the team may only require a refresher course in group dynamics. At some point this will become a normal way of doing business that the people assigned to a team can work out on the fly.

23.2   Specialty engineering in requirements analysis

IPPT activity at any one moment is a function of program phase, product item characteristics and applicable technology (software versus hardware, for example), and team position on the development progression. Normally, IPPTs will not be assigned until the product entity structure has been established by early program work accomplished by the PIT. A project in its earliest phases should begin with the PIT growing out of a proposal or marketing experience. This team should accomplish early program work and accomplish the planning work for later program phases including defining teams corresponding to major product system entities. Once IPPTs are identified and staffed, they should follow a predictable progression outlined in this chapter.

23.2.1   Parent team requirements development

The author believes that each team should be presented with a portion of the IMP/IMS and a specification crafted by the PIT or superior IPPT when the team is formed. If a specification is not offered at formation, the IPPT must implement an organized requirements analysis process to define item requirements derived from the information supplied to them by the PIT. In any case, an IPPT must develop specifications for lower-tier entities subordinate to the item their team is formed around. If the IPPT has subordinate teams, it should develop the specifications for the items they assign development responsibility for to IPPTs.

Figure 23.4 illustrates one view of a generic requirements analysis process to a concurrent development environment employing a traditional structured development process for a single specification that can iteratively and cyclically be applied for a whole program. This process has to be going on within the PIT for the system specification and first layer of entities to which teams will be assigned. Upon formation, the IPPT must first study the requirements in the specification provided by the PIT. As the concept development work continues for that item, the IPPT must extend the requirements analysis for any subordinate entities. Note that this includes not only product requirements but requirements for the production process, logistics support, quality implementation, and operations activities in task F4113C.

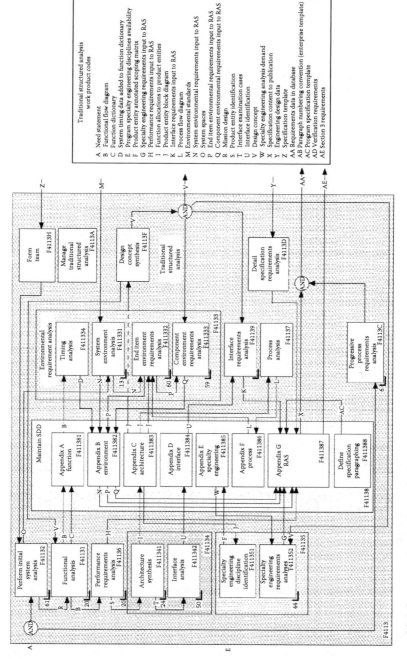

Figure 23.4 Item requirements analysis process.

Specialty engineering disciplines that must participate in the analysis for a particular item are identified by the PIT in task F411351 using a specialty engineering scoping matrix that coordinates disciplines versus product entities. Each specialty engineer with a call for that item specification must apply their modeling approach to produce one or more requirements from their specialized perspective. All of this requirements work must take place concurrently with a sharing of the resultant information taking advantage of the communications resources described in Chapter seven. A concurrent development bond formed by intense conversation must exist between these tasks leading to a coordinated set of requirements.

23.2.2   Child team requirements development

If the parent team does not provide a specification to its child teams, the parent team should provide them with the following:

1. An identification of the product entity superior to the item for which they are responsible and a specification for it.
2. A list of functions allocated to the item assigned.
3. Identification of required external interfaces in list, n-square diagram, or schematic block diagram form.
4. System environmental requirements if the item is an end item immediately subordinate to the system plus a system use profile with the item mapped to the process steps of the use profile or some other means to help the team understand the environmental requirements imposed on the item. If the item assigned is a component, they should be supplied a copy of the requirements defined by an end item zoning analysis.
5. A list of allocated specialty engineering disciplines.
6. A list of required customer applicable documents that must be complied with.

During the item requirements analysis process, someone must coordinate and integrate the results. This work could be performed by the team leader or by a system engineer selected by the team leader depending on the magnitude of the process. The integration work should entail:

1. Identification of conflicts in specialty requirements like high reliability and maintainability figures and snow load on a hot tin roof.
2. The item requirements must be checked for traceability to parent requirements.
3. Identification of a verification approach for each requirement and coordination of these approaches with analysis and test planning activity.

4. Review for unnecessary requirements and exclude the same.
5. Review for completeness and add other needed requirements.

While the team leader or appointed system engineer integrates the requirements within the team (intra-team cross-functional co-product integration), the PIT must monitor the evolving requirements set for the item to assure they are compliant with system requirements. This includes integrating the item verification needs with an integrated system test plan and other system-level activities. The PIT reliability person should evaluate the item reliability requirement for compliance with the allocated failure rate, the PIT mass properties engineer checks the item mass properties requirement for compliance with allocated weight, and the PIT life cycle cost analyst checks item compliance with the cost allocation. These are all cases of inter-team cross-product and cofunctional integration.

23.2.3   Specialty engineering requirements
identification responsibility aid

We must have a foolproof way to communicate to the several specialty engineering analysts that they have a specific requirements analysis task to perform. We must have a way to place a clear demand upon them to provide a specific service. One method for doing this is to use a design constraints scoping matrix (DCSM) encouraged in AFSCM 375 series. The matrix, an example of which is illustrated in Figure 23.5a, renamed a specialty engineering scoping matrix, correlates the different engineering

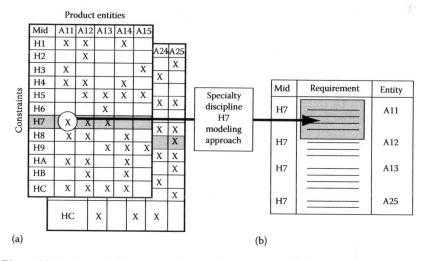

(a)   (b)

Figure 23.5 (a) Specialty engineering scoping matrix and (b) requirements analysis sheet (RAS).

specialties on the left margin (using the author's modeling ID or MID designation approach) with the product entities across the top of the matrix. An appropriate symbol in a matrix intersection means that the responsible engineering specialty discipline (left margin) must identify one or more constraints for that discipline against that product entity (top margin).

The example identifies components using product ID codes along the top to designate system entities, but WBS could be used instead of, or in addition to, these codes. Obviously, one matrix for a complete system may become completely unwieldy because of the number of elements in the system. This can easily be handled through multiple pages each with the same vertical axis defining the constraint categories. This also allows us to use this approach in a progressive way as the product entity diagram expands through the application of structured decomposition.

We place a symbol in each intersection to denote whether or not the constraint category applies to that indicated product entity. We can use a simple "X" to indicate yes and a blank to indicate no. In this case, the specialty engineer responsible for a given specialty constraint category simply responds by writing one or more specialty engineering requirements against each item marked "X" in his or her row.

But, we may also use the matrix to capture a more complex reporting mechanism in addition to tasking the specialty engineering analysts. The system engineer on the PIT could first mark up the matrix with "Xs" as indicated above. The matrix (some subset of the total matrix) is then passed around to the specialty engineers on an IPPT for review. Each specialty group annotates the matrix to indicate the anticipated impact or difficulty using one of the codes listed below and returns the marked-up matrix to the system engineer for integration and database update (this action can be performed more efficiently in a fully automated, online fashion, of course).

X The constraint applies to this element. No difficulty is foreseen in implementing the constraint.

Y The constraint applies to this element. Additional study or analysis is required to determine the impact of implementation.

Z The constraint applies to this element. A serious impact is foreseen in implementing the constraint in terms of cost, schedule, and/or performance risk.

— The constraint has been reviewed against this element and found not to apply.

A blank (no entry) indicates that the analysis related to this intersection has not been completed.

23.2.4   Requirements capture

The engineering specialty analyst responds to the specialty engineering scoping matrix by applying his or her specialized tools and procedures to identify design constraints for which they are responsible, often by reference to applicable documents appropriate to the contract. These documents must have been screened earlier (preferably in the proposal phase) for any needed tailoring and the tailoring agreed upon with the customer (see Chapter nine). Alternatively, the specialty analyst may appeal to a system mathematical model to allocate a requirement value from parent to child through each level and branch of the product entity structure. This response could occur within the environment of any of the three basic requirements analysis strategies.

The freestyle approach: In the freestyle strategy, the analyst responsible for a specialty discipline simply ensures that he or she writes one or more constraints against each product entity and provides them to the appropriate principal engineers. This could be in the form of a tabular printout of product entities versus requirement values, an applicable document referencing requirement statement that applies to all product entities, or a series of statements with blanks that are filled from a tabular listing as a function of product entity structure.

The cloning approach: In the cloning strategy, the specialty analyst might provide the person responsible for preparing boilerplates with the generic information that applies across all of the elements that might use a particular boilerplate and a tabular list that contains specific values for each product entity. Each principal engineer using the boilerplate would simply fill in the value from the table for that item into the blank provided in the boilerplate text.

The flowdown strategy: The flowdown approach is especially appropriate to many specialty engineering disciplines such as reliability and maintainability. It is accomplished by partitioning the requirement value for a parent item into component values for each child item. No matter what strategy is used, flowdown should be used as the detailed mechanism for numerically valued specialty requirements.

The structured strategy: In the structured approach, each specialty engineering analyst would be required to define one or more specific design constraints for each intersection marked with a letter character on the specialty engineering scoping matrix, as in the other cases. The difference is that they would be required to respond in a definite time frame for given product entities on a special form provided for that purpose. Figure 23.6 illustrates a design constraints identification form (DCIF) appropriate for an analysis process implemented with paper years ago as described in AFSCM 375. This form could be used directly in a manual

Analyst	Date		LSA reference		Action	
DCC code	DCC name				Rev	Date
Functional requirements information				Allocation information		
Line item	Design constraint statement	Source reference	CRIT	Arch ID		STAT
	Note: This form is only of historical interest, it has been replaced by a RAS implemented in a computer database system					

Figure 23.6 DCIF.

implementation of design constraints identification or as a computer data entry form. Within this structured approach, many specialty disciplines would apply the flowdown strategy as their means of determining lower-tier requirements. Ideally, the organization would employ a computer requirements application into which the requirements could be written directly by the analysts. Figure 23.5b shows a view of a requirements analysis sheet (RAS) reflecting the content of a requirements database into which this data would be entered. The specialty engineering MID in Figure 23.5a coordinates the specialty engineering requirements with the models from which they were derived just as performance requirements are mapped to the functions (such as F4732) from which they are derived, interface requirements are mapped to the interfaces illustrated on n-square intersections or lines on schematic block diagrams (such as interface I132), and an environmental requirement is mapped to a hostile (such as QH1), natural, noncooperative, or self-induced environmental requirement using a standard, three-dimensional service use profile, or zoning model from which it is derived. For more on this traceability approach refer to *System Requirements Analysis* (Grady, 2005).

23.2.5   Requirements integration

The product entity principal engineer or IPPT should be held accountable for integrating the effects of the specialty constraints. The principal engineer will normally be a system engineer at the highest system levels, possibly an IPPT leader at intermediate system levels between segment and subsystem or a subsystem or component design engineer at lower system levels. All of these principal engineers will have different degrees of skill and experience with specialty engineering integration and some may need support from system engineers.

There are two alternative applications of a system engineering function, or PIT, in the integration process. First, they could be used to audit the performance of the principal engineers in this process and provide feedback to principal engineers about their conclusions. Second, system engineers could be assigned to teams to do the integration work for the principal engineers. The particular program team must evaluate the skills of its personnel and select an appropriate approach for their situation.

In addition, the system engineering community should be tasked with coordinating the specialty engineering requirements analysis activity to include (1) ensure all required disciplines have access to budget to perform their task, (2) monitor that each required discipline has qualified personnel assigned to specialty requirements tasks, (3) verify that each discipline is responding in a timely way to the specialty engineering scoping matrix content, and (4) ensure that the specialty requirements integration process is working. This requires energetic pursuit on the part of the specialty engineers and eager interaction on the part of the design engineers. All too often, both parties tend to withdraw from this interface. What is required is that both plunge across the abyss and engage in real teamwork.

Particular interest must be focused on identifying and resolving issues corresponding to constraints that are identified on the DSCM by a letter "Z" where the specialty engineering scoping matrix is used.

The DCSM and DCIF are relics of a past century, of course, when main frame computers were used to capture requirements and generate specification. In the 1980s, this application was pushed aside by the emerging engineering workstation, and PC industry matched up with servers and networks. Today, each of the specialty engineers would likely enter their requirements directly into the common requirements database. But the DCSM, renamed a specialty engineering scoping matrix, remains a useful artifact for system engineers to put order and discipline into the process of identifying specialty engineering requirements work responsibilities.

23.2.6 Specialty constraints communication

Many specialty engineering requirements are contained in often extensive applicable documents. These are imported into the program through a specialty engineering requirements statement such as "Pressure vessel design shall conform to MIL-STD-1522A." This simple statement places a constraint on the responsible designer to comply with the many requirements in Sections 4, 5, and 6 of that document to the extent that they apply to pressure vessels if they are not tailored out.

One of the most difficult tasks in system engineering is the efficient communication of the meaning of the contents of the pile of applicable documents referenced in specialty engineering requirements statements to the designers. Designers who have a great deal of experience will have acquired a good understanding of many of these requirements, but not all. Recent college graduates will have little experience with the content of any of these documents or the concept of importing requirements through the referencing mechanism.

Each engineering specialty discipline must work assertively and interactively with the design community to ensure that the designers understand the specialty requirements and their consequences. This interaction could take the form of any combination of three principal initiatives on the part of each specialty discipline: (1) specialty checklist, (2) person-to-person discussion, and (3) organized interaction meetings. This interaction could also occur through participation in trade studies and engineering review boards. This very difficult process may eventually be replaced by integrating specialty requirements into computer-aided design packages.

23.2.6.1   Checklist approach

In the checklist approach, each specialty discipline that invokes an applicable document containing a voluminous list of design constraints must translate that listing into a checklist focused on the specific product line and element, if possible. The checklist must be formatted such that the designer can clearly and rapidly understand the important attributes that must be satisfied. This checklist must be made available to each design principal engineer affected by the requirements. The checklists could be communicated with printed material or via a networked computer system.

Each engineering specialty creates a specialty checklist that is responsive to the applicable document(s) called out for that discipline. The checklist simply lists each specialty requirement contained in an applicable document and provides a space for each item to be checked off. A specialty checklist may have several columns for different kinds of elements (AGE, flight vehicle, etc.). The checklists should be created during the early period of the program when the top-level system elements are being synthesized.

Specialty requirements often apply to all system elements across several layers of the product entity structure but can best be dealt with in association with assembly or component-level design work. The checklists must be available by the time it is necessary to identify requirements for component-level elements that will yield to detailed design.

If the checklist approach is applied to all specialty disciplines, there are many disciplines on a program, there are many requirements per discipline, and all of the specialty disciplines use the checklist approach effectively, the design engineers can easily become completely overloaded.

It is, however, important to make every reasonable effort to help the design engineers understand the complete set of requirements that must be complied with. One element of that assistance could be a joint checklist peer review. This is an organized review of all of the specialty checklists by the specialty disciplines. A representative from each specialty discipline reads all of the checklists of the other disciplines with an eye for spotting any conflicts with their own checklist.

One way to implement this review is to meet for 1 h per day for "n" days (where n equals the number of specialty disciplines on the program).

Each day, a different one of the specialty disciplines should act as the host, providing copies of his or her checklist, explaining the content, and reacting to critical comment. A PIT system engineer should set up these meetings and oversee them as necessary to ensure issues are clearly resolved. The resultant joint checklist can then be briefed to some or all of the designers with follow-up interaction on a personal basis by the specialty engineers.

Another checklist simplification approach is for each specialty to partition their aggregate checklist into subsets that provide all of the constraints related to particular kinds of elements common to the company product line or particular program elements. This may eliminate many checklist items from the list that a designer interested only, for example, in valves must use. This becomes very important when you consider that a designer may have to deal with checklists from 10 or more specialties.

By now, you should completely understand how the system development process became so difficult through progressive specialization without concurrent improvements in integration of the specialty engineering views.

23.2.6.2   Individual person-to-person

In the person-to-person approach, each specialty discipline interacts on a one-on-one basis with each designer, with or without benefit of a checklist, to help them understand their requirements. If checklists are used, these conversations can focus on the checklist items and whether or not the designer understands the listed items. Where checklists are not used, the specialty engineer must develop a feeling for whether the designer understands the raw content of the applicable document or allocated specialty requirement.

In these discussions, the specialty engineers may be able to offer advice about which potential design solutions under consideration would have the better result in satisfying their requirements. The specialty engineer may also give the designer some examples from similar programs about how these requirements were satisfied.

23.2.6.3   Organized interaction meetings

The designers and specialty engineering representatives could be brought together periodically for informal discussions about current problems and shared insights. This may be most useful when a development team must respond to a specialty engineering discipline very uncommon to their experience. For example, if a private aircraft company won a U.S. Army contract to provide an air vehicle that could survive in a nuclear battlefield situation, they would have to master some very uncommon design solutions. A periodic meeting may be very effective in this situation until the designers have come to understand appropriate design approaches. These meetings may be more a matter of instruction at first eventually turning into effective technical exchanges.

23.3 Specialty engineering in design

23.3.1 Concurrent preliminary design development

Given that the team has developed a set of requirements either developed by the parent team as a condition of forming the team or the new team has done the requirements work themselves, the new team must accomplish a concept development activity to identify a concept with an acceptable risk profile that they can gain parent team approval for in concert with the related requirements.

The IPPT is now ready to begin preliminary design leading to sketches, layout drawings, validated requirements, and analysis reports for the product. At the same time, the product design engineers are leading the product design effort, a design is concurrently being crafted for each of the other development channels: manufacturing process and facilities, tooling, procurement and material, quality, test, operations, and logistics. The design agents in each of these cases must actively interact to evolve mutually compatible concepts that evolve from and survive frequent concurrent team meetings where each party must expose the features of his or her concept and answer questions from the other design agents and specialty engineers.

Figure 23.7 illustrates the preliminary concurrent design process. It is composed of three major steps: a concurrent concept development, the selection of a preferred design, and the documentation of that design. The team members join to synthesize the requirements they developed in the previous requirements analysis work. Alternative product, operations, logistics, and manufacturing design concepts are developed creatively by team members and mutually evaluated through a trade study process and the optimum concept set selected. Throughout the preliminary design phase, the teams may find it useful to apply the interim common database (ICDB) information concept to capture development data for definition of the current baseline and communication across the team members and teams. Teams review ICDB information in meeting rooms employing the real-time projection review and update technique and program information directly from the DIG for internal reviews. The whole team is physically collocated and they are able to carry on easy cross-functional conversations within this space. Their work space includes at least one large wall upon which they may place materials they find useful to stimulate a synthesis of their individual team members' ideas. This preliminary design is documented to the degree necessary and subjected to a review by management at a PDR.

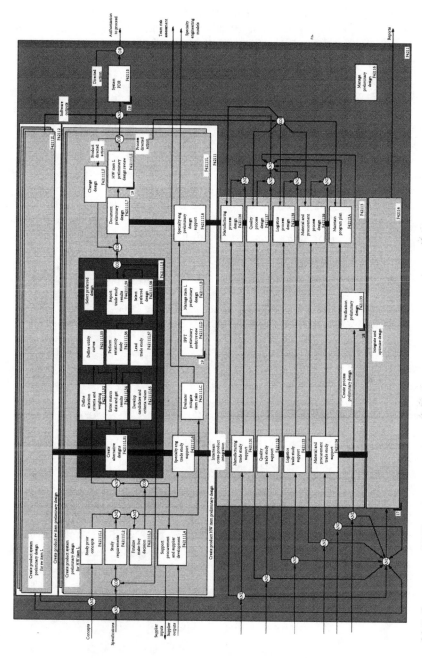

Figure 23.7 Concurrent IPPT preliminary design process.

23.3.2   Concurrent detailed design development

In the detailed design phase, the IPPT develops the selected design concepts concurrently into a complete set of product engineering drawings, detailed manufacturing and quality inspection planning, and test procedures. This phase is characterized by the most intense product integration activity in the complete development process. Figure 23.8 offers a diagrammatic view of the process.

There will be several of these activities going on simultaneously, of course, recognizing that the development process proceeds from the top down. If the system is very complex, this process may be in work by five top-level IPPT immediately subordinate to the PIT. Each of these teams may have two or more subordinate teams at work on lower-tier entities and some of those may have sub-teams assigned.

Note the continued dependence on the concurrent development bond running through the work that results in the design. This includes the specialty engineering work focused on helping the design members of the

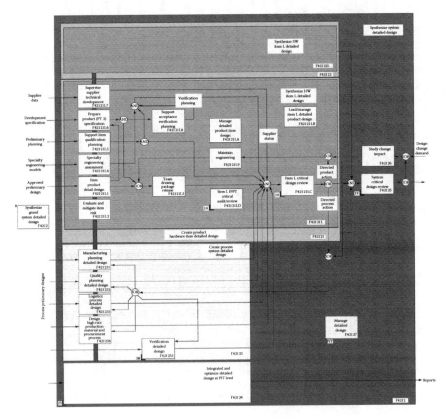

Figure 23.8 Concurrent IPPT detailed design process.

team to understand the requirements they included in the item specification. This will be a period of potential stress between the specialty engineers and the design engineers. The team will be committed to the design and to the extent that specialty engineers discover the design in jeopardy of failing to satisfy their requirements, tough choices may have to be made that will be argued powerfully from two or more perspectives. System engineers will have many opportunities to apply their skills in integration and optimization.

23.3.3   Decision support

Throughout the preliminary and detailed design activity, it will be necessary for team leaders and program managers to make difficult design decisions. Specialty engineers will be called on to interact with the design development process to participate in trade studies, design reviews, and engineering review boards. In these forums, the specialty engineer is responsible to ensure that the preferred solution offered or the decision arrived at in the selection among alternatives has properly taken into consideration the relative valuation offered by the specialty engineer and any specialty concerns offered by them.

23.4   Specialty design assessment

From the 1950s through the 1980s, many companies allowed growing specialized engineering design, analysis, and specialty disciplines to form departments and build walls such that development work progressed in a serial fashion. Specifically, specialty groups were allowed to review finished drawings without concurrent interaction during the creative concept and design development process. This phenomena is called stovepipe engineering by some people.

As discussed earlier, the specialty and analysis groups must be employed in a concurrent way in order to create around the creative design engineer the effect of a single, complete all-seeing engineer effective in initially designing-in all product specialty engineering requirements simultaneous with the design of the product manufacturing and employment process. This requires the formation of effective, physically collocated teams of design, specialty, and analysis engineers and their manufacturing, logistics, and operational specialists.

Even in an effective concurrent design environment, however, we should not fail to independently and formally assess the design for requirements compliance. If a project organizes by IPPTs, this can be done through cross-IPPT assessment of each other's designs to preserve a degree of independence or by a system-level specialty engineering function within the PIT. The results from this assessment should be captured in some enduring way so that it may be used as specialty requirements verification at a customer-mandated functional configuration audit (FCA). If checklists are used, a set of signed-off checklists may be adequate for this purpose. This checklist could have one column for pre-design interaction and post-design assessment.

23.4.1   Noncompliance identification

During the design process, the engineering specialists must study the preliminary sketches and detailed engineering drawings offered for check and release to assess the degree of compliance with their requirements. If a checklist was used as a means to inform the design engineer, the specialty engineer should use the same checklist as a systematic aid in the assessment.

The specialty engineer must have some understandable rationale for a conclusion of noncompliance. It is not enough for the specialty engineer to simply conclude that a design fails to meet the requirements. The designer must be offered an understandable statement of the reason in terms of the requirement. A checklist will help in this regard where the specialty engineer may refer to a specific checklist item and describe the design characteristic that is at fault.

It is less costly for the specialty engineers to identify noncompliance issues before design release than after, and the earlier the better. But, situations do arise where it is necessary to view several drawings and associated analyses before the specialty engineer can form an educated opinion about compliance. In these cases, some drawings may have been released in the process of waiting for the others needed for the analysis. This is an unavoidable problem the severity of which can be reduced if the specialty engineer gives these kinds of compliance assessments top priority when the complete set of data does become available.

23.4.2   Noncompliance correction

When the specialty engineer finds a design that fails to comply with a valid specialty requirement, that engineer must first try to resolve the issue with the designer. It will be helpful to refer to the specific checklist item or documented requirement and the specific way in which the design fails to satisfy the requirement in this conversation. The experienced specialty engineer can also offer fairly specific ways the problem could be overcome in the design. The specialty engineer should recognize that the designer may very well feel protective and defensive about the results of his or her creativity. A little tact mixed with assertiveness will generally be repaid in this conversation.

If the direct approach to the designer fails to achieve a satisfactory solution, the specialty engineer must bring the noncompliance issue to the attention of the designer's supervisor, manager, or the chief engineer. Where the specialty discipline is on the drawing signoff list, recognition of a problem is guaranteed, but correction is not. Where the engineer is not on the drawing signoff list, he or she must ensure that the noncompliant issue does not pass unnoticed. A program issue system can be useful in such cases. A program issue is a special kind of action item that must be resolved by the designer of the element against which the issue is written. Like all engineering problems, the engineering review board is the court of last resort for these issues.

23.5   *Engineering specialty activities*

Refer to the subsequent chapters of this part for coverage of the work efforts of the individual specialty disciplines. Not every one of the disciplines covered here will be required on any one program. During the program planning process, we must understand which specialty disciplines will be required as a function of the nature of the system, the maturity of the related technology, and the degree of correlation with past solutions to this or related problems.

The SOW or IMP for a given program should include an identification of the specialty disciplines required. It is possible that a program may conclude that one or more specialty engineering disciplines must be applied even though a customer has not specifically called for it in order to ensure the design is adequate for the requirements defined in the customer's system specification. The specialty engineering scoping matrix is a good place to start for the system engineer during a proposal effort to identify the specialty disciplines needed for each entity identified in the evolving product entity structure.

chapter twenty-four

Reliability

24.1 Reliability overview

Reliability is a measure of the probability that an item or system will continue to function for a specific duration and under prescribed conditions. It is measured in terms of the probability of success (1 minus the probability of failure), the mean time between failures in hours, or the failure rate in failures per unit of time (units, thousands, or millions of hours, for example). The reliability engineer allocates the overall system reliability figure to lower-tier items in the product entity structure forming a reliability math model. The design team fashions a design that satisfies this allocated figure that is verified by assessing the reliability of the parts and computing the resultant predicted reliability figure for the item as a function of the way the parts are connected and used. Part and component reliability figures are commonly extracted from reference documents listing proven reliability figures for specific kinds of components.

Reliability engineers commonly also perform a failure modes effects and criticality analysis (FMECA), which identifies every way that every part can fail and the consequences of such failures. This work can be begun at the system level and expanded as far down as the piece part level if time and money are available and the risks from failure while in operation are sufficiently dire. This data should be the basis for many other specialty engineering activities since it defines every failure that can occur. Therefore, it should be the basis for maintenance responses needed, built-in test design, spares determination, as well as be part of the safety hazard analysis stimulus.

FMECA can be applied to processes accomplished by personnel as well as product entities. We often apply a reliability of 1 to the two most fault-prone elements of systems—people and software—because it is so hard to do otherwise. In practice, it is very common for people to be at the root of failures. Failure modes applied to processes can be a very productive way to ensure good system performance for the customer and this is often the task of human factors and system safety engineers.

24.2 Reliability modeling and allocation

During the requirements definition period on a program, the principal function of the reliability engineers is to define reliability requirements.

This process starts at the top with a system availability figure of merit that must be partitioned into a reliability and maintainability components each of which can then be allocated down through the product entity structure. In order to do this work efficiently, the reliability engineer must construct a reliability math model of the system product entity structure that recognizes how the system will be employed and to what degree redundancy is introduced into the design.

The system reliability figure must be decomposed into the end item reliability figures as a function of the mission intended defining the principal operational end item reliability figure and reliability figures for other end items in the system. Figure 24.1 illustrates a simple end item model where all of the parts of the end item are characterized by failure of the end item through any single part failure. Granted, some failures may be more severe than others and we take that up under FMECA shortly, but any failure in this case will cause end item failure.

If the end item reliability figure were $R = 0.98$, we could first convert that into a failure rate or mean time between failure (MTBF) figure. Failure rates are especially easy to work with in the model because we can use simple addition rather than more complex probabilistic relationships necessary in working with reliability or MTBF numbers. To make the conversion, we need to know the amount of time the end item will be exposed to failure potential, that is, the length of a typical mission. Let us say that user information has defined the mission as being 3 h long. Then the relationship $R = e^{-\lambda t}$ can be solved for lambda (λ) which is the end item failure rate. If one were interested in MTBF, then it is simply the inverse of λ, but failure rate is so much easier to work with. In this case, the failure rate is $\lambda = -\ln R / t = 0.0067$, where $\ln R$ is the natural log of R.

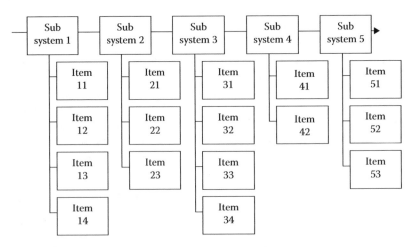

Figure 24.1 Typical end item reliability math model.

The reliability engineer must now partition this figure into allocations for each of the five subsystems where the sum of these allocations equals 0.0067 unless one was applying a margin account that acts as an imaginary sixth architecture item. In this example, we may allocate the end item failure rate as follows based on an understanding of the complexity of and technology maturity being implemented in the subsystems:

Subsystem	Failure rate
1	0.0017
2	0.0006
3	0.0018
4	0.0004
5	0.0022
Total	0.0067

This process can be continued down to the items in the subsystems and item circuit boards and parts on those circuit boards to produce a complete system reliability math model. At each system level of indenture, the figures in the math model should pass into the item specifications as the reliability requirement. The reliability engineers should then be working with the design engineers on the teams to help them understand what that reliability number means and how they might have a chance of satisfying it in their design work.

Integration work is appropriate across the reliability and maintainability math models because the specialty engineers doing this work may not coordinate their work satisfactorily. If a reliability engineer allocates a relatively high reliability figure to item 21, for example, and the maintainability engineer allocates a low mean time to repair figure (often simply a remove and replace time) to that same item, the pair of engineers have failed to select an optimum pair of figures for the item. Since the item has a high reliability and will not fail very often, there is no need to provide for a rapid repair time. To optimize the reliability and maintainability figures toward satisfying the availability figures, one should have an opposite relationship between reliability and maintainability numbers for any one item. That is, low remove and replace times should be associated with high failure rate items. The maintainability engineer could be assigned the responsibility to monitor these figures for imbalance because the maintainability formula includes failure rate in it whereas the reliability formula does not include the maintainability figure.

24.3   Failure modes effects and criticality analysis

If failure of the system being developed will result in death or serious injury of friendly personnel (weapons systems are supposed to kill the enemy but not those who operate them) or a significant economic loss, the customer will be vitally interested in preventing those failures or controlling them so as to minimize adverse effects. A model to determine, in a very exhaustive fashion, how systems can fail has been developed, called failure modes effects analysis (FMEA). The FMEA can be accomplished at the system level, the unit level, or the part level. Therefore, one can perform the FMEA from the top down as the product entity structure is determined on an unprecedented development in accordance with the techniques covered in *System Requirements Analysis* (Grady, 1993, 2005).

The FMEA is accomplished by considering each element of the system in an organized fashion, determining how it can fail (its failure modes), and the consequences of those failures. Where the effects are very serious, remedies are sought in the design to make the product more reliable. Some techniques that are commonly considered are (1) replace the part type with another that does not have that particular failure mode or with one that will fail with a lower frequency, (2) provide for operator announcement that the part is at risk of failing, (3) replace the part as a time change preventive maintenance step before it is in danger of failing, (4) apply redundancy so the design is failure tolerant, or (5) derate the part such that it is exposed to less severe stresses than it is designed to handle.

In this analysis, we are interested in effects of failure of each entity evaluated since some failures are more serious than others. Some failures may cause the system to be destroyed (as in an aircraft crash) whereas others may degrade the mission capability of the system or create only a nuisance. Thus, we consider each failure from an effects perspective and judge how critical it is that the failure be avoided. The FMECA process is applied during the design process, and where the reliability engineer discovers a failure that causes serious adverse consequences of failure, he or she should encourage the design engineers to see an alternative solution or consider other alternatives to control these adverse consequences.

FMECA may be accomplished as a team effort by reliability and safety because the results feed both analyses. Where the consequences of a failure expose a system safety hazard, the safety engineers should participate in the resolution to ensure the hazard is actually mitigated adequately. One could make an argument that a maintainability person should be involved as well since they have a role to play in responding to failures in terms of support equipment needs. Depending on how one is organized to accomplish the complete sustainability analysis, this latter role could be a logistics engineer. The point is that there are several ways to cut up the budget for the FMECA task.

24.4 Reliability analysis and prediction

Throughout the design process, the reliability engineer should have been supporting the design engineers by assessing alternative solutions from a reliability perspective. This work might include participation on trade study teams. As the design solutions begin to become available, the reliability engineer must begin the formal process of assessing the design for compliance with the requirements previously inserted into the item specifications from the reliability math model. This work is accomplished through analysis of the design solution by the reliability engineer for failure potential and inserting the predicted reliability figures into formulas to determine if the predicted values are at least as good as the allocated numbers for the same items. This work may start at the circuit board level for electronic equipment where the reliability engineer counts the number of each different kind of component and applies a nominal failure rate figure from a reference source. If derating has been used in the design of an electronic assembly by operating the device less aggressively or providing suitable heat transfer features in the design to channel heat away from the device, for example, the reliability engineer would claim this derating in the predicted figures. It may happen that the designer has effectively derated 15 out of 28 cases of the use of a particular kind of electronic part so the reliability engineer may not be able to use the same reliability prediction for every application of that part type. In this case, 15 parts would have a failure rate of perhaps 0.000003 and 13 identical parts might have to be assigned a predicted failure rate of 0.000005.

The predicted failure rates based on the actual design solution are assembled up through the reliability math model compared at each item level with the requirements value included through allocation in the specifications. Some item designs will be found to be well within their allocated values and some may fail to satisfy them. There are several alternatives available to the reliability engineer and the program where failure to satisfy the reliability allocation is discovered. The reader will recall that the reliability numbers were assigned to the items based on a best guess and it is possible that the actual relative degree of difficulty of satisfying the reliability figures at one level of the architecture may have been very different from what had been assumed by the reliability engineer during allocation. There is nothing evil in reallocating the failure rates such that the current design actually satisfies those different allocations so long as the higher-tier reliability figures are still being satisfied. Where reallocation is applied, the math model and the specification figures should both be updated so that the models reflect the program realities.

As an alternative to what appears to be a very difficult or impossible reallocation approach, the reliability engineer could appeal to the team or program manager for any available reliability margin or for an outright change in the required reliability that could go all the way up to a customer decision. These requests should be avoided if possible but the reliability reduction might be acceptable if it was accompanied with a significant reduction in cost. In the margin consumption case, if there is available reliability margin and there are no other design areas in a similar risk condition, it may be possible to justify margin assignment. If there is no remaining reliability margin, all is not necessarily lost to the reliability engineer. If there is available mass margin, it might permit the design engineer to acquire it and apply it toward a larger heat sink that derates critical parts enough to satisfy the allocated reliability. Alternatively, if there is remaining design to cost margin it might be applied to support a better class of components that could also solve the reliability problem.

In some cases it may be impossible to solve such problems through reallocation or consumption of available margin and it may be necessary to change the design to improve item reliability. The reliability engineer can help the design engineer in this process by calculating alternative reliability figures for candidate solutions or identify which candidate has the best chance of satisfying required reliability. Often in these situations, the problem is not as simple as the isolated case of failure to satisfy reliability. Rather, the case is often that the design has multiple problems and the team is trying to hit the most promising case in the aggregate.

24.5   Other design support functions

Where a significant portion of the product design is procured from suppliers in accordance with procurement specifications, those procurement specifications must include a reliability requirement coordinated with the system reliability math model and the supplier design managed to achieve that figure. This may require the participation of a reliability engineer in the supplier design reviews to assess the degree of confidence that the program should have about the supplier's chance of success.

Reliability people often participate in and even manage the failure reporting and corrective action system (FRACAS) on a program. This system is primarily intended to support the manufacturing process to evaluate hardware entities reported as damaged in production or failing to satisfy acceptance criteria. However, on a large program employing a full rate production phase preceded by engineering shop or low rate production fabrication of dedicated qualification items, FRACAS can be very effectively applied in the qualification process to assess any reported failure to satisfy qualification criteria. A unit may fail to satisfy a particular qualification requirement either because it failed due to an inadequate design or because a random failure has occurred. It is important to determine which is the case so that unnecessary re-design is not entered into with its attendance delays and cost. The early implementation of FRACAS also has the added benefit of wringing out any problems before the program enters full rate production.

It is important that the program develop a design reference mission as early in the program as possible. This mission defines a list of environmental parameters and their corresponding values over the period of one mission. The problem is that environmental requirements are analog quantities and that there are an infinite number of values to each one of them as well as an infinite number of possible combinations of traces of them over a mission. We would like to control the cost of the qualification process by defining a single or a few possible combinations of value traces and will probably need an agreement with the customer on what those are. Once this is agreed upon, advance planning can begin on the environmental chambers that will be required for unit qualification. The reliability engineer should check the environmental parameter values, ranges, and rates of change to ensure that they are consistent with the conditions called out for the reliability numbers. Otherwise, failures may occur in qualification testing that suggest a lower reliability than actually achieved.

24.6   Reliability verification

The system engineering approach to system development calls for clear definition of requirements followed by an attempt to synthesize those requirements into a design solution, and followed by a proof process to determine whether or not those requirements were satisfied by the design. Reliability must participate in the third step of this process as well as the first tow as described above. Four methods of verification are generally identified: test, analysis, demonstration, and examination. Of these, testing and analysis are common reliability verification methods. The analysis method is covered in the section on reliability prediction. The reliability engineer assesses the actual design solution for the reliability that it should have based on its design features and the reliability of the parts used. The prediction data is summarized in tables and reported upon along with related text telling whether or not the allocated reliability figures were achieved supported by the detailed analytical data.

Reliability testing is a much more complex undertaking. In order to follow this pathway, there must be several dedicated units reserved just for reliability testing and this cost must have been foreseen during the original cost estimating for the program. One should not plan on using any qualification item as a deliverable product because in the process of qualification, commonly, some life will have been extracted from the item, so these reliability test units must be dedicated units above and beyond the units needed for delivery. It is also not reasonable to assume that you can use the reliability test units for other qualification purposes or vice versa. It would generally be advantageous to first prove the design adequate in environmental and operational qualification testing before entering reliability testing because you would then be able to assume that any failures that occur in reliability testing are based on a random failure rather than a design flaw.

Reliability verification testing, to be successful from the developer's perspective, must prove that the design satisfies the reliability requirement. In that reliability is a probabilistically defined parameter, it stands to reason that it will have to be proven through a statistically valid process. If the item under tests has a reliability requirement expressed as an MTBF of 200 h, it is obvious that testing will have to take place for at least 200 h. There will be cases where the item fails after 300 h and others where it fails after 82 h. We have to test the unit or units long enough to derive a sufficient number of failure data points to qualify statistically. Several factors are involved in this number and a reliability engineer can calculate the number of failures and hours necessary.

24.7   Field reliability data

Field reliability data collection can be collected during system test and evaluation. Subsequent to deployment of the system operationally, the user may have their own failure reporting system operated by user maintenance personnel. When a failure occurs in such a system, the maintenance people have to complete paper of computer forms and the data is eventually entered into a database used to track the consumption of resources and identify failure trends that require more detailed attention. It is very difficult to collect accurate failure data in the field. Those who have the responsibility do not connect the utility of the mass of data collected from each maintenance event with the need for accuracy in the entries they make into the same system when they have to report a failure.

Where the system is operated by the producing enterprise, the enterprise may have to operate the field failure reporting system using its own forms or reporting media either because the ultimate customer may require it in the contract or because the enterprise understands the utility of the data. The author, while a field engineer for Teledyne Ryan Aeronautical, led a small team of engineers for over a year and a half operating and maintaining a loran-guided unmanned reconnaissance aircraft team out of a base in Thailand running flights into North Vietnam. He had been a field engineer for 6 or 7 years at the time and had never relished the expectation that he was to fill out a Ryan trouble report on each failure at his location and send it to the company in the weekly report. However, he realized that he had an unusual opportunity to create a believable record of failures that could be used as a basis for recommendations for changes in the maintenance practices then in effect. At the time, every vehicle that flew a mission with a flight time of about an hour and a half would have to endure several hours of checkout to ensure it was in a mission-ready condition for its next mission. The vehicles all had a great deal more maintenance time on them than flight time.

Over a one and a half year period, the author reported each failure and the probable cause. He also tried to determine whether or not the failure would have caused a mission failure or degradation and whether or not the fault would have been detected prior to launch if the vehicle had not been subjected to system tests using a console in the hangar. A few years later, a Tactical Air Command drone unit was looking for ways to deploy from Davis Mounthan AFB in Tucson, Arizona, to Europe without placing a burden on over-taxed air transport in the event of a Russian invasion of Western Europe. The author produced the data previously collected that revealed that only one aircraft over a one and a half year period might have failed to return in the event system test was limited to DC-130 ground preflight and captive carry checkout. The unit tested the concept on an operation out of Hurlburt Field, Florida, and concluded that they could fly themselves into the war with the initial launch on the way in and subsequent launches based on between-flights checkout on the DC-130 until transport was available to support the unit. Unfortunately, TAC was burdened by the development cost of new manned aircraft, there was no active hostility under way that might result in a desire to minimize pilot loss, and the unmanned aircraft of the time required a substantial maintenance response. The unmanned group was disbanded. With the resurrection of interest in the employment of uninhabited combat air vehicles, these same issues may return to the fore.

chapter twenty-five

Maintainability

25.1 Maintainability overview

Maintainability is a probabilistic statement of the time it will require to repair a failure. This can be stated in terms of the remove and replace time, total repair time, or other parameters. The maintainability engineer allocates system-level repair time to items in the system and tracks design team performance in responding to these allocations. As design alternatives are evaluated, the maintainability engineer looks for features that will encourage or deter maintenance actions. When the design is firm, the maintainability engineer applies a process model to determine resources needed to support all maintenance actions and the time required to perform them. These time values are plugged into a maintainability prediction formula and the results compared with required or allocated values in the specification.

25.2 Maintainability modeling and allocation

Time is the metric of importance in maintenance. How long does it take to return a failed item to a fully performing state? There are many components of time in the logistics support process contributing to this repair time. A user can tolerate some cases where it takes a considerable time to execute repairs so long as the average or mean time to repair (MTTR) is reasonably low on a time scale of importance to the user. MIL-STD-721 offered an excellent breakdown of time for use in this situation. Figure 25.1 illustrates these times defined in the standard.

The time often used for contractually controlling the product design development process is mean corrective maintenance time (M_{ct}) and then often only the mean time to remove and replace (remove and replace with like item in Figure 25.1). The rationale for this is that these are figures for which a development contractor can easily and properly be held accountable. The contractor most often cannot control the logistics delay times, for example. If the contractor fails to adequately spare the program despite customer willingness to fund adequate spares for the logistics situation they have described, then a customer could hold the contractor responsible for subsequent delay times, perhaps, if they found in practice that delays in maintenance were not related to any inefficiencies on the part of their personnel.

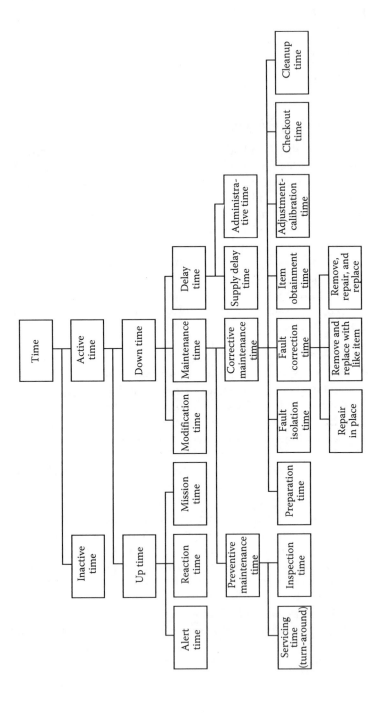

Figure 25.1 Logistics time relationships.

As in starting the reliability math model, one should convert the system availability figure into top-level reliability and maintainability figures of merit. The top-level maintainability figure might be, for example, an M_{ct} of 0.5 h. That is, on the average, it will require no more than 30 min to repair any failure. To ensure that the system that is developed actually has this attribute when it is delivered, a maintainability engineer will have to determine appropriate remove and replace times for every maintenance significant item. So, the first task is to identify every maintenance significant item, that is, every item that can be repaired consistent with a particular maintenance policy. Every item on this list will have to have a maintenance time assigned.

In determining what items should be on the list of maintenance significant items we should use some common sense as well as logic. There may be entities in the system that could conceivably be removed and replaced while in an operational condition, but the design that might be required to assure that the system post replacement was free of any faults would be very costly compared to removal of the end item to a maintenance environment. So, the things on this list may require a particular status of the end item in terms of its operations and maintenance cycle. For, example, if it were determined while a liquid-fueled space transport vehicle was on the pad undergoing the launch sequence when it was determined that there was a fault in one of the booster engines, should the system be configured such that the engine could be replaced on the pad or would it be acceptable to unstack the payload and other stages and remove the booster stage to a maintenance environment where the engine could be removed under better circumstances and all of the interfaces carefully verified during the installation of the new engine?

If the system being developed has no clear precedent, it should be developed from the top down applying some form of structured analysis followed by concept development work for the items in the system architecture responsive to the requirements defined for the item. The maintenance analysis process can be accomplished first based on these concepts and refined based on preliminary design work, if necessary. As each branch and layer of the architecture is exposed in the structured analysis, parent item M_{ct} can be flowed down to the child items on that branch and those figures plugged into a formula to confirm that the relationship is correct. This work is commonly accomplished using an equation applying a normal distribution that permits a fairly accurate assignment. An equation applying a log normal distribution is encouraged for the prediction process discussed next because the resultant maintenance times have been shown to be a better match with reality.

The formula for M_{ct} includes estimates of item maintenance times and failure rates. Thus, the maintainability engineer can be asked to act as an integration agent relative to these factors. If a reliability engineer happens to assign a high reliability figure to item x and the maintainability engineer independently assigns a low remove and replace time for the same item, then an opportunity has been lost. It serves no useful purpose to apply low maintenance times to items with high reliability. It is better to apply low remove and replace times to items with less reliability that will fail more frequently resulting in a higher availability.

25.3   *Maintainability analysis and prediction*

As the design concepts mature, maintainability engineers can start assessing that design to determine if it satisfies maintainability requirements previously allocated to the items in the system. This work must be done as early as possible during the design effort so that if problems are identified, there will be enough budget remaining to correct the problem. If done too early, on the other hand, subsequent design changes driven for reasons other than maintainability may require iteration of the maintainability analysis. Ideally, the design people on the item teams would be able to estimate the degree of stability of their designs encouraging the earliest implementation of an affordable specialty engineering assessment process.

The maintainability prediction process for an item can be accomplished by identifying every maintenance action that may be required, predicting how long it would take to accomplish each of these tasks, and plug these values into a formula to derive a mean. The corrective maintenance responses could be based on a failure modes effects and criticality analysis (FMECA) that are supposed to cover every way an item can fail. The corrective action should be based on removal and replacement of a logistics critical item (LCI) that is coordinated with the spares planned. This process can be applied to end items where the LCIs are components as well as at the shop level where the LCIs are modules in those components.

25.4   *Maintainability verification*

MTTR times are commonly verified by technicians and mechanics being asked to repair the product entities with no fore knowledge of the nature of the fault. This is a demonstration process where someone must time the performance of the repair task. To encourage credibility, each fault can be dealt with by different technicians and a mean time collected. Obviously, this method of verification can be quite costly and require a considerable amount of time. The predicted remove and replace times could alternatively be accepted as verification of design compliance.

chapter twenty-six

Availability and RAM integration

26.1   Availability overview

Availability is a measure of the probability that the system will be available for use at a point in time. It is measured in terms of one of several particular combinations of system reliability and maintainability figures. Availability is a very high-level system measure and does not have a great deal to offer as a figure of merit at the lower tiers of the system below the end item level. By observing life cycle cost (LCC), availability, and operational effectiveness figures of merit, system engineers and the program manager can gain an excellent assessment of overall product development success.

26.2   Availability measures

Several different availability measures have been identified and applied on programs. Inherent or achieved availability can be applied by the contractor where the contractor has no control over the operational environment. Operational availability is more appropriate for assessment of equipment in a realistic operational environment.

26.2.1   Inherent availability

Inherent availability is defined as the probability that a system or equipment, when used under stated conditions in an ideal support environment (i.e., readily available tools, spares, maintenance personnel, etc.), will operate satisfactorily at any point in time as required. It excludes preventive or scheduled maintenance actions, logistics delay time, and administrative delay time, and is expressed as

$$A_i = \frac{\text{MTBF}}{\text{MTBF} + M_{ct}}$$

where
   MTBF is mean time between failures
   M_{ct} is the mean time to repair or corrective maintenance time

This is essentially the ratio of up time to total time. Only corrective maintenance is considered in this form of availability.

26.2.2   *Achieved availability*

Achieved availability is defined as the probability that a system or equipment, when used under stated conditions in an ideal support environment, will operate satisfactorily at any point in time preventive maintenance is included. The equation for it is

$$A_a = \frac{\text{MTBM}}{\text{MTBM} + M}$$

where
MTBM is the mean time between maintenance
M is the mean active maintenance time

Both corrective and preventive maintenance are considered in this availability figure.

26.2.3   *Operational availability*

Operational availability is defined as the probability that a system or equipment, when used under stated conditions in an actual operational environment, will operate satisfactorily when called upon. The equation for it is

$$A_o = \frac{\text{MTBM}}{\text{MTBM} + \text{MDT}}$$

where
MTBM is as described above
MDT is the mean maintenance down time

26.3   *Availability integrity*

If an availability figure is employed in a contract, the number must of course be coordinated with the reliability and maintainability numbers also employed.

26.4   *Availability verification*

Availability could conceivably be verified during OT&E and early operational use by collecting actual reliability and maintenance factors observed in the field. Customers most often wish to have their requirements verified prior to delivery, however. So, the availability figures can be derived from figures that fall out of reliability and maintainability analytical or test-based verification work.

chapter twenty-seven

Logistics engineering

27.1 Supportability and integrated logistics support

The logistics specialty engineering activity seeks to identify features that will result in optimum supportability in terms of maintenance (testing, servicing, handling, etc.), spares provisioning, support equipment, personnel staffing and training, and technical publications. A more current name for this area of interest is system sustainment as it is referred to on the common process diagram in *System Management: Planning, Enterprise Identity, and Deployment* (Grady, in press) and in Figure 27.1 extracted from the common process diagram included in the management book. The whole life cycle consists of system development and system employment. System employment consists of system sustainment and use system, the latter involving the operational employment of the system to achieve the ends suggested in the original system need statement.

Sustainment combined with system use covers the whole life cycle from the customer's perspective starting with delivery or deployment through disposal at the end of system life. Often, disposal will not result in a total destruction of residual system elements, rather a partitioning of them into several sets: those parts that are of no use whatsoever that can be sold for scrap or junked, those that may have continuing value with no further action and may be retained in a new system to replace the old one, and those parts that can be modified to provide adequate service in the new replacement system. The latter two classes may be combined with new elements and the aggregate will be suitable for satisfying a new need that the current system can no longer satisfy.

Logistics engineering applied during development deals with the engineering of good sustainment features into the system during the design process. Commonly, logistics includes many of the other specialties as subsets such as reliability, maintainability, maintenance engineering, spares provisioning, and even operations. This process must be based on the items in the system as well as the process intended for operation and maintenance. Therefore, integrated logistics support (ILS) needs a definition of the architecture of the system and a process diagram through which the system and its elements will flow in normal operation and maintenance. Each step in the process flow at some level of indenture

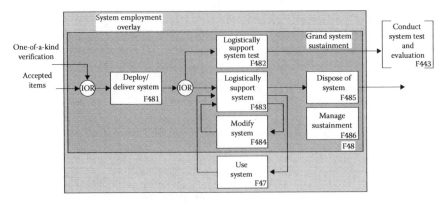

Figure 27.1 System employment overlay.

is analyzed at some level of architecture indenture to determine all of the related logistics consequences.

Systems have two kinds of maintenance that must be applied: preventative and corrective. Preventative maintenance seeks to take actions that prevent failure during operational use of the product. Corrective maintenance focuses on repairing things that have failed. The preventative maintenance can be analyzed effectively using the process orientation and corrective using the architecture orientation. Most systems have some major reusable element that encourages a cyclical use trace, and the preventative maintenance focus can be linked to this cyclical sequence in terms of pre- and post-mission maintenance actions. The corrective focus must inject into the normal use cycle whenever symptoms of failure are observed. So, part of the preventative maintenance focus must expose failures to easy observation such that faults can be acted upon in a corrective fashion. At the same time, preventive maintenance is intended to encourage a fault-free period of performance where the required reliability figure can be expected to be realized.

27.2   Maintenance engineering

Perfect systems that will never fail cannot be built in a physical sense because, among other reasons, a system with a reliability of 1.0 will have an infinite cost. Systems consisting of millions of parts cannot possibly exclude all possible faults. The author used to marvel at how the low-budget unmanned reconnaissance aircraft systems he worked on in the 1960s and 1970s would hold together for the few hours they operated resulting in successful flight and good photo coverage of target areas defended by the most intense antiair warfare capability that had been seen on Earth at the time. These operational systems consisted of unmanned aircraft containing fairly complex systems, DC-130 launch and control aircraft, ground control stations, and CH3C midair retrieval helicopters. On one day in 1969, for example, three unmanned reconnaissance aircraft overflew the Hanoi Hilton through intense anti-aircraft fire and were successfully midair-recovered with good film coverage. This was thought of at the time as an unprecedented performance, not some routine expected result. These machines were controlled by relay-diode discrete logic and analog flight control systems.

These same systems were extremely difficult to maintain. Prior to the advent of a remote control officer (RCO) aboard the launch aircraft, after flight there was no pilot squawk sheet listing the real or imaginary problems with the vehicle in flight other than what the ground RCO might have observed on the 120 mile inbound flight after the unmanned aircraft's command radio had been reactivated by on-board program assuming it made it through what was then a very formidable air defense system. The avionics and mechanical technicians had to completely inspect the vehicle visually and with special support equipment. It often required a whole day and sometimes two to post flight a vehicle. Preflight preparations then required many hours to make the aircraft ready for the next mission. At the end of the line of this series of unmanned reconnaissance aircraft, the Ryan model 147SD(L) or USAF AQM34M(L), the undermanned Ryan crew operating this system out of Utapoa, Thailand in 1973 had only one man turning the birds around in the evening in the hangar, Gerry Boyd. Gerry found a way that he could hook up all of the support equipment at one time to the vehicle and run one continuous test rather than stop and change the test equipment connections several times as covered in the technical data. The author, the Teledyne Ryan Aeronautical team leader at the time, found Gerry in the hangar one evening stepping on edges of the test equipment (so as not to disturb the panel settings or damage the equipment) laid all over the floor to get between the bird and checkout console panels. There was so much special support equipment surrounding the vehicle there was no room for the technician to walk on the floor. This was not an ideal maintenance design but probably about the best that could be done at the time.

Another model being developed at that time was outfitted with a checkout console containing a digital computer that worked with an on-board digital computer to stimulate and respond to a testing process controlled by software running in these two computers. A computer-controlled airspeed and altitude signal generator was controlled by the console computer to simulate points in the airspeed altitude envelope to a vehicle-mounted pitot tube hood. Signals being monitored were routed to a console digital voltmeter that relayed the reading to the console computer for comparison with what the reading should be. A computer-controlled transponder issued computer-generated commands and monitored vehicle downlink data.

Neither the AQM-34M(L) program nor its predecessors had been funded to cover logistics support analysis (LSA) during development. From time to time, a new vehicle would be added to the mix and support equipment modified as needed with some new equipment added for special systems. Ideally, the system development team should be funded to perform an analysis of the planned maintenance concept as the preliminary design progresses with maintenance analyst feedback to the design team encouraging alternative design approaches that will enhance maintenance ease where necessary. The best tool for this analysis is process analysis where the analyst creates a flow diagram of the intended process and analyzes each block of that diagram for needed support equipment, tools, procedures, materials, and talent. In any case where the resources available are found to be in conflict with those needed, the supporting resources should be changed or the design of the article needing maintenance is changed to preclude the need.

Maintainability, as discussed in Chapter twenty-five, identifies repair times and assesses designs for compliance. Maintenance engineering synthesizes a set of maintenance actions that when implemented by properly trained personnel will result in a system ready for use as needed. Given a maintainability figure of 30 min mean repair time (corrective maintenance), the design team must develop a design that can be repaired on the average within 30 min. The first step in this process is to identify all of the things that can fail, identify the level at which all of these things may be replaced, and the maintenance consequences of removal and replacement of them.

The analyst can make a list of all of the items that may be removed and replaced as a corrective maintenance action and, hopefully, this will include everything that can fail in the item at some level of indenture. The failure modes effects and criticality analysis (FMECA) should be the basis for identifying all of the possible failure modes, but many of these failures will be repetitive relative to items in the system. Now, the analyst must determine how long it will take to remove and replace the item that caused the failure. This can be done analytically, of course, the analyst using his or her imagination to go through the process recoding times for each step. While doing this, the analyst can jot down the steps, make a list of tools, consumables, and common and special support equipment needed either in notes reported upon in a report or via computer tool entries as part of a LSA. Depending on the contractual requirements, remove and replace time may have to be combined with diagnostic, repair, and confirmation times. Some customers would add spares delay times and other factors. See Chapter twenty-five for commonly identified times often called for.

The mean repair time can be computed and compared with the required time and appropriate action taken. If this work can be accomplished during the preliminary design, there will be time to interact within the responsible team and change the design to reduce repair times if the analysis has shown the average to be too high. If the analyst waits to study the design during detailed design or drawing release, the responsible design engineer may very well truthfully claim that there is no money left to accomplish the desired changes. You must use models to be able to accomplish this work during a time when funding remain available.

There may also be requirements for turnaround time and preventive maintenance time. These tasks can similarly be analyzed at some level through use of the imagination and models fairly early in the design effort. There are clear linkages between this work and the work needed to determine what tools and support equipment are needed, the procedures that should be captured in technical data, and the identification of potential safety hazards and candidates for human engineering improvement.

The Teledyne Ryan Aeronautical AQM-34 series of unmanned reconnaissance aircraft operated by the Strategic Air Command (SAC) in Southeast Asia during the Vietnam War, referred to earlier in this chapter, was initially deployed with a full Ryan crew because there had not been time to create an Air Force organization and train the personnel between the time the government decision was made to deploy and the time the system was supposed to be in place operationally in 1965. The system went right from flight test, not yet complete, to operational employment with no careful logistics analysis. As a result, the initial Ryan crew had to find ways to supplement formally provided support equipment, some of it factory test equipment, with special aids that eventually were passed on to the Air Force personnel when they started coming on board.

During the author's last trip to Southeast Asia on this system as the team leader of the AQM-34M(L) deployment in 1973, he had a lot of time on his hands so he made an attempt to identify every piece of equipment that was needed to properly maintain the system. In the process, he identified 30-odd items that were not identified anywhere in the formal technical data for the system. Eventually, the customer purchased a technical order change that identified all of this equipment and told how to locally manufacture it.

27.3   Spares

A spare is an element of a whole (end item) that can be replaced to change the state of that end item from being defective to working or operable. Spares are provided to permit the using organization to repair end items when they fail through parts replacement. The beginning of the spares story is FMEA that tells us everything that can fail at some level of indenture. We wish to spare things that can fail but we can spare them at different levels of indenture. Given a whole airplane, it would generally not be a good idea to spare it at the airplane or end item level. It is true that in World War II one could truly say that many airplanes were replaced at the end item level due to high loss rates of the B-17s flying bombing runs over Europe from Briton. These aircrafts were not considered spares but replacements for expended items.

Material supplied to replace that expended is in a different logistics class than spares. Some examples of this class for your automobile are gasoline, filters, engine oil, transmission fluid, tires, brake pads and fluid, and windshield washer fluid. These items are replaced as they are consumed in normal operation in what is called servicing the product. It is understood that they will be consumed. Spares are used to replace things that are not intended to fail or be consumed in normal operation of the end items. However, some parts, if they are sufficiently low in cost, may be treated as throw-away items resulting in a one-way supply process.

The spares selected must be compatible with the other maintenance resources selected for the system so all of these questions must be addressed as a whole. The support equipment must be able to support identification, diagnosis, and isolation of faults to the items spared. The technicians and mechanics must be trained to use the support equipment properly to find the faults at the level spared. The technical data supplied must focus on how the system works and the maintenance actions that correspond to the spares.

Our maintenance concept may entail more than one level and each level may require a different level of spares. For example, let us assume that a three-level maintenance concept is used where line replaceable units are replaced at the line maintenance level, shop replaceable units are replaced at the shop level, and depot replaceable units are replaced at the depot level. Obviously, each level will need a different set of spares. Since technicians and mechanics capable of replacing level 3 (depot) items would have to be qualified to replace items at levels 2 (shop) and 1 (line), which would entail simpler work, the spares for these three levels would commonly appear in the form of telescoping sets as suggested by Figure 27.2. Systems tend to be designed in a modular fashion such that major and minor pieces of end items may be removed and replaced easily and this is a major part of the logistics and maintainability work purpose, to encourage this kind of design structure.

The spares analysis should identify the right level of replacement of all replaceable elements and thus identify line replaceable, shop replaceable, and depot replaceable items, which must in turn be stocked at the right places to be effective. The spares situation is also influenced by inventory and shipment plans to be applied in placing and shipping the spares. Both of these decisions are, as well, related to the geographical stage upon which the system will play. Spares can be derived through use of hangar queens and cannibalization. This is not a desirable method because it is

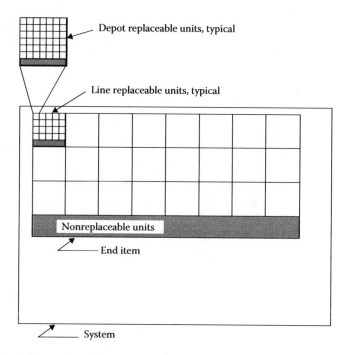

Figure 27.2 Spares distribution by levels.

generally good only for one failure of any one kind and it deprives the operating organization of one operational asset.

The quantity of spares of any one kind also have to be determined. This is primarily a question of anticipated failure rates determined in the reliability analysis and the intensity of use. All of these spares factors must be considered and integrated into the final answer of the total number of spares to be purchased based on a planned distribution of them over the geographical space within which the system will be employed.

27.4   Personnel and training

Some systems are operated and maintained by a user element of the customer. An example of this would be the F/A-16 fighter aircraft. General Dynamics originally built these machines prior to Boeing purchasing the developing division, but they were operated and maintained by the U.S Air Force. Similarly, Boeing builds 777 commercial transport aircraft operated by American and United Airlines, among others. On the other hand, Boeing built a Delta IV space transport vehicle and also provided the people at Vandenberg, California, and Cape Canaveral, Florida, launch sites to maintain and operate the vehicle. Some systems are maintained and operated by the building contractor and others by a customer. Generally, when the builder staffs the operational organization, the people will need less training unless these people are hired off the street for that purpose rather than staffing the team from engineering and manufacturing personnel with some experience with the system acquired during the development process.

It is necessary to identify all of the jobs that are necessary in operation and maintenance of the new system. Then we must determine how many people must be trained for each job. This will require a clear understanding of how the system will be employed: (1) how many end items are coordinated with one operating unit, (2) how many operating units, and (3) the operational tempo of the units. Training requirements for people being prepared for a given job are a function of two questions: (1) what are the skills and knowledge that people will have who are attracted to maintenance tasks on this program and (2) what are the needed skills and knowledge? The difference is the training requirement. It is necessary to determine what skills must be covered and how they will be packaged into courses.

The total training requirement for a particular job must be partitioned into one or more courses recognizing possible different levels of experience. Each course to be prepared must have appropriate goals established, the methods of instruction (on-the-job, lecture, demonstration, laboratory) selected to satisfy those goals. Finally, qualified instructors or training material builders must build the class materials.

Each course should be piloted with a test class possibly consisting of experienced people who can comment authoritatively about the quality of the course. After adjustments, the class can be rolled out in accordance with the training schedule.

27.5   Technical data

Technical data needs can be determined, like so much of the logistics story, by building the physical model of system maintenance. All of the needed work can be categorized into line and off-line work when in the former case you are working on the end item and in the latter case you are working on some part of the end item removed from the end item. Line maintenance work should be discouraged whenever possible because it ties up the complete end item often precluding other work. Some online work will require that all other work cease such as repairs to the fuel tanks, fueling, ordinance upload, and installed engine test. Some line maintenance tasks may be accomplished together with others in parallel. The latter is advantageous because they tend to reduce the total span time of the needed maintenance work.

Given that all line maintenance tasks have been identified at some level of granularity, we must then describe in some detail how each task will be accomplished. In each case, an effort should be made to make it possible for a single person to do a job alone. Otherwise, a team task may have to be described. There are cases where one should require a two-man team concept even though it is physically and mentally possible for one person to accomplish the task. For example, when dealing with nuclear weapons, two people must cooperate to employ the weapon. This is based on the redundancy principle. One person could fail to follow the tech data and prematurely release a nuclear weapon.

The analyst doing the task analysis must simply run through the process of doing the work in his or her imagination, jotting down each step in the process and also noting the support equipment, tools, and materials used. The time of each act should also be tracked such that maintenance times can be collected. Sketches should be included where necessary to make it perfectly clear what has to be done and how. Commonly, notes, cautions, and warnings are woven into the text to alert the technician or mechanic to potential problems that should be avoided.

Once the data is believed to be complete, it should be validated by someone from the developing organization trying to perform the work as documented. The person doing this work must be very careful not to let their knowledge of the system fill in any blanks missing in the data, rather they must mindlessly follow the written word. Problem identified must, of course, be corrected. Then the data should be verified by the user of the system ideally at a user site. Clearly, this verification of the technical data must be timed to occur after the initial product elements have been delivered to the user. The author would sometimes draw the assignment as a field engineer to support tech data verification at field sites like Davis Mounthan AFB in Tuscon, Arizona. It often occurred that the verification process would discover problems with the technical data that had not been found in the validation process. These differences could often be traced to the validation process having been accomplished by engineers with long product experience while the verification process was accomplished by airmen with relatively little experience. This might be instructive for the validation process suggesting that perhaps we should employ people with little experience for the validation work. In the extreme case, we could let Julie the mail room clerk have a go at it.

27.6 Testability, integrated diagnostics, and built-in test

A system will generally require some means of determining whether it is in an operable condition under certain circumstances. This specialty determines an optimum way to accomplish this end through identification of an integrated view of product and support equipment features and capabilities that together will assure effective testing. The process should be driven by the FMECA results that should reveal every possible failure. A specification might require 92% built-in test (BIT) effectiveness. This could be verified by showing that 92% of all faults listed on the FMECA will show up on BIT. A system might be built using BIT for all line checkout but rely on special support equipment for shop and depot-level maintenance.

Clearly, a system in total determines its health. This outcome can be achieved either by locating all of the health monitoring capability in the operational product or separating the product into operational and maintenance components. The former solution, in the extreme case, eliminates the need for separate support equipment but at the price of adding weight and complexity to the operational equipment with the possible result that operational payload performance may suffer. Also, this solution may result in added cost because each operational asset needs the support equipment included within itself rather than supplying operating organizations with some smaller number of test equipment items.

Thus, the question is how do we partition operational and support equipment? This decision should not be made in isolation from the decision about how maintenance will be performed and at what levels it will be performed. Clearly, all of the logistical factors must be coordinated.

27.7   Support equipment

Support equipment is used to evaluate mission equipment status, to prepare mission equipment for use, and repair mission equipment subsequent to identification of faults. This support equipment falls into two subsets. Support equipment developed specifically to maintain a particular system is referred to as special support equipment. Equipment that can be applied in a more general way to many systems is referred to as general support equipment. It is possible that a program might have to develop special support equipment to accomplish support equipment functions relative to other special support equipment, but this is not the ideal case. Ideally, it would be possible to maintain special support equipment using only general support equipment such as voltmeters, signal generators, and o'scopes.

As the operational equipment design concepts become clear, decisions must be made regarding how it will be verified that the operational equipment is in a fully mission-ready state. The range of possibilities includes no support equipment because the operational equipment includes a full complement of BIT equipment. The other extreme has no BIT capability requiring all test equipment needs to be satisfied by support equipment. There exist many combinations in between these extremes.

Support equipment includes other kinds of equipment besides test equipment, of course. Often, it is necessary to provide special handling equipment for lifting, hoisting, and moving mission equipment. It may also include servicing equipment to charge batteries, bleed hydraulic brakes, or add environmental system fluid. Different kinds of support equipment are required for the three layers of maintenance often employed in systems: depot, intermediate, and operator. At the operator level, generally, line replaceable items are simply removed and replaced, sent to a shop where they are checked out, repaired, and made ready for reissue. At the intermediate shop level, shop replaceable units are exchanged and defective ones shipped to depot or discarded locally.

The author drew an assignment as the electronics chief of a U.S. Marine transport helicopter squadron in the late 1950s that was based in Oppama, Japan. The squadron was supported by a headquarters squadron, which was responsible for intermediate (shop) maintenance, and his squadron was only responsible for operator or line maintenance that involved identifying the faulted units and taking them to the headquarters squadron shop. The radio equipment in the HRS helicopters at the time was of an ancient World War II vintage including AN/ARC-1 UHF and AN/ARC-5 HF radios. His squadron shop possessed only a PSM-6 multimeter and a UHF wattmeter.

The squadron often deployed on maneuvers such that it was not supported by the headquarters squadron shop. The previous shop chiefs had built a wooden box that mounted all of the equipment racks from the helicopter and wired them up so that the shop could take a work bench with them on these deployments. The author had the shop technicians reduce the mass of this device that we were not supposed to possess to a small metal box with wire harnesses and racks. The Marines were experimenting with the use of helicopters for vertical envelopment at the time with only a single World War II jeep carrier acting as a landing platform helicopter so the squadron often had to operate from the fan tail of an LSD using the empty well deck for a hangar hoisting the helicopters between levels with the LSD deck edge crane. In other cases, the squadron might operate from a dirt strip scraped out of some farmland in the Philippines with a shop provided by a pyramidal tent full of wind blown dirt when the helicopters turned up.

On one operation, the unit accomplished a maneuver in the Philippines, boarded ship on an Essex class carrier in a panic joining an invasion force for Sumatra to rescue Americans holed up in the Embassy that was solved prior to the operation by President Sucarrno backing down, but the carrier was diverted to Tricomalee, Ceylon for flood relief duty. At the end of that activity, the unit was sent back into a maneuver in the Philippines following a brief stay at Clark Air Force Base where a headquarters team finally replaced the AN/ARC-1 radios with AN/ARC-27 radios.

During this whole operation stretched out over several months, the squadron never lost a moment of planned flying time due to radio malfunction despite operating with essentially a single level logistics system. Some other factors also highlighted that despite the best-laid plans of logistic support during development, reality may intrude and supply some really bizarre twists in the way units actually have to operate. When trying to leave the base in Japan, a sailor opened the wrong valve on an LSD tied up in Yokosuka, Japan delaying the squadron's boarding by several days while they cut a hole in the well deck, removed the flooded generators, dried them out, reinstalled them, and patched the hole in the well deck. When the squadron finally got everyone together in the maintenance shack at the flight line the night the repairs were completed, mostly every one was nearly drunk including the maintenance officer and one of the crew chiefs who argued over who was going to fly the crew chief's helicopter aboard ship in the morning. The corporal who had been up and down the ranks a few times won the argument with a coin flip and did so remaining a corporal this time. The line chief said that the maneuver would be short so that no one had to take rifles or 782 gear (pack, ammo belt, etc.) aboard. While planning the amphibious operation on Sumatra later in this extended operation, it was decided that a radio tower had to go ashore right away to control the landing of the ground support aircraft after their initial flight providing support for the invasion force. The author was selected to take a radio jeep in. He reminded his electronics officer that he had no rifle so the officer agreed to loan the author his pistol. Then it was discovered that it was not so easy to get a radio jeep into an LCVP from a carrier and there was not a AN/MRC-35 jeep aboard anyway. Some other work around was elected. The question was, of course, why the tower was required so early in the operation. It turned out that the Douglas AD attack aircraft squadrons that had been loaded aboard by crane at Cubie Point in the Philippines only had one pilot who had ever taken off from a carrier and none who had ever landed on one. Apparently, it had been decided that anyone could make a take off but that they had better land on a ground strip. So this may have set a record, if it had ever occurred, for the shortest amphibious airfield capture of all time in that the AD aircraft had to provide close support for the landing force that had to capture the airfield before the close support aircraft fueled out.

Now, you say that these silly problems from a past that included vacuum tubes and reciprocating engines could not happen today with our lighting-fast logistics system. Do not be so sure. Military operations often have to be thrown together more quickly than good sense would allow and the military, especially the Marines, loathes the phrase, "We can't do this." To the extent that a system can be designed to possess flexibility in using it will find a higher probability of delighted users under conditions of stress.

27.8 Transportability, mobility, and portability

Some elements of the system may require some degree of physical mobility. This specialty discipline determines the character of the needed mobility and defines ways to satisfy these needs. Weight, size, and interface simplicity issues are chief among the design features that will lead to good or bad transportability. The item may require transportability by particular kinds of vehicles not part of the product operating in or on a particular medium (air flight, sea surface, or ground surface, for example) or have mobility features built into the product itself.

Small radios have been developed that can be backpacked into combat supporting unit commanders in their need to coordinate action up and down the chain of command. These units have to be lightweight while having a combination of transmitter power and receiver sensitivity that encourage success in communication over the range needed. Large four-engine aircraft have also been developed with the responsibility to support battlefield commanders with communications and battlefield status and trends. Both of these systems entail an ability to move the physical resources about geographically but on a different scale.

27.9 Packaging and shipping

MIL-STD-961E the DoD standard for specifications, encourages that Section 5 of specifications dealing with packaging and shipping should be a void in that these are contractual matters. The author believes this is bad advice or direction to the contractor, but we must do what our clients require. The reason the author would encourage strong coverage in the item specifications for this area is long experience revealing that often this is the most stressful part of the life cycle for many products. A guidance set, for example, will seldom in its life cycle be treated any more roughly than when it is being shipped not installed.

The packaging and shipping design problem is to so contain the product that stresses expected in transportation of the product will not degrade product capabilities. The stresses can come from several environmental factors like moisture, shock, vibration, and temperature. One way to predict what these forces will be long before anything must be shipped is to send an instrumented box on the planned routes the product will have to follow and read out the results in terms of peak stresses. The instruments are very reasonably priced. Moisture and temperature stickers can be purchased that turn colors when certain extremes are reached, and by placing a series of these on or in a package it is possible to determine the peak value accurately enough for specification purposes.

There are some stresses for which it is, of course, impossible to prepare and to do so would be a case of overdesign. The U.S. Air Force unit the author was serving as a field engineer in Vietnam in the 1960s eagerly awaited receipt from the home station in the States of a Doppler test set for use in testing unmanned reconnaissance aircraft Doppler navigation equipment. When the test set was received, it had a big hole in the case and front panel where it had been pierced by the tine of a fork lift truck.

Good shipping and packing equipment sometimes provides unintended benefits. This same Air Force unit flew low and high altitude unmanned photo reconnaissance aircraft. Sometimes a contrail suppressant was used in high altitude birds to make it more difficult for hostile forces to visually spot the vehicle over flying at above 60,000 ft. The suppressant tanks were shipped over to the operational location in special crates that were supposed to be returned containing depleted tanks for refilling. The Air Force troops found that they could ship whiskey and other materials back to the States in these containers avoiding interference by the customs people. No one was interested in opening these containers with all of the wicked warnings posted on them and besides they entered the country via a military base.

chapter twenty-eight

Safety, human engineering, security, and environmental impact

28.1 System safety and health hazards

The safety engineer identifies safety requirements based on customer needs and cooperates with design and analysis personnel to understand and identify safety hazards to life, health, and property value. The principal approach is to build a model of system use in cooperation with the systems, maintainability, and logistics personnel, and to examine this process model for conditions that can cause hazards to develop. The product is evaluated for ways to prevent these conditions from ever developing, or ways to control the risks when they do occur. A safety hazard analysis develops essentially a hazard list that should really be part of the program risk list. Ways are found to eliminate or control each hazard in terms of its anticipated frequency of occurrence or severity of occurrence.

Commonly, the safety engineers use essentially the same criteria used for other risk identification work. Safety hazards are measured in terms of the probability of occurrence and the severity of the occurrence. Criteria are established to numerically assign these factors to particular hazards and all hazards with a numerical figure above a certain value must be mitigated. The mitigation process seeks to find ways to lower the probability of occurrence and/or severity of impact and may entail design changes, procedural prevention, or safety placarding, for example. Figure 28.1 is a variation on a figure provided MIL-STD-882B on the categorizing of safety hazards. The safety hazard assigned smaller hazard indexes for more serious hazards, which makes it very difficult to combine hazard indexes to form an aggregate program hazard index and track it over time with energy applied to reduce that index by reducing the probability of occurrence and seriousness of occurrence of those hazards.

Figure 28.2, also motivated by MIL-STD-882B (but with the sense reversed), provides guidance for the safety engineer performing system analysis in the selection of the best category match for the hazards observed. Clearly, this is not a completely objective process because the criteria are broad enough to permit two analysts to come up with two different

Frequency of occurrence	Consequences of occurrence			
	4 Catastrophic	3 Critical	2 Marginal	1 Negligible
5 Frequent	20	15	10	5
4 Probable	16	12	8	4
3 Occasional	12	9	6	3
2 Remote	8	6	4	2
1 Improbable	4	3	2	1

Hazard risk index and hazard criteria

20–10	Unacceptable	6–4	Acceptable with review
9–7	Undesirable	3–1	Acceptable without review

Figure 28.1 Hazard risk assessment criteria.

Hazard categories		Meaning
4	Catastrophic	Death or system loss
3	Critical	Severe injury, minor occupational illness, or major damage
2	Marginal	Minor injury, minor occupational illness, or major damage
1	Negligible	Less than minor injury, occupational illness, or system damage

(a)

Hazard probability		Meaning
5	Frequent	Likely to occur frequently
4	Probable	Will occur several times in item life
3	Occasional	Likely to occur sometime in item life
2	Remote	Unlikely but possible to occur
1	Improbable	So unlikely that it can be disregarded

(b)

Figure 28.2 (a) Hazard categories and (b) probabilities.

categories for the same situation. But, when the guidance is applied by qualified safety engineers and their work is subject to review, discussion, and approval, the process has the effect of objectivity. Clearly, one could fabricate a model with any number of categories on the two axes.

28.2   *Hazard analysis and reporting*

Designers are seldom the best people to depend on to spot unsafe situations with the product for which they have design responsibility. Designers tend to have a very positive attitude about the results of their

Hazard id	Hazard name	CAT	Freq	Index	
201	Fuel supply line leak	3	3	9	
202	Engine mechanical failure	4	4	18	
203	RPM indication failure	2	3	6	

Figure 28.3 Program safety hazard list.

labor and commonly cannot see faults in it. A safety engineer should try to maintain a detached feeling about the evolving product design. If the safety engineer identifies too tightly with the design, he or she may become part of the problem and fall victim to overidentification with the design. While safety engineers should be depended upon as the principal hazard identifiers, they should maintain contact with the engineers doing maintainability and reliability work because people in these disciplines often observe the cause of potentially hazardous situations that may be missed by the safety engineer.

One source of potential safety hazards is realized by evaluating failures of system elements. Thus, the failure modes effects and criticality analysis (FMECA) results should indicate any adverse effects from each possible failure in the system. Ideally, everyone on a program would be empowered to identify risks of all types including safety risks most often called safety hazards. All of them identified by anyone should be added to a list and the probability of occurrence and severity of impact determined. An example of such a list, which may include many more fields, is included in Figure 28.3. The hazards on the list should be reviewed periodically by team and program management with safety engineering report of current status. Ideally, the aggregate safety risk should be descending over time and if it is not, then management must take action to cause it to do so.

Using the MIL-STD 882 index figures rather than their reversed form shown in Figure 28.1, the reader can see that a decreasing safety problem would be measured by an increasing aggregate hazard index. The problem with this metric is that initially before safety hazard analysis has begun, the program will have a zero score suggesting that everything is unbelievably perfect. As hazards are identified they may have low index values (high safety hazard) with increasing numbers of hazards resulting in a display that does not intuitively communicate meaning. The author would prefer to reverse the sense of index scoring as in Figure 28.1 such that an aggregate score would increase both from increasing numbers of hazards and higher safety concerns for each one. The data shown in Figure 28.4 could then provide an overall program metric for reporting safety program status. Early in the program the number of hazards starts to mount and many of them are serious hazards measured with a high

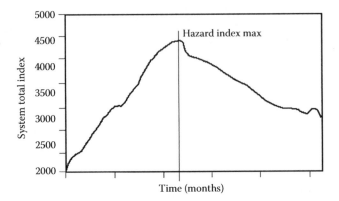

Figure 28.4 Program aggregate safety hazard metric.

index value. At some point in the program, one would pass the peak hazard index value suggesting that the number of hazards has stopped rising rapidly and the safety engineers are being successful at reducing the hazard index of those that have been identified.

The safety engineer must participate in ongoing design work at the system and team level and offer his or her opinion on whether or not current design intent is consistent with safety requirements and to work with the design people to make it so when a problem is identified. This work is often very difficult for the specialty engineer because of the designer's positive attitude about the design and their perception that the specialty engineer is very negative in outlook. Safety engineers must be tenacious and determined to remain focused on the job and avoid being drawn into a personality and ego battle. Team leaders and program managers must also be alert for symptoms of this problem and stamp it out quickly.

The first priority should be to resolve those safety hazards that have the highest index numbers (lowest if using the index from MIL-STD-882B). Three common ways to mitigate safety hazards is to (1) redesign to preclude the hazardous feature or outcome, (2) redesign to give warning of a hazardous condition, or (3) establish procedures to prevent occurrence of the conditions necessary to trigger the hazard condition. After the action is taken we can adjust the safety index to a lower number reflecting a reduced probability of occurrence or reduced frequency of occurrence.

28.3 Human engineering

The human engineer seeks to ensure that design features reflect human capabilities with respect to recognition of critical conditions and ease of actions that must be taken to operate and maintain system items. Commonly, the system will have to be designed for a certain population percentile where the physical measurements, strength, and sensory capabilities are known. The human factors engineer obviously must be very familiar with real human capabilities. Military and commercial standards for human performance are available and one must be careful to match the standard used in analysis with the actual population that will operate and maintain the system. If one chose a military standard as the basis for the human factors for a system intended for commercial use, it may be found in actual system use that many purchasers had difficulty using the system because the military standard focuses on people in good physical condition in their prime.

In addition to ensuring that system features compare favorably with human physical and sensory capabilities, human engineers must also ensure that system features are consistent with human cognitive processing, perception, problem solving, decision making, and memory capabilities. It is a difficult task to assess the human work load of a fighter pilot where the pilot must maintain appropriate flight conditions, maintain situational awareness, remain alert for hostile aircraft and weapons systems presence and threats, monitor aircraft and weapon systems readiness and status. It is entirely possible that the engineers may provide adequate indication of all conditions of interest but not recognize the time limits working on the pilot.

Early unmanned aircraft remote pilot equipment was heavy on the engineering and weak on the human factors. One could say the same about manned aircraft instrument panels, no doubt. The unmanned aircraft used by the USAF in the Vietnam War were remotely controlled using 8 analog dials and 16 on–off lights. The design of each new air vehicle would include a different combination of analog and discrete parameters with some of the same parameters possibly assigned to different meters or lights. Different models might also have different scaling factors. Perhaps it was not so strange that USAF engineers often attracted to electronic warfare and navigator roles generally made better remote pilots than did former actual pilots who had difficulty not being in the air and often experienced anger at having been dropped from a flight status for medical reasons.

In one antisubmarine warfare computing system the author worked on, the human factors engineer's contributions were limited to ensuring that there was an indent on the operator's panel to prevent a coffee cup from sliding around the panel and insistence that a panel have a light green color. In another system, that the author had to repair in the Marines, air control officers working in a darkened air control center would often place their coffee cup on top of the communications selector switch protective panel. Often the coffee cup would be knocked over and coffee would drain down into a set of 32 intercom push buttons eventually causing them to jam completely. In a ground launch unmanned aircraft system that had been converted from an air launch system, a red light on the launch control panel called the Launch No Go light was illuminated throughout the launch countdown. The problem was resolved by changing the light to a green bezel and renaming the light the Launch Go light. The light had no significance in the ground launch mode actually.

The aileron linkage for an aircraft included a turnbuckle to permit the center point to be adjusted to cause aileron actuator electrical zero to align with mechanical zero and streamlined ailerons on both wings. The turnbuckle was threaded incorrectly such that the mechanic had to remove the actuator linkage to make the adjustment.

The HRS helicopter used by the Marines in Korea up through the late 1950s, when it was replaced by the HUS, had a very practical but harrowing method for ensuring that the rotor blades were in good vertical alignment, that is, they all remained in the same plane as the rotor head spun around the hub. The mechanic would mark the tip of each of the three blades with a different grease pencil color. A pilot or maintenance man would fire up the engine, engage the rotor, and adjust engine speed for a particular rotor head speed. While the blades whipped through the air, two maintenance men holding a blade tracking pole would edge the canvas flag on the pole into contact with the blade tips so as to hear three blade hits (whap, whap, whap), at which time they would pull the flag clear. The mechanics would then measure the distances between the marks and if necessary stop the rotor and adjust the appropriate rotor linkage. This process would be repeated until flag hit distances were within limits. Crew chiefs in these helicopters also used a foolproof method for determining their helicopter was developing mechanical problems. Depending on whether the problem was in the engine, main rotor, or tail rotor, they would feel it as different parts of their body would vibrate while sitting in the troop compartment with the helicopter flying. This trouble-shooting feature was acquired at no cost to the tax payer, one would hope.

During World War II, there were no weapons systems employing electronic digital computers. A World War II vintage mechanical analog computer used for antisubmarine warfare was bolted to a bulkhead and the technician gained access to it by unbolting the front panel and folding it down for access to the insides for cleaning, repairs, and alignment. Once the lid was bolted back, the solution would change from that observed with the lid open. The technician had to know how that particular serial number acted so as to pre-adjust it such that the solution came in correctly when secured. These kinds of problems seldom occur today because the mathematics is being accomplished in digital computers that are not adversely influenced by similar stresses.

Many design problems may be introduced into a design that could be discovered by any number of different specialty discipline engineers following the discipline methods. No matter who discovers a problem in a design, the program should declare victory and move on to resolve the problem.

28.4   System security

The system security discipline seeks to identify security risks to the system and identify ways that the system can avoid these risks. The threats to the system must first be listed and then each one of these threats must be dealt with in the design to preclude their occurrence from having an adverse effect on system operation. The driving analysis commonly produces a threat report identifying the ways that a hostile force can interfere with system security. These security threats might be categorized as physical, information, etc., and coordinated with a risk assessment metric similar to the safety measure involving two considerations: the probability of occurrence of a threat and the degree of difficulty if it occurs. A security index could be similarly formed.

During the time this book was being written system security had been elevated to occupy a very important role because of the unique threat posed by middle eastern fanatics who would attack a target by blowing themselves up in a fashion similar to Japanese kamikaze pilots but a lot less detectable. In addition, our military has become very dependent on systems that require that the computers within them that are linked via radio into very complex networks employing GPS and Internet all subject to attack not only on the nodes explosively but on the networks and their computers with possible false data and viruses of all kinds.

28.5   Environmental impact

The system must operate within a prescribed environmental definition. The system and the environment will interact in certain ways and the goal is to design the product system such that it can sustain operation within the extremes of environmental stresses anticipated while minimizing the adverse impact of the system on its environment. The latter is accomplished by understanding the interface between the system and the environment in terms of all materials and energy that are exchanged across this interface. Each of these interfaces is studied for ways to reduce environmental impact. Military systems often, by their very nature, result in serious environmental damage and for years there was no restriction against this impact. Now, DoD must generally respect environmental impact restrictions though it is unavoidable in these systems that when they are employed in anger, they will inevitably result in a terrific impact. The worst-case example of this is the use of a nuclear weapon.

One of the longest-term and most serious environmental impacts has motivated the nuclear weapons development clean up. At the time that work started on atomic bomb development we were not fully informed about the problem and even if we were the immediate nature of the need for the weapon, if it were even possible to build such a thing, probably would have mitigated against developing and operating a concurrent cleanup program.

Now the Department of Energy has developed a comprehensive solution to the problem, but the location chosen for the storage of the post-processing residual materials that will offer a hazard for a considerable period of time is predictably being challenged by those living in that area. So, there will often be not only technical problems of considerable difficulty to deal with in solving these kinds of problems but political ones as well. One of the hottest political issues on Earth in 2007 was global warming. One side claims that man is the cause of the Earth average temperature increasing 1° in 100 years while people of another persuasion offer that the explanation is clear from the history of our planet's temperature passing through hot and cold cycles for a very long time and the current rise being coordinated in some way with a similar rise in temperature on Mars that we have not been able to influence in any way. This argument comes down to economics, which is an important part of any discussion about environmental impact.

There are development, operating life, and disposal components to environmental impact that ought to be considered. Some companies formally doing business in California moved out of the state rather than pay the cost of complying with very tough laws on environmental impact related to the use of solvents, plating, fiberglass manufacturing, and painting. Teledyne Ryan Aeronautical and General Dynamics Convair for many years operated facilities at Lindberg Field airport in San Diego with drainage systems emptying into San Diego bay. With all of the naval ships in the bay discharging sewerage and whatever else into the bay, this drainage might not have been all that easy to detect. But, as the ships cleaned up their act these companies often were cited for polluting the bay.

It cannot be denied that military systems do pollute during their useful life. These systems blow things up and in the process release toxic materials into the air and water. Airplanes sit on the tarmac leaking fuel. The author worked on one target system that burned inhibited red fuming nitric acid (IRFNA) and rubber in its hybrid rocket engine. Upon return from flight it was necessary to drain any residual IRFNA from the vehicle onto a concrete deck that drained into a hole in the ground filled with sea shells that were intended to neutralize the acid before seeping into the local water table. One large open space park in San Diego, California, had to undergo a cleanup operation that lasted several years after a child had been killed by unexploded ordnance left in place from a World War II military training site.

Many space transport development efforts have been begun with a serious interest in avoiding further clutter in space by requiring that anything that is placed in orbit must be eventually removed from orbit by the system that put it there. Most often this requirement has been deleted early in the effort because of the added cost to lift the desired payloads to space while preserving some propellant to de-orbit the payload at some point in its life. Oddly, some weapons of war have found energetic environmental supporters of late. Some complete naval ships have been sunk to form the equivalent of a reef that evolves as a teeming home to sea life.

chapter twenty-nine

Parts, materials, and processes engineering

29.1   PMP engineering overview

This book is generally focused on the development of the higher levels of the system, but all systems are composed of parts and materials and they should be manufactured in accordance with written standard processes. If discipline is not introduced into the development of designs, the engineering drawings will call for a wide range of parts and materials and refer to many specialized manufacturing processes. A program parts, materials, and processes (PMP) activity seeks to introduce a degree of standardization in selection of parts and materials by the designers and use of manufacturing processes. Figure 29.1 illustrates this selection process that applies to every part, material, and process called out by the designers on drawings. Obviously, this requires a mountain of work that can most easily be sustained through computer-aided design (CAD) automatic evaluation of all PMP called for by all of the designers against those listed in the standard.

29.2   Parts engineering

The process entails lists of PMP from which the designers may select. If a designer finds that there is nothing on the list to satisfy their design need, they have to request addition of new entities to the list or evolve a design that can work with the existing standards. PMP lists can be coordinated with CAD making selection a click away.

The role of parts engineering is to assure the use of parts qualified for the application and to standardize on the fewest possible number of different parts. This is accomplished by development of a standard parts list from which designers may select parts. Any suggestions for additions to the list by designers are reviewed to determine if a suitable part has already been identified or an existing listing can be applied to the new application. Some parts may require a company parts specification written by someone from parts engineering and someone from the requesting team. The parts list should be a responsibility of the program integration team (PIT) when the program is organized as discussed in this book.

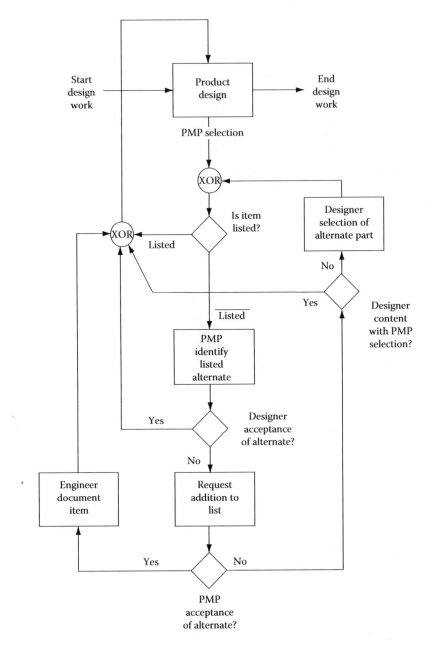

Figure 29.1 PMP selection process.

In an enterprise organized as a matrix, the parts engineering department, often in an electronics design department, should have the enterprise resource responsibility in this area and may develop a generic enterprise parts list that all programs either use in the generic form or tailor for their own needs. There are some manufacturing advantages in an enterprise able to standardize it use of parts across all programs. They can standardize more narrowly on tooling and manufacturing skills. They can also employ robotics more extensively.

29.3   Materials engineering

Materials engineering, often combined as a materials and processes engineering under a mechanical engineering department, seeks to standardize on a minimum number of materials qualified for the product. Some materials may require company specifications written to characterize them where adequate definition does not exist elsewhere. Designers must select materials from the standard list as in parts.

29.4   Process engineering

Engineering should define a set of standard manufacturing processes appropriate to manipulation of the materials called for in the design. These standard processes are then called for on design drawings rather than each engineer writing up a process unique to their drawing. The same logic applied to parts and materials applies here as well. These processes have to be consistent with manufacturing capabilities so should not be developed without manufacturing cooperation or review.

29.5   Contamination control

Contamination control seeks to limit contamination of the product during manufacture by any foreign material generally defined in terms of particles larger than a certain size. Control is exercised by requirements for processing within areas qualified for a prescribed level of cleanliness and special transportation and handling process instructions. In the manufacture of an aircraft fuel system normal work performed can result in contamination fairly easily and processes must be developed to prevent it, identify any contamination occurrences, and remove any contamination identified.

Aircraft fuel tanks in wing structures may be contaminated by broken rivets, sealant compounds adrift, and manufacturing materials like rags and tools undetected during the manufacturing process that can later become lodged in check valves, foul filters, or are ingested in an operating engine, Manufacturing personnel should be trained in their tasks and ways found to motivate them to want to do good work so as to reduce contamination events. Quality inspections can be performed at the right manufacturing stages to look for contaminating materials. The design can recognize that contamination may occur and include features that will prevent the contamination from introducing a fuel flow blockage. Tools used in the manufacturing process can be placed on a board with tool outlines indicating where the tools should be located if not in use. If a job is finished and there is a tool missing from the board it is possible that the tool is in the product. This might also be a good idea for hospitals to make sure all of the instruments are outside of the patient after the operation.

29.6   Application to software

Parts standards have been effectively used on development programs for many years. These ideas have not penetrated the software world pervasively, however. Some software developers worry about a proliferation of solutions to common problems within different parts of a software product in development and claim that earlier methods of problem space modeling like flow charts and modern structured analysis, both based on a functional view of the problem space, expose the program to the potential for different teams to separately invent different solutions to similar problems. This concern was, of course, precisely the basis for the evolution of interest in PMP standards in engineering. Parts need not be restricted to hardware parts in this process. Adding software parts to these standards will also enable the reusability of software entities that software people are very supportive of.

Such a solution may also relieve software developers of an excessive, and the author believes misplaced, concern over the application of top-down development using dynamic modeling (and particularly functionally based dynamic modeling) as the lead analytical approach over static modeling. All of the early object-oriented perspectives encouraged that the analyst first discover the objects, the static entities of which the software would be composed and then analyze these objects internally for their dynamic behavior using behavioral and functional modeling. This is exactly opposite to the approach that system engineers have applied for decades following the form follows function notion of the architect Louis Sullivan.

The development of unified modeling language (UML) has corrected that problem encouraging the analyst to first model the problem space using use cases that expose the needed system functionality from an inside–outside perspective revealing benefits that actors (outside entities) will receive from the system followed by a dynamic analysis of those use cases applying some combination of activity diagrams (much like the flow charts of old), sequence diagrams, state charts, and communication diagrams and in the process one exposes lower-tier entities that will require dynamic analysis. This is precisely the sequence engaged in with functional analysis identifying entities through allocation of functionality to them—form follows function.

On the hardware side, it is entirely possible that a team assigned the responsibility for an entity identified through functional analysis and possessing a specification filled with requirements derived through an application of traditional structured analysis may decide to synthesize the requirements in the form of a particular hardware design solution. Simultaneously, another team in the same or another company associated with the first company in the development may have a similar problem and solve it differently. The ultimate customer would generally benefit from an application of common solutions to identical problems throughout the system they will have to live with for years. So, to the extent that system integration and optimization can be effectively applied across the teams, no matter the different company situation, the better for the customer. Adding software parts to the program parts, materials, and process standards can have a beneficial effect. Eventually, perhaps this can be added to the software modeling tools as it has been added by some to CAD preventing the use of particular parts called out that are not on the standard list.

In the past, hardware dominated system engineers have been unequipped to investigate the software side of a hardware–software divide much less the two sides of a software–software interface to discover integration and optimization opportunities. It is precisely this problem that causes the author great anticipation in the evolution of a combined modeling capability where system engineers can master UML and system modeling language (SysML) and observe the intentions of the teams on two sides of a team boundary situation. Part of the benefit to emergence of a coordinated pair of modeling approaches is the general use of a similar development sequence. During the time when early OOA was popular with software engineers believing in identifying static entities first and system engineers implementing Sullivan's idea of form follows function, there was good reason to expect significant difficulty in either side understanding the actions of the other.

Will this solution be welcomed by software engineers? There will be no cheering from them if the software engineers have anything in common with their hardware brethren. Many engineers prefer full freedom to express themselves and dislike any successful effort to constrain their prerogatives. However, enterprises are in business to solve problems for customers and run programs to do so. All of us populating those programs have to realize that we cannot possibly understand the complete problem presented to a program nor devise single handedly a solution even to parts of the whole. We all have to cooperate in ways that individually we may not find fulfilling but the effect is to solve a problem that only a collection of people can attack successfully thus advancing the work of the whole team.

chapter thirty

Other specialty disciplines

30.1　How many can there be?

We have already named and described several of the specialty engineering disciplines. We will name a few more in this chapter. It is likely that some have been omitted. But how far can this go? The reality is that as man's knowledge base continues to expand, and it will continue to do so, design engineers will be forced to specialize more finely to stay up with the latest technology and knowledge base in their profession that will continue to reduce in breadth while increasing in depth. Some engineering disciplines will shred out more than they have already. New specialty disciplines will form to relieve the design engineers and perhaps other specialty disciplines of some of their burden. Another reality is that the task of the system engineer will continue to change as well with two or more layers being formed recognizing people sometimes called architects and others a little more focused on a particular domain like software or telecommunications.

30.2　Survivability and vulnerability

This may include nuclear, biological, and chemical analyses as a function of the threats posed by a hostile force. The threats may be covered directly in the system specification or in a classified document referenced in the specification. The effects of these agents is defined for the benefit of design teams and design alternatives reviewed for compliance with recognized effective solutions to the problems posed by the agents.

　　The threats to the product may not be entirely the result of hostile actions. For example, a nuclear armed cruise missile en route to the target may have to fly in close proximity to other missile detonations causing intense electromagnetic effects in the structure. One of the most complex problems the author ever experienced in development entailed exactly this problem. The missile was overweight and everyone on the team understood weight and how to eliminate it. The airframe design engineer for a major part of the fuselage could not get the nuclear survivability and vulnerability engineer to approve his design after trimming considerable weight out of his design. In a meeting called to understand the problem, the nuclear survivability and vulnerability engineer explained in broken English that a near burst could cause such currents to flow in that part of the airframe as to result in structural problems under maneuver. Convinced that mission deconfliction could not alone solve the problem of near burst, the program had to accept an overweight condition. This was a case of specialty engineering conflict where everyone understood the one discipline but no one understood the other. The solution to a weight problem is to have less mass that was exactly the wrong path for the other discipline. We tend to manage what we understand better than what we do not.

30.3   Electromagnetic compatibility

On systems that include sensitive electronic circuits that could be upset by strong electromagnetic fields or that are capable of generating such fields that could interfere with other systems, the system should be studied for sources of interference and any that are identified should be corrected to within the levels prescribed by law, program requirements, good engineering judgment, or recognized standards. System susceptibility to outside electromagnetic interference should be studied as well. Engineers skilled in this field can spot problems from a study of bonding, shielding, and grounding design features. These system stresses are examples of a noncooperative environmental influence.

Electronic warfare signals intended to deceive, mask, overload, or otherwise interfere with a system may appear to be examples of electromagnetic incompatibility, but these kinds of signals are generated and directed at the system with hostile intent and should be treated as hostile stresses.

30.4   Radio frequency spectrum management

Systems that include radio frequency emitters, such as radio or radar transmitters, must have their frequency assignments coordinated with available spectrum assignments by one or more controlling organizations. If the system being developed includes no emitters in that it is a receive-only system, the frequencies may not have to be cleared by a controlling agency but the frequencies used will have to be coordinated with the emitters intended to be received.

In any system using the radio spectrum, we should seek to minimize dependency on radio signals and minimize the bandwidth and number of frequencies used consistent with other system factors like cost and operability. This is a very difficult order in systems that must interact via radio to complete communication pathways and control loops that weapons systems depend on.

30.5   Electrostatic discharge

Systems operating in the atmosphere are susceptible to a build up of electrostatic charge that, if allowed to reach a high potential relative to the charge in the surrounding atmosphere, can have a detrimental effect on sensitive on-board electrical equipment. This work may be extended to include lightning effects. Efforts are made to bleed off any charge on the structure by providing low impedance paths from the surface to the surrounding medium generally in the form of sharp points at trailing edges of thin structures like wings. These static dischargers must have a low resistance path from the metallic fuselage. A discharger base may have to be bonded to the structure using a conductive adhesive, for example, and conductivity tested during manufacturing processes.

Unless protected in some way from human contact, these sharp points can be a safety hazard leading to potential conflicts between the needs of two specialty camps. Static dischargers exist that are effective and have no exposed sharp points but can still be a hazard to physical eye injury.

30.6   Producibility

It is possible to design a product such that it is either very easy to manufacture (or produce) or it is very difficult to produce. Producibility seeks to bring about the former condition by close and cooperative work on the part of manufacturing engineers with the design engineers to encourage the use of processes already clearly characterized, simple design features, and materials for which the organization's manufacturing capability has mastered the machining and working.

30.7   Operability

Operability, in concert with human engineering, seeks to optimize the ease of operation of the system and the operational effectiveness of the system. A process diagram can be used as a basis for evaluating the steps needed to operate the system where the operational technique is fairly simple. Human activities needed to operate the system in each of these steps are studied and operability features offered to the development team. The principal issue for operability is the man–machine interface for the principal operator, such as the aircraft pilot, ship's captain, or train engineer, but the interest can range to mission planning and determination of mission settings. In more complex situations flow diagrams with swim lanes marked out for the different entities involved in the control pattern can be very useful. Some people refer to these as operator sequence diagrams.

30.8   Design to cost and life cycle cost

Design to cost (DTC) is a technique to encourage cost-conscious behavior in the development team toward the end that the product development cost targets are met. A system development cost number is first identified and this cost is then allocated down through the hierarchy based on anticipated development difficulty. Margins may be assigned to introduce slack cost figures that can be reallocated as problems develop in design. The DTC allocations should be managed by PIT but the responsible IPPTs should be accountable for satisfying them.

Life cycle cost (LCC) combines DTC with cost figures representing recurring manufacturing, operations, maintenance, and disposal costs. It seeks to define the true total system cost over its entire life cycle. This is commonly used as a principal decision-making figure of merit in early program phases.

It is interesting that MIL-STD-490A specifically rejected the inclusion of a cost requirement in a specification because it was said to be a contractual matter. As the military moved toward cost as an independent variable and real cost consciousness following the loss of the Soviet Union as a hostile partner, affordability was recognized as a programmatic consideration. MIL-STD-961E into which the function of MIL-STD-490A was folded does not include cost as a paragraph title but at least it does not prohibit a cost requirement.

30.9   Value engineering

Value engineering is a structured method for finding ways to improve a product after it has entered a production status, though some people apply it during the development process. Ideally, it involves the employment of cross-functional teams in the performance of trade studies focused on selection of the most cost-effective solutions to production problems. This process may be married to a preplanned product improvement program for the purpose of determining the precise way that the preplanned improvements will be implemented. The value engineer intent is to increase customer value either by decreasing product unit cost or increasing customer benefits from the product.

30.10   Operations research or analysis

The system development effort often begins through analyses conducted formally or informally by users. These analyses may not be heavily mathematical or be based on scientific principles, but they produce useful qualitative insights into user needs. Often the content of user documentation falls into this pattern. Thus the requirements they include in early documentation may reflect the users needs, but they may not be quantified because those preparing the document were not skilled in applying some useful mathematical tools. Many of these tools evolved from the operations research or analysis field. Some would argue that the whole system engineering profession evolved from operations research in companies like Thompson Ramo Wooldridge (TRW).

This book does not cover this rich field of work but a system engineer would be wise to develop competency in its methods. The techniques of operations analysis (OA) are especially helpful in early program work to understand quantitatively the relationships between key parameters of interest and determine optimum selections for these variables. The general approach in OA problems is to first formulate the problem. This is seldom simple because, as noted above, the customer commonly will not have a lot of knowledge about the problem they seek to solve. Secondly, the analyst will construct a mathematical model of the system under study. The OA field possesses several modeling templates that a person untrained in the field will commonly have difficulty seeing the relationship between the problem to be solved and any of these templates. But once an operations analyst makes the connection we would all say, "Well, of course." Thus, for a system engineer who wants to improve their understanding of this field might speed up their rise to competency by first reading about these techniques and then by finding an older engineer probably employed in a pre-design group to help them understand how to apply these techniques on a few specific problems they are likely to encounter in that company.

The third step is to derive a solution to the problem using the mathematical model. The common view of this construct is a box with inputs, outputs, and something going on in the box that can be explained by a collection of equations. These equations may be simple arithmetic, or algebraic, or follow the rules of the calculus, or differential equations involving discrete or analog real or imaginary values.

The reader will perhaps recall that in Chapter eighteen we recommended that once a trade study solution had been derived the decision matrix should be tested for sensitivity based on the logic that at the time we accomplish the trade study we are not always fully informed about the reality of the problem we face and we may not have had the time and money to fully examine the problem to the depth we might have preferred. So it is in OA problems as well. For these same reasons one would test the solution derived rather than simply accepting the first run answer. It is interesting, for example, to observe whether or not the model applied will predict the effects of alternative courses of action. In evaluating the model and its performance, it is useful to have two people in a conversation with the analyst. One of these persons would ideally be someone who is at least as well qualified as the analyst but someone who was not involved in the analysis. The other person should be unfamiliar with the

OA field but skilled in asking dumb questions. This second person could be a system engineer or program manager, for example. If the model and its working can survive their questions it has a good chance of providing good service. To the extent that the model applied produced a good answer for the specific problem applied and this same kind of problem occurs frequently, there may be an advantage in further developing the model into a template that can be applied more easily by any member of a team.

The techniques operations analysts apply include linear programming, network analysis including program evaluation and review technique (PERT) and critical path method (CPM) used in program planning by many enterprises, dynamic programming, integer programming, nonlinear programming, Markov chains, game theory, and queuing theory. The mathematical programming techniques apply systems of equations in a stylized fashion to encourage identification and selection of an optimum result. The operations analysts have been so successful in applying some of these techniques that many of them have leaked out from their profession and are in common use by others. Some would argue that reliability began life as an OA technique. It now boasts a functional department with that name. LCC also followed that pattern. PERT and CPM have been claimed by schedulers. Simulation has become a routine part of any new development effort. This effect has in many organizations resulted in the demise of an OA group to the loss of organizations that have consciously or otherwise allowed that to happen. In these cases, the functional system engineering department would do well to encourage some of its members to master OA techniques that have not been spun off into other departments.

30.11　Other specialty engineering disciplines

We can continue almost indefinitely with our list of specialty disciplines, but in the interest of preserving our forests, let us try to move on to other pursuits. You should understand by this point that we have specialized with very fine granularity in the engineering community for reasons covered earlier related to individual human knowledge capacity limitations and the continuing expansion of our knowledge base. The challenge for the system engineering community is to provide programs with effective methods within an environment conducive for integration of the work of these many specialists toward program goals.

The reality is that the people doing their work within these specialty disciplines appear to the design engineers to be very negative thinkers finding ways that the design engineers' perfect design can fail and kill people. As a result, if the team does not consciously deal with this potential, the specialty engineers may be treated with such hostility that they become ineffective. Team leaders, system engineers, and program managers must take action to prevent hostility from ever developing, ideally, but certainly cutting it off if it does appear. This can easily be done through team conversations that have the effect of making it clear that these attitudes are common but unfounded and if permitted will seriously damage the potential for team success. To the extent that team loyalty can replace individual ego drive the team will be successful.

chapter thirty-one

System analysis disciplines

31.1   Specialty engineering differences

The system analysis disciplines differ from those grouped into a specialty engineering collection by the author in several ways. First, they tend to have been in existence for a longer period of time and have become more fully accepted by the design engineering community than the specialty disciplines. They tend to employ engineering mathematics as in differential equations and the calculus rather than probabilities, which is not all that easy to master, but nonmathematicians can be trained to perform the needed manipulations fairly easily. Specialty engineering disciplines have to appeal to qualitative and subjective judgments often rather than deductive logical and mathematically quantified ones. The key point is that design engineers tend to accept people doing the work of system analysis as engineers but commonly do not have the same attitude toward those doing specialty engineering work. Thus, the latter may have more trouble contributing to concurrent engineering activities and may require team leadership and system engineering support more frequently than system analysts.

It should be understood that each of these disciplines is a mature domain with a large literary basis available to us all. In this chapter, we will only give a very light overview of what in some cases is a tremendously difficult undertaking on a program. The good news is that people doing this kind of work will respond like any one else when asked to explain something about their work. They will bend your ear as long as you stay in one place.

31.2   Mass properties

The mass properties engineer is responsible for ensuring that the design falls within weight and center of gravity (CG) constraints established by them for the product. The principal method involves allocation of available weight to system elements and monitoring the design process to see that responsible teams and designers remain true to their allocations. An indentured weights table is established that lists all of the system elements and their weights with subtotals and grand total. Weight margins may be established to protect the project from weight growth problems and provide for the management of difficult weight issues as the design matures.

The mass properties engineer must also compute the CG of elements where this is a critical parameter. In maintenance situations, this data may be required not only for a whole end item (hoisting and lifting, for example) but for various conditions where the item is incomplete as in assembly, disassembly, and maintenance activities.

31.3 Space engineering and packaging

Mass properties people may also be responsible for space allocation for the many things that must be installed within the physical confines of the system end items. Unfortunately, space is not always handled in a systematic way in the development of systems resulting in two or more things sharing a common space or inefficient use of available space and environmental control resources. Some units in a system end item may have to be controlled in terms of their form factor so as to fit into an available space.

An engineer at a military aircraft builder shared a story with the author about the development of an aircraft where the contractor had been badgered for quite some time by the customer to implement effective system engineering practices. The company finally implemented integrated teams that seemed to work well for a period of time. After awhile, there appeared symptoms that the teams were loosing their effectiveness. In one case, they found that a black box was overheating for no apparent reason. Further investigation led to the knowledge that some holes in the structure had become filled with wiring bundles and plumbing in a series of modifications and those holes had originally been sized to convey cooling air to the now overheating black box. In that case, design engineers had become leaders of the integrated teams (not an uncommon or necessarily bad happening) and had permitted the specialty engineers to be so badly treated that many of the specialty engineers had stopped going to the meetings. Here was a case of space utilization run amuck in what was intended to be an effective concurrent engineering environment.

Space can be treated exactly like weight or any other allocable commodity. One of the most tedious tasks in an evolving design job is to make sure that two objects do not occupy the same space. This is an area where computer-aided design (CAD) specialty engineering support can be fairly easily implemented. Three-dimensional CAD can be set up to detect interference between parts or to ensure minimum clearances are respected. One should not stop with static clearances but include dynamic clearances. As acceleration, vibration, and shock forces are applied to or by the product, items may move relatively whereas they appear to have adequate static clearance. This is especially true, of course, where one or more elements are shock isolated.

31.4 Guidance analysis

If the system includes an element that must move from one point to another with a degree of path precision, engineers in this discipline provide requirements that encourage the needed accuracy and evaluate design features to assure that those requirements will be satisfied. Guidance error budgets may be assigned to the guidance set, installation and alignment accuracy, and any separate sensing instruments involved.

31.5 Structural dynamics and stress analysis

This discipline determines the needed strength of structures under static and dynamics conditions under all system conditions. This should include manufacturing in terms of the structural tooling points where the structure will be physically secured during structure buildup. It should include any shipping or transport cases as well as normal operations. Computer tools are used to model the structure and support structural design personnel in selection of materials and design concepts.

At one time this work was done with slide rules and mechanical calculators. It would take many weeks to complete an analysis during which the design would proceed only to find part way through the design that it was inadequate. Today, computer software has radically sped up the analysis process and it can be accomplished based on CAD inputs when using the right software applications.

31.6 Aerodynamics

If the system involves movement of an element through the atmosphere at speed, it may require an aerodynamics analysis and/or wind tunnel testing to assure that its shape minimizes drag and offers adequate lift to balance weight under all conditions of flight. Similar considerations may be necessary where the fluid medium is water instead of air.

While the most exposed work of aerodynamicists entails airframe shapes, very specialized and difficult tasks exist in engine intake design, external stores separation, and ram air power pathways. The wind tunnel is the principal laboratory for the aerodynamicist. Instrumented models and full-sized shapes can be used as a function of the vehicle and available tunnel sizes. These models must be built early in the design effort and someone must be careful to ensure that the shapes remain faithful to the design that can change over time.

31.7   Thermodynamics and thermal analysis

Heat sources and sinks are identified and the resultant temperature of items in time are determined. These engineers are involved in positioning and mounting of items for thermal control and elements involved in altering the environment within which items are located.

A very common environmental problem in systems is thermal in nature and these problems are often solved through design of an active or passive environmental control system. This may entail simply routing external cooling air over the items wanted cooled or entail blowers and active refrigeration equipment. In other cases the problem may be that the item is not warm enough, in which case heaters are called for. Heat exchangers are not uncommon in thermal control systems where heat in one fluid medium is transferred to another fluid. In one case the author can recall the guidance set was cooled by pumping fuel through it that also had a beneficial effect on the fuel. But, who in their right mind would first offer this solution to a program team—to pump flammable fluids through an electrical box?

In today's electronics intense systems, cooling the black boxes becomes a major problem. This flows down all the way to printed circuit boards and integrated circuits. Those familiar with the insides of their personal computers will know that the processor chip in their computer very likely has a fan assembled right onto it. Circuit board design is another area where CAD can be programmed to support the designer directly. CAD used for circuit board design commonly includes parts data that can include thermal characteristics. When the designer calls these parts onto the circuit board design, the dimensional and thermal relationships can be analyzed by the computer yielding a go-no go conclusion. This is a tremendous improvement over human engineers performing this tedious work freeing them for the thinking work that humans do so much better than computers.

31.8   Anything missing?

With all of the disciplines described in this part of the book, can there be any problem that could possibly fail to register on some specialty engineer on a fully staffed program? The author while working as the team leader for a contractor team operating and maintaining an AQM-34M(L) unmanned photo reconnaissance aircraft during the Vietnam War ran across numerous design and manufacturing problems in the field that, in retrospect, he could not trace to a failure of any particular specialty engineering discipline. A case in point occurred when it was discovered that low altitude photo reconnaissance birds returning from flight over North Vietnam would often fuel out short of the recovery site at Nakhon

Phanom, Thailand on the Laos side of the Mekong river when they were flown with no external fuel tanks but had plenty of fuel when the flight called for external tanks.

Now you might first say, "Well of course, dummy, you have less fuel without external tanks." But external tanks were used or not based on the planned mission range so this was not a factor. This meant that the recovery helicopter with an armed party had to go across the river and retrieve the bird and its on-board camera film when the bird fueled out short. This team would most often find that the local population was attacking the bird with machetes to pull off anything of potential value in the local economy that was just about anything.

The fact that the birds flew as planned when flying with external tanks confused the analysis for some time. Finally, the author and a team mechanic, Gordon Allen, inspected the fuel tank one evening on a bird returned from flight that day having come down the other side of the river. They discovered that there were no vent holes in the top of one of the fuselage bulkheads. This fuselage station was where Ryan Aeronautical had modified its BQM-34A (Ryan model 124) target drone design to satisfy an Air Force requirement for an unmanned reconnaissance aircraft with added range. A 30 in. tank section had been added to the fuselage at this point. The added fuselage bulkhead stiffener did have the needed fuel vent holes, but the original design of what was the model 124 target drone fuel tank aft bulkhead was not retrofitted for reconnaissance vehicle use to provide vent holes. So when the model 124 design was added to the new Ryan model 147 added tank section, the result was no vent hole between the original target drone tank section and the added section. These birds were intended to be air launched full of fuel from a DC-130 launch aircraft that used a captive refueling capability. The fuel input was from a DC-130 pylon-mounted quick disconnect mounted up forward in the original drone tank and the vent was located in the added model 147 fuselage extension. The vent action was interrupted by the failure to put vent holes in the original model 124 bulkhead. This design had been flying for 8 years at the time and this problem had gone undetected throughout that period.

This explained why clean winged birds sometimes fueled out short but did not explain why birds with external fuel flew as planned. The team scrutinized the whole fueling process at its base in Utaphao, Thailand and throughout the captive mission and discovered that often when they fueled the external tanks, the tanks were installed on uploaded drones thus were in a nose down attitude rather than level as required in technical data. The technical data was built on the assumption that the external fuel tanks would be fueled with the bird setting in an upload trailer that had leveling capability and that the tank would be in a level attitude. The tank was to be filled until fuel was up to the aft fill port lip.

In the interest of time, the Ryan team and Air Force troops would fuel the external tanks on the flight line with the bird uploaded on the DC-130 when the refuel truck would come around for the DC-130. This meant that the external tanks were being overfilled, which compensated for the underfilling of the main tank on external fuel tank missions. The birds were simply fueled from the mother aircraft through the captive flight refueling capability and the Ryan team built a battery-powered kluge that they used to open the drone fuel dump valve connected to a hose placed in the wing tank fill port.

The wing tank overfilling problem was apparently tolerated by some margin in the system capability allowing for added fuel mass during launch and early flight. Clearly the vehicles were being launched in the wrong fuel loading configuration with fuselage fuel tank underfilled and wing fuel tanks overfilled. Once the tanks were drilled properly for the vent hole and tech data was followed for filling the pods, planned fuel margin proved adequate on all flights.

This was a failure in structural and fuel system design at a time when a customer with a very urgent need wanted delivery of a vehicle. Very likely, on the initial flight test birds, which were built up by a team of very experienced technicians, mechanics, and engineers, these main tank vent holes were bored but the engineering was redlined and not completed in the formal engineering documentation to retrofit the original bulkhead design and manufacturing apparently did not pick up on the problem. This is only one small example of such a problem. It is the kind of problem that a system engineer should spot in the process of thinking through system interface relationships in coordination with operational mission employment. There is simply no one else who covers this kind of complexity.

This kind of problem can also fall into a different pattern. The initial manufacturing crew may understand the problem and accomplish needed tasks even though they are not clearly identified in the engineering or manufacturing instructions because they were there when the problem was exposed in early development. Over time the original manufacturing people may retire or take new assignments leaving people who actually follow a flawed plan. The user begins to see an unexplained problem in operational use of these products manufactured after the personnel changeover.

part six

Concurrent and post-design process synthesis

chapter thirty-two

Procurement and material integration

32.1 Customer choices

A customer has the first choice of suppliers, of course, in that it must pick its contractors to deliver the system. They could conceivably select one contractor to deliver the compete system. If the development of a new unprecedented system offers significant alternatives for which different contractors have exhibited potential for success in these different approaches, the customer could conceivably procure systems from two or more contractors and require them to compete their systems in some kind of operational evaluation with the winner to receive the contract. If the system is quite complex, multiple contracts may be awarded for major elements of the system often called segments. In the latter case, it will be necessary for several contractors to cooperate in accordance with special associate contracting clauses in their contracts to jointly develop the interfaces between their segments. In this chapter, we are focused on one contractor's program for whatever portion of the system being developed that they have received a contract for, and its dealings with its suppliers.

32.2 Procurement as an early team member

This book has outlined a three-step process for system development with the understanding that all three steps must take place within an infrastructure of good management. Step two of that process, synthesis, is composed of three steps itself. Step two in the synthesis process is to acquire the material and finished goods required to manufacture the deliverable components and end items defined in the engineering drawings and contract. However, this process should have started much earlier than the release of the drawings. During the requirements analysis work, procurement should be represented on the team that makes the product entity structuring decisions. The fundamental decision is whether to make or buy a given item, but if it is to be purchased, it is necessary to gain confidence that acceptable suppliers exit. A procurement person will have that information or can research available suppliers quickly.

Procurement people will also be aware of the product scope that various suppliers will be capable of providing. For example, would the program be better off purchasing the components of a radio system (receiver, transmitter, power supply, modulator, antenna, and coax) from different suppliers or the whole integrated system from one supplier? These decisions also have to be coordinated with enterprise management so as to respect an intended retention of enterprise technical skills base. The higher in the food chain the enterprise decides to make its purchases, the higher will be the product entity structure boundary below which it will be difficult to retain a lower-tier technical skills base.

32.3   Purchasing modes

As the product entity structure matures, identifying specific pieces of equipment and software that are selected as purchased items, we have to make a decision on exactly how the program will acquire them. The first of two basic modes entail contracting with a supplier to develop the product so as to be compliant with the content of a procurement specification that we publish. Alternatively, we could complete the specification and accomplish the design work as well and elect to purchase items built to our engineering. A third alternative is to purchase items off the shelf from suppliers that sell the items to many customers.

32.3.1   Buy to specification

If the buying organization does not possess the staff that has mastered the knowledge or technologies needed to execute a good design to satisfy the functionality associated with an item or it does not have the numbers of people needed due to other commitments during the period when the design must be accomplished, the item may be a good candidate for purchase in response to a specification prepared by the buyer. There are a few fields where it is fairly common place for the seller to prepare the specification (guidance sets and engines, for example) but it is a good business practice in general for the buyer to prepare the specification.

It is possible that a buyer could contract for the design but elect to manufacture the item in house. This is an odd combination that could conceivably be brought about by a buyer that has the capacity and experience to accomplish the manufacturing job but is swamped at a critical time with other engineering demands precluding the design job. In the more normal case where one company is going to complete both the design and manufacturing work, one or more companies are asked to bid on the design and manufacturing job. A selection is accomplished by the buyer, where a single supplier is not the case, to select the company that will accomplish the work. A contract will include the procurement specification and a means to control the work to be accomplished by the supplier. The latter may include design reviews, periodic reports, visits, and financial reports of progress developed on an earned value computer system.

It often happens that the buyer has some difficulty preparing a good procurement specification in a timely way to support the procurement process. Part of the problem is that a procurement specification commonly includes both the development (performance or part 1) and product (detail or part 2) content and the latter requires some knowledge of the design solution. A good way to relieve the program of this problem is to let a contract in two steps. The first step would involve the design of the product that could include qualification verification through the conduct of some kind of functional configuration audit. The second part would involve manufacture and acceptance of the product. The buyer would prepare the procurement specification part 1 prior to the contract award and the supplier could be tasked with preparing the part 2 during the tail end of the first portion of the contract after the design is understood. Either the buyer or the supplier could prepare the part 2 specification, but since it is intended to drive the development of the acceptance test plan that the supplier will accomplish, it probably would be better for the supplier to do it.

If the article that is being contracted for is a long lead item and fraught with technology concerns, the procurement process becomes more critical. In one program that the author can recall, the customer's system specification required more accurate guidance than had been commonly called for on unmanned vehicles. This is the contract that motivated Figure 16.1 titled the advancing wave. The program had to develop the procurement specification for the guidance set before the requirements for the complete avionics system had been clearly understood forcing the team to plunge to a lower tier before they were ready at the immediate parent level. The problem was more serious because of those bidding only one looked to have a chance of meeting the requirements, but it was risky while another had a concept that might not satisfy the customer requirements but at least they could provide a guidance set on schedule to satisfy early flight test needs where guidance accuracy was not critical. As a result, the program elected to trade a performance and schedule risk for a cost risk and let two guidance set contracts. The hope was that a guidance set from one or the other would become available in time to support long-range guidance accuracy flight tests. The author at the time already had the mental image of guidance set program managers gifted with the grin of the Cheshire cat when told that there were changes necessary in their product and this scenario was guaranteed to produce that result.

In general, the engineering drawings for the item will be the property of the supplier unless the contract has included the cost of ownership of the engineering package. The buyer must weigh the risks of possible non-availability of this product in the event their customer executes an option for 100 more units that include the supplier item. If there is a danger that the supplier may not respond favorably to such an option, purchase of the drawing package might be warranted. An option is to put the engineering and manufacturing package in escrow such that if the supplier refuses to enter into a contract for additional items in some time frame, the rights flow to the buyer. The price for this clause in a contract is commonly less costly than an outright purchase of the rights to the package.

The author believes in this case it is better to include qualification in the procurement package, but some companies prefer to accomplish the qualification work themselves. In any case, the procurement specification content should be the basis for the qualification tasks. To make the buyer responsible for the qualification process makes the process exceedingly difficult to manage with too many cross currents between the engineering staffs in the two companies. Better to focus the buyer's energy on management of the supplier accomplishing the qualification work including analysis and testing. This would also, in the author's opinion, include building the qualification test articles because it gives the buyer organization the knowledge it can use profitably in perfecting its manufacturing, acceptance, and quality assurance plans from the experience of building the qualification test article(s) in an engineering laboratory.

32.3.2   *Buy to drawings*

When buying to a drawing package, the supplier must develop a manufacturing, acceptance, and quality assurance plan that will result in delivery of a satisfactory product. The author believes that the qualification work should be accomplished by the owner of the engineering, which in this case is the buyer. However, the supplier should be required to participate or at least observe the building of the qualification unit(s) in order to encourage that the qualification item is built and quality assurance accomplished in a fashion as close to the final manufacturing process as possible. It is important that the supplier be able to say truthfully at the functional configuration audit that the buyer should feel confident that the qualification units built in an engineering shop would produce the same performance as those passing through the final manufacturing process, especially if that process is under the control of a supplier.

In this case, the formal configuration management baseline under the control of the seller in accordance with rules stipulated in the contract will include not only a specification but the drawing package as well and the contract must deal with the problem of supplier-defined engineering changes that might reduce manufacturing cost or risk.

32.3.3 Buy off the shelf

The U.S. Department of Defense (DoD) has popularized the acronym COTS meaning commercial off the shelf, but regardless of the nature of the product line being of a military or commercial nature, the developer has to deal with the special problems of off-the-shelf purchase. The understanding commonly is that the supplier will not make significant changes in their profitable product to satisfy all of the needs of the buyer. COTS is a subclass of what DoD refers to as nondevelopment items (NDI) and for the past several years they have shown a preference for these kinds of items to the maximum extent possible in the systems they take delivery of.

Many years ago, while employed at Teledyne Ryan Aeronautical long before the COTS acronym was invented, the author recalls that Techtronics, a great test equipment supplier, would tell Ryan procurement and engineering people who tried to push a procurement specification on them to simply back their truck up to the loading dock and the company could buy as many o'scopes as desired. In some cases, Ryan simply added its own cover sheet to a supplier's specification qualifying for what was called at the time under MIL-STD-490A, a type C (commercial) specification.

When you buy off the shelf you will have limited control over the product received. The supplier can change the configuration based on his internal engineering work to improve the product, availability of parts and materials, and the desires of large purchasers of the product like Wal-Mart, for example. Wal-Mart might be selling 300,000 a year across the country and that dwarfs the 25 you need on your $3,000,000,000 program. But your program price tag will not impress the supplier to the extent that the 25 articles does. Your lack of control over the product can, of course, play havoc with your responsibilities for logistics support for your total product. You may be able to purchase some kind of maintenance agreement for the item and it may have to be spared at the level you purchase it with you acting as the depot with returned items funneled through the agent selected by your supplier.

As noted in Chapter twenty-one, the use of COTS can be worked into a spiral development scenario for a military customer quite well. The spiral sequence offers much to the development of systems where the customer can be presented with a capability that reflects what they are hoping to acquire very early. This capability may not necessarily be fully ready for use in a combat environment so long as it functions the way the final system will. During the earliest program spirals, COTS items can be depended upon to function well so long as they are not subjected to environmental stress in the extreme. As the program moves through early spirals, it may become clear that the design of the COTS item is actually adequate for the application and it is low enough in cost that it can be simply replaced from time to time to satisfy the logistics support requirements. Another alternative is that the customer, impressed with the performance of the COTS item in the system, may be willing grant a waiver for the environmental requirements that it cannot satisfy. A third alternative is that you can create an altered item that modifies the COTS item to sustain the environmental stresses. If none of these routes is satisfactory on some later program spiral, the COTS items may simply have to be replaced with a new development item but the COTS item will have served a useful purpose to have provided a realistic capability very early in the development effort and have clearly defined the functionality needed by the item whatever its source.

32.4   Procurement cycle need for engineering support

32.4.1   Contract help

Procurement and data management will need help from the program office and the responsible engineering team or principal engineer on that team to help define the documentation that the buyer will require that the supplier provide during the development period. At the prime contract level between the buyer and their customer this data is often referred to as being on a contract data requirements list (CDRL). At the supplier level it is referred to by many as the supplier data requirements list (SDRL) with customer replaced by the word supplier. The intent is the same in both cases. The buyer wishes to review and possibly approve particular documentation that controls the development of the product that they wish to take deliver of. In the engineering area, this will most often include the following:

Specification (if the supplier develops it)	Reliability and maintainability report
Some key engineering drawings	Safety hazard analysis report
Preliminary design review package	Security hazard analysis report
Long lead list	Qualification verification plans and procedures
Critical design review package	Qualification verification reports

Engineers making the request for this data must ensure that there is enough cost and schedule coverage in their task estimate to permit them to review this data when it starts flowing in and they are very busy with their own work. Procurement and the program office are depending on engineering people being a part of the review of supplier data because they are the only ones who can understand the technical problems they may be facing.

It is a good idea to include on the SDRL a data accession list that will inform the buyer of all of the program documentation created that is not on the SDRL and provide a means for the buyer to obtain copies of any document listed there. This is commonly implemented by charging a fee per page provided. It does also involve some administrative costs for the supplier to maintain and provide the list periodically that will be an added expense.

32.4.2   Source selection and contract award

It turns out that the trade study approach covered in Chapter eighteen is very effective in supporting a source selection process in procurement. It may be necessary for a system engineer on the PIT to provide a quick 2 h briefing on trade studies and provide a template for performing them. That engineer could also review some of the first ones created to ensure that the selection boards have understood the intent. These boards will especially need help in understanding the concept of a value system through which the selection process is accomplished.

32.4.3   Monitoring the supplier

Data management people on the program will require the help of engineering people as the flood of documents start flowing in from the suppliers. Ideally, data management would have engineers identified as the principals for each data item and these would be routed to the principals via internal e-mail with a suspense date for when data management needs a response. It is very common for engineers to try to escape this responsibility because they are too busy. They should have thought of that when they were responding to data management request for identification of documentation required from the suppliers. The engineers simply must respond to these requests from data management in a timely way. The whole program is depending on them to do so.

Engineers may be recruited to attend design reviews at the suppliers so as to advise the program office attendee of their conclusions of supplier technical progress and credibility of risk reports presented. Ideally, each purchased item would have a technical person identified in the buyer program who would be carrying on a conversation from time to time during the period leading up to a major design review with reports to the responsible team leader or program manager, as appropriate where a concern developed.

In the case where the supplier has developed a design in response to a buyer-provided specification and the buyer is to accomplish the qualification work, engineers will be required to review and approve qualification verification plans and procedures. If there are very critical tests to be performed that could influence overall program schedule and cost, especially if the test item is destroyed in the process of accomplishing the test, procurement would be wise to have an engineer witness the test and especially the preparations for it before the button is pressed. Engineers should also participate on functional configuration audit teams to review the qualification reports to determine if the supplier's design satisfied the requirements in the procurement specification.

32.5   What is so different about suppliers?

Whether an item is to be developed by a supplier or an in-house team, a program manager would be wise to treat all of these suppliers as teams under contract. The outside suppliers will all have contracts through which they can be managed and controlled. The internal teams can be held to the content of the IMP/IMS and the specification for the item for which they are responsible. But there may be some elements missing in the internal "contract" that could be covered in the program plan in the form of some expectations the program manager has for all in-house teams. A cursory review of the language used in vendor contracts might lead to identification of the right language to be used and a contracts person could be helpful in adjusting this language to the in-house situation.

chapter thirty-three

Manufacturing process design and integration

33.1   Process overview

In Chapter thirty-two, we dealt with those parts of the system that were going to be purchased from other organizations. For all other entities in the system, we will have to manufacture them. Even if we purchase everything in the system, it will have to be assembled. Many suppliers of large and complex systems have found that they are very inefficient manufactures of lower-tier entities, in fact have difficulty competing with outside suppliers, and they seek out suppliers who have specialized in particular kinds of products that they need for final assembly of the end items of their system. Some of these large enterprises have purchased companies that have become divisions that do supply many of the items that the big system divisions require. This is true of aerospace and the auto industries.

During the time that the product requirements analysis work and related system concept work is being accomplished, manufacturing engineering should be funded to participate in that work as well as identify requirements for the manufacturing process that are consistent with the make-buy decisions. During the product preliminary and detail design processes, manufacturing should be interacting with the design teams to encourage recognition of product requirements driven by manufacturing needs as well as evaluating maturing design features for their producibility aspects offering critical input on the appropriate teams in an attempt to bring about an optimum solution from a product and process perspective. Actually, the manufacturing people may not have such a high viewpoint and may only be focused on their anticipated manufacturing difficulties. That is not to denigrate the manufacturing folks because it is what all specialists are doing in the give and take of the system development process.

The system engineers are being paid, however, to approach the development work from a higher plane and should be sensitive to symptoms of suboptimization and work to push the optimization level higher if possible forcing a change of the engineering design and/or the manufacturing process. In order to be able to gain insight into both the engineering and manufacturing requirements and designs, the system engineer should have some understanding of what the manufacturing engineers are doing during the time the product is being defined.

The purpose of this chapter is to provide an overview of that work. The manufacturing planning process can be partitioned into the parts discussed in this chapter: (1) process definition, (2) manufacturing instructions development, (3) kitting material development, (4) tooling, (5) personnel and training, and (6) facilities. A system engineer should, however, take opportunities to become familiar with the manufacturing capabilities of their employer including walking through the facilities and talking to people. It will often not be possible to stop and engage manufacturing people in a discussion of their work, but a manufacturing supervisor can probably either explain what is going on or arrange time for a discussion with workers on a break.

In 2006, the author presented an INCOSE tutorial for the Crossroads of America Chapter covering the Indiana and Illinois area at the Columbus, Indiana engine plant of Cummins. The classroom had one wall entirely of glass that overlooked a manufacturing space on the ground floor below. One of the Cummins employees told him that the space had earlier been the production floor for a particular diesel engine. The author asked about the long row of truck and trailer parking spaces just outside the door to this facility and was told that at the time those engines were being built at this facility (not that long ago at the time), any truck driver could pull up and with very little interference walk onto the manufacturing floor and discuss the engines being built with those putting them together. With the increased interest in security, finding a similar situation even for employees of the same company will not be all that common.

33.1.1 Process definition

As the product entity structure expands and the items that are going to be manufactured or assembled by the program are selected, each of those items will require a manufacturing process defined for it. This includes identification of a finite number of steps accomplished at particular workstations within specific facilities. It is possible that some of these workstations will be in plants that are some distance from the final assembly facility. It is necessary to identify all of the materials that will be required at each workstation in terms of product elements that will become a final part of the product such as (1) raw materials or finished goods, (2) processing materials consumed in the process of doing work such as lubricants and solder, (3) tooling that will be used to hold or act upon product materials, (4) manufacturing aids (maids) that support the production process, (5) possible production line apparatus, (6) the kinds of people who will be needed to do the work, and (7) possible facility improvements involving modification work performed by people in the building trades.

This process could be defined in a simple flow diagram or IDEF-0 diagram. The latter was actually developed to chart manufacturing processes. The original meaning of the acronym was Integrated Computer-Aided Manufacturing Definition language. It provides not only for flow from one step to another but the flow of supporting resources and process controls. A schedule of the steps charted is also required, of course. Alternatively, a PERT or CPM chart could capture process and schedule definition. For each block on the flow diagram we need to define the parts of the process ticked off in the previous paragraph.

The work of planning a manufacturing process has a lot in common with logistics support analysis where engineers have to define the maintenance process and all of the resources that are required in order to accomplish those steps.

33.1.2   Manufacturing instructions development

The work that must be accomplished at each workstation must be written in simple language with clear meaning. Product items and other materials used must be clearly identified and how they are processed from entry into the step through completion. If the work entails the use of any hazardous materials like acids the instructions must provide adequate warnings and cautions to encourage safe completion of the work. In these cases, it may also be necessary to provide special protective clothing and equipment for the manufacturing personnel.

Today, the manufacturing instructions can be presented to the workers from handheld devices or on screens conveniently located for good visibility. Ideally, this same means of communicating the instructions could be used to gain insights into the design from CAD.

33.1.3   Kitting material development

The material received from suppliers and that manufactured at lower tiers in our own company will have to be related to the manufacturing process steps we have laid out in our manufacturing plan. It may be the responsibility of manufacturing to do this or the procurement organization, but someone must do it and it will probably be easier to manage the work if manufacturing is responsible. The manufacturing process will have been partitioned into some finite number of stations where work is accomplished. For each of these stations, it is necessary to have a clear description for the work that must be accomplished at the station in manufacturing planning plus a clear identification of the input material needed and output material condition. The complete collection of input material is often said to be kitted relative to the workstations. In a production line situation, the kitting work, obviously, has to stay ahead of the production line demand for material to preclude stoppages of the line.

A kit for a particular workstation must include all of the product material as well as any consumables needed. This includes the attaching hardware like screws, rivets, and cotter pins. If the station work entails attaching plumbing it might require some form of thread lubricant or sealing compound as well as safety wire to preclude the b-nuts loosening in operation due to vibration. Commonly, this smaller material is enclosed in plastic bags and larger entities wrapped in some kind of protective wrapping all of which must be disposed off in some fashion or collected for reuse. It could be returned to the kitting station in the container that was used to move the kit to the workstation as shown in Figure 33.1.

The packaging of all product material including attaching part may seem like an unnecessary burden when they could be supplied as open stock in bins. Figure 33.1 offers a caution for open stock available on the production line. The Teledyne Ryan Aeronautical unmanned reconnaissance vehicles *ca.* 1970 had been developed from the BQM-34 target vehicle by cutting the fuselage into two and adding a 30 in. fuel tank section, extending the nose to contain a large camera, and other system changes. The extension of the fuselage required installation of a 1/4 in. thick stiffening strap down the bottom of the fuel tank that was the main portion of the fuselage. The jet engine was contained in a nacelle below the fuel tank that was attached to the fuselage at four points. The two aft hangers for the nacelle were pinned to fittings mounted on the underside of the fuel tank.

Because of the 1/4 in. thickness of stiffening strap on the reconnaissance birds (model 147) this fitting had to be 1/4 in. shorter than the comparable part on the target drone (model 124). It happened that the model 124 and 147 fuselages were built up using similar tooling on the same manufacturing station and the loose parts required in the processes were contained in loose parts bins, as illustrated in Figure 33.2. It is not known how many times this happened, but it is known to have happened once because the vehicle that had this problem failed to meet an acceptance test requirement that limited the size of a step in adjacent fuselage panels. The problem was traced to the use of a 124 part for one nacelle hanger and a 147 part for the other. As a result, the whole fuselage had a twist in it because these were two of four basic tooling points upon which the whole fuselage was built. In this case, it was decided that the bird could be passed with body putty applied to clear the fuselage step though the author cannot recall if the government inspectors were told the cause of the step.

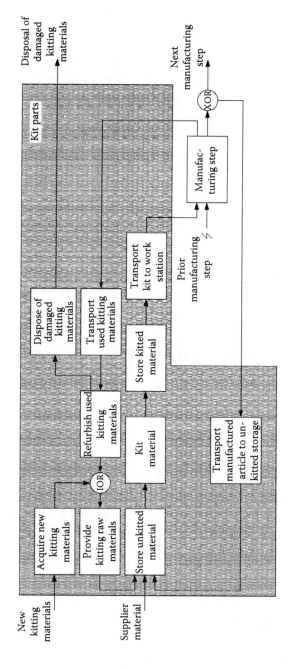

Figure 33.1 Kitting process flow.

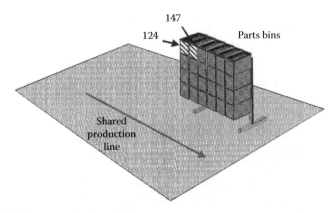

Figure 33.2 Loose parts bin problems.

In building the manufacturing planning, we should strive for the fewest cases of re-kitting as possible. Ideally, the kitting process would be accomplished at the lowest levels of assembly with intermediate material flowing direct from one station to another. But, each station will likely require a kit of attaching parts.

33.1.4 Tooling

Tooling falls into the same two categories as support equipment in the field. It includes general and peculiar tooling. The former category includes lathes, milling machines, metal brakes, and so forth. Computer-controlled milling machines qualify for the general category, but program peculiar software can drive the machines to accomplish fantastically complex cuts uniquely related to a particular program.

The peculiar tooling includes the program peculiar software for all manufacturing machines as well as machines made specifically for the program. This can be as simple as special cutting tools for use in general purpose milling machines and as complex as special equipment to wind the coils of a superconducting magnet. If a program requires castings, the tooling may call for program peculiar casting forms. One area of potential conflict is the tooling design relative to the product design. The tooling design should be reviewed by the IPPT responsible for the design of the related product entity.

33.1.5 Personnel and training

The work instructions, facilities, and tooling all have to be consistent with the kinds of people that will be recruited to perform the work. Ideally, the people who will be employed in these jobs will already be employees of the company that will be drawn from other programs that may be winding down from a manufacturing perspective. It may, however, be necessary to hire new people into the company to completely staff the force. The abilities and knowledge these people will be required to have mastered must be compared with the people who will be assigned and from that a training requirement derived. The method for accomplishing this training could include classroom instruction or on-the-job training conducted by people who are determined to be ready to accomplish program work. The latter could involve special training aids or other equipment similar to that required on the new program.

33.1.6   Facilities

The enterprise will have certain facilities available to it at the beginning of the program we are considering in this chapter. Those facilities will have certain loading as a function of other ongoing programs. Either they will be adequate to satisfy the aggregate (existing and added program) manufacturing capability or they will not be. If they will not be adequate, we may consider purchasing some items to print as discussed in Chapter thirty-two more energetically than if facilities are available. Alternatively, the enterprise may have to make an investment in new facilities buying and modifying existing structures or building new ones for use by the new or some existing program.

The design of the facilities will have to include adequate architectural and engineering features and local services like electrical power, communications, sewerage services, trash pickup, water supply, and gas service. It is conceivable that other services will have to be provided like periodic gaseous nitrogen service, water treatment services, and compressed air.

33.2   Integrated product development during production

Integrated product development teams are very effective up through first article test. Beyond this point, factory production problems can best be managed through one or more production-oriented teams organized around associate contractors, physical plants, and major suppliers. A given company's production facilities may, at any one time, be involved in production on several programs. None of the programs want another program's IPPTs adversely influencing the production of their product. At the same time, each program would hope to influence the production of its own product. A company's production resources have limits so they must be managed in the best interests of the overall company by company functional management with consideration for each program's needs.

Throughout the product development activity on each development program within a company, we have to find some way for the PIT production representatives to interact with each other and the EIT to integrate their aggregate production needs for the good of the company. Note that this is a case of cross-product, co-function integration at the highest level within the company. It is a case of cross-program integration not included within our integration spaces definition in Chapter three. Throughout this period, the first level of integrated development is on the product while the second level is on function (production). Manufacturing personnel must serve on the IPPT and PIT during development where their responsibility is to ensure that manufacturing needs are satisfied in the product design and the production process requirements and designs are consistent with the product features. These teams are primarily focused on engineering product development with strong participation by people representing the process views. Throughout the work up through critical design review and possibly functional configuration audit, it is appropriate for the teams to be led by people with an engineering background.

As each program passes through the first article milestone, their IPPT focus should join the production-oriented team structure organized about physical plants. This places integration by production facility at the first level and program integration at the second level. In both cases, both kinds of integration are needed but primacy should shift as you move through first article. The IPPT may be disbanded except to the extent that one or more of them may be needed to work design changes and new variants concurrently with production of previously defined product. Where they are not needed, a PIT nucleus may be all that is required to support the needs of the production-oriented teams. Some companies call this continuing engineering support of the production process liaison engineering. No matter the title, it can be carried on by production-facility-oriented teams with program membership.

Figure 33.3 illustrates a suggested view of this relationship in transition. The program managers for three programs each manage their program during development through a PIT and several IPPT. Each team has production membership that is coordinating with the management of appropriate production facilities throughout the development process and production functional management. The reason for this transition of teaming responsibilities to production is that it is easy to manage the production process through physical facilities. They are run by specific managers and staffed by specific teams of people who can be easily motivated within the context of their facility. They have specific material inputs and product outputs clearly defined by the responsible programs. Metrics can easily be focused on this structure. This arrangement also fits perfectly with the management of subcontractors and vendors because that is precisely the

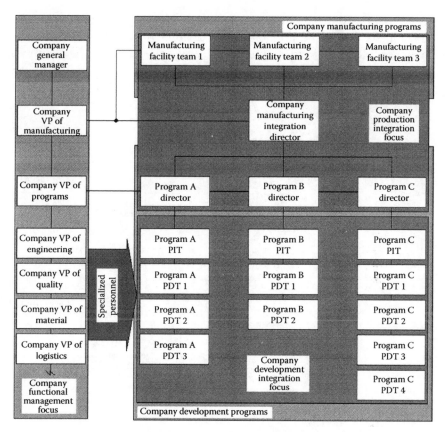

Figure 33.3 Integrated development transition to production.

way you have to manage them. All production, whether internal or external, is then managed through essentially the same basic process.

The production facilities should form integrated teams that include representatives from all of the programs served. These teams need not be structured exactly like the IPPT used in development, but they should include all specialties that are involved in the production process including engineering. Ideally, some people from the IPPT and PIT should transition to the production teams as each program matures to production status so that continuity is assured. The production teams use the first article production process in concert with IPPT support to gain needed experience for subsequent rate production.

When it develops that the production facility 1 team concludes that it could reduce production span time of the Program A product by 3 weeks by making a design change, the decision on whether to implement the change must be considered against several criteria. The Program A

configuration control board or PIT must determine the cost/benefit relationship based on the anticipated production run. Programs B and C
must verify that their needs will not be adversely influenced by changes
planned. Production management must determine the overall production effects of allowing facility 1 to reduce span time by 3 weeks. It may

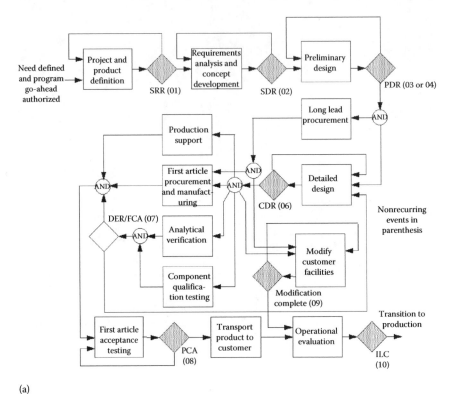

(a)

(b)

Figure 33.4 Team correlation with overall program flow: (a) nonrecurring process flow (PDT responsibilities), and (b) recurring process flow (production and
operations team responsibilities).

develop that the product output of facility 1 will only have to be stored in facilities that do not now exist for the whole period of time saved because other facilities cannot take advantage of the change. The time saved may be an overall detriment to the company's interests.

The techniques of operations analysis/research offer these production teams powerful tools to solve many of the problems they face. Queuing theory, linear programming, the transportation problem, and inventory models in particular are very useful.

Figure 33.4 illustrates, from a production process perspective, the suggested break point in responsibilities for IPPT and production teams. IPPT should work development issues through first article inspection, as shown in Figure 33.4a. Thereafter, the production teams should be responsible for all work done at their facility for the steps illustrated in Figure 33.4b.

The suggested team transform from IPPT to production-oriented teams also fits in effectively in companies where the development and production facilities are geographically displaced. In these cases, it is very difficult to extend IPPT effectiveness beyond the development phase.

chapter thirty-four

Quality influences

34.1 What is quality?

Juran's Quality Handbook (Juran, 1999) offers many definitions of the word quality. For the purposes of this book, quality is accepted as a freedom from faults in a system that satisfies the requirements for which it was designed. We would all prefer products with good quality but like all good things, quality does cost money. So, the question is how much quality can we afford? Generally, money spent on reliability and other system qualities will result in better quality as well as lower life cycle cost. Clearly, product quality approaching perfection is both unaffordable as well as impossible since cost approaches infinity as quality approaches perfection, as suggested in Figure 34.1.

A good argument can be made that spending more on quality in development can cause a reduction in life cycle cost because the largest element of life cycle cost is operations, and maintenance extended over the life of the system could be reduced through a design that encourages a lower maintenance response that may very well cost more in development to achieve. In Figure 34.2a, low spending on development could result in a higher life cycle cost than the higher spending on development, as shown in Figure 34.2b.

34.2 Product quality components

34.2.1 Product quality

The product that results from the development process has matured through a very orderly process that has been created specifically to produce a quality product. Requirements are defined that have been subjected to review and approval as properly defining the problem to be solved. A design effort will have been conducted in response to a carefully thought through plan. The design work would have been accomplished in accordance with the plan and a set of enterprise best practices. A set of components would have been produced and subjected to a thorough verification process proving that the design satisfied the requirements that drove the design. Planned product quality will have been established through this organized process. The question is to what extent were the program plan and enterprise best practices complied with during the

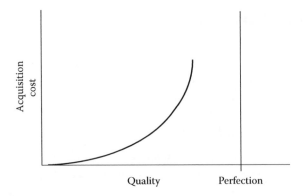

Figure 34.1 Quality versus cost.

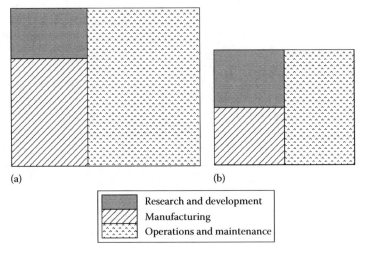

Figure 34.2 Early spending effects: (a) low R&D spending and (b) high R&D spending.

development effort. Therefore, a large part of the quality assurance (QA) effort is focused on processes.

34.2.2   Process quality

The manufacturing work must be planned in concert with how QA is going to interact with that process. It must be determined under what conditions is QA going to apply process controls. This can be overlaid upon the planned manufacturing steps in the manufacturing planning. In each step, we must determine what QA will evaluate and how it will be done.

Ideally, it would not be necessary to stop manufacturing activity in order to apply the QA activity but that may be necessary in some cases. In some processes that take place in manufacturing steps, it may not be possible for QA to inspect some aspects of the work performed because the work prevents access to the product features that should be evaluated. In these cases, it may be necessary to change the order of QA versus manufacturing work. In other cases, it may be necessary to inspect in real time with work performance. It may develop that QA and logistics share a common problem of access. It may not be possible to inspect and it may not be possible to repair some aspect of the product due to its structure and planned assembly order. Ideally, these cases would be discovered as early in the design process as possible so an alternative design solution can be formulated and implemented so as not to interfere with the program schedule. QA, like manufacturing, and procurement should have early membership in the PIT and appropriate the IPPTs.

QA has come a long way in American industry over the past 50 years. It is now an independent organizational entity reporting very high up in the management structure. It is no longer under the direction of manufacturing that it was originally created to support. But, QA has one more hurdle to leap to take its proper place in industry. If the application of QA to manufacturing was such a good idea, why would it not also be a good idea to apply it to engineering and program management, that is, to the whole enterprise. In the author's view, some parts of many enterprises have been reasonably successful in guarding against having their processes evaluated by QA. The two candidates suggested for this dubious honor are program management and engineering. Some enterprises have elected to apply integrated capability maturity model integration (CMMI) developed by the Software Engineering Institute of Carnegie Melon University to assess their maturity that includes an assessment of the extent to which the enterprise actually follows its practices. The integration model to date covers system engineering, software engineering, and concurrent engineering activities. This can have a positive effect on the performance of the enterprise on programs so long as the ranking the enterprise receives causes internal work to improve their performance so as to attain a higher ranking.

However, the CMMI approach would be so much more effective if it was made the responsibility of the QA organization rather than some CMMI guru. QA in companies has years of experience running process controls and can apply them to engineering and program management without skipping a beat.

QA should be responsible for assessing whole organizational performance on all programs in accordance with the procedures prepared by all functional departments to control the performance of their people on programs. The results of that assessment should be provided to the EIT that should be the overall process owner for the enterprise. The EIT should evaluate all of these inputs and prioritize the process changes suggested selecting a pathway from the current process definition to a new baseline. It is not necessary to establish new gurus for every new management initiative. It is much more satisfying to integrate good new ideas into any existing successful enterprise definition. These ideas will be further developed in *System Management: Planning, Enterprise Identity, and Deployment* (Grady, in press).

chapter thirty-five

Post development*

35.1 System employment overview

The life cycle for any system consists of two major activities: (1) system development and (2) system employment. This book covers a portion of the development activity but this chapter rounds out the system life cycle providing an overview of employment.

This process begins with the development of the base of operations for the system in one or more locations and the shipment of accepted material from the manufacturers to the operational facilities and the staffing of those facilities.

35.2 Operational test and evaluation

A special test organization or the planned initial operating organization may be tasked with the conduct of the operational test and evaluation (OT&E) that is conducted to evaluate the system from a user perspective. At this time, the system will have passed a development test and evaluation (DT&E) conducted by the contractor that developed the system. This process evaluated the system relative to the contract for its development including the system specification content that was part of that contract.

It is possible for a system to pass DT&E yet fail OT&E if the program has been acquired based on the wrong requirements, or the requirements have changed significantly during the time the development work was occurring and the contract not updated for those changes. It is not uncommon in DoD work that the acquisition process cannot provide adequate funding to permit fully satisfying the user requirements in one procurement. In these cases, the development process must reject some of the user requirements or scale back on the performance to some extent. It is possible if the development program has been badly managed that the user may evaluate to system against their complete set of requirements even though the procurement did not cover the complete set.

On a program where the user has been very aggressive in pursuing the acquisition agent to acquire the capability offered by the new system at the earliest possible date, system test and evaluation may have been

* The material in this chapter is from Grady, J., *System Synthesis Management*, CRC Press, Boca Raton, FL, Chapter 4, pp. 203–228, 1994. Used with permission.

approved before all of the fixes determined necessary in qualification testing have been completed, resulting in DT&E entry with less than full capability available. During the DT&E, some of these fixes will have been completed and kits installed in the equipment involved in DT&E allowing completion of some of the full capability testing. During DT&E, additional problems may be uncovered that will not have been fixed in the product expedited through the low rate manufacturing process to encourage the earliest entry into OT&E. As a result, OT&E is entered with a system that cannot satisfy all of the system requirements. At some time later, the modification kits become available to bring the system up to full capability so that OT&E may be completed.

On the other hand, the OT&E may uncover some real problems that may not have been found in DT&E and changes are necessary for these problems as well. Eventually, all of the changes will have been completed and OT&E completed. It is entirely possible that even though the system passed DT&E some negative OT&E results will be accepted resulting in additional changes paid by the acquisition agent in that the system had been shown to satisfy the contract. During this period, the contractor may have some difficulty supporting OT&E from a spares perspective especially with changes flowing through the system for the reasons discussed. Modification kits will very likely be flowing through the logistics system either requiring items to be modified in the field or be shipped back to the depot (the contractor early on) for modification.

The contractor often will have some difficulty following what is happening in OT&E because it is conducted by the user in accordance with a plan prepared by the user. Two ways the contractor can gain some insight are (1) by requesting an observer status or (2) through an on-going field engineering support arrangement. Throughout the 1970s and 1980s, for example, Teledyne Ryan Aeronautical maintained field engineering support contracts through Air Force logistics Command for service at Davis Mounthan Air Force Base in Tucson, Arizona where the Strategic Air Command and Tactical Air Command maintained home base units for their unmanned aircraft units as well as at operating locations on bases in South Korea, South Vietnam, and Thailand. The technical representatives at Davis Mounthan often were aboard the DC-130 flights while OT&E missions were launched providing technical support and could provide the company with accurate data about the performance of the product.

Field engineers operate under a very difficult set of loyalties. They are paid, of course, by their company but they are serving with people who they are depending upon for information. If it gets back to the unit that a tech rep has passed information back to their company that has created an embarrassment for the unit or people in it, the tech rep will get closed out of the flow of information. The author recalls a case where a DC-130 returned to base in Bein Hoa, South Vietnam with a missing vehicle that was known to have been uploaded after flight and recovery at Da Nang, South Vietnam. During debrief, the launch control officer (LCO) on that side of the 130 reported that he did not know why but during the return flight the bird had simply fallen off the pylon suggesting a bomb rack failure or human error in attaching the vehicle to the bomb rack. The author walked by the LCO hut subsequently that evening and had a beer with the LCO who told him that he had fallen asleep with his feet up on the launch console desk surface and the heel of his boot had apparently operated the locking mechanism on the launch switch. The author was able to get this information back to the company and a witch hunt was averted into an evaluation of the bomb racks on many aircraft like those on the DC-130 without poisoning the career of the Captain or himself. On several other cases similar attempts to keep the company informed about the truth resulted in serious problems for the author in the field.

35.3   Initial operational capability

When a predetermined criteria has been satisfied, the system will be declared to have arrived at an initial operational capability (IOC). For an aircraft, this criteria may include one squadron or one group equipped, staffed, and trained. This same unit may have accomplished the OT&E operations.

All of the pieces of the logistics support package have to have come together in the period leading up to declaring the OT&E milestone satisfied. The following list is phrased and motivated by a military scenario, but similar capabilities would be required for commercial activities as well:

1. A full complement of system end items will have been delivered and initially placed in a condition such that they are ready for an assigned mission.
2. The unit will be staffed by people new to the system some of whom will have to have completed special system training.
3. Initial training is complete for the staff. Some positions may require some form of certification and those cycles also have to be complete.
4. The spares at field and depot level will be adequate to support sustained operations.
5. All needed support equipment will be in place. This includes equipment supplied under contract plus any customer-furnished common support equipment. It also includes the special and common tools needed by mechanics and technician. Some support equipment may require calibration and everything of this kind should be within its calibration period.
6. The facilities are complete to the level needed to support sustained operations. On a DoD program, this may have been the

responsibility of the Army Corp of Engineers and a collection of private contractors.

7. The technical data for operations, maintenance, and supply must have completed validation at the contractor and verification at the IOC unit and changes from these experiences introduced into the copies available at the unit.

8. An effective depot capability must be in place. This will likely initially be at the contractor. Eventually, the responsibility may transition to a logistics organization liked to the operating unit. During initial depot support work, the contractor may employ the same equipment it used in the manufacture and acceptance testing of the product to support depot repairs, but in preparation for movement of this function to a customer organization, it may be necessary to develop a set of depot test equipment. Another alternative is the customer could let a contract for depot support covering a considerable period of time. The latter is especially true for depot support of software.

9. The IOC unit will have completed all administrative processes for being a tenant on a base. This may require housing and food service; fire and police protection; electrical power and facility communications services; water, sewer, and trash services; and establishing a clear chain of command with the next higher command.

35.4　Operations and maintenance

The most costly phase of life cycle cost is the period when the user is conducting operations and maintenance (O&M) on the system throughout its useful life. It is the span of time that this is ongoing that causes it to be so costly. This figure is staggering for an aircraft like a B-52 but its cost relative to its service has probably been a great bargain because it has stretched over a period of nearly 50 years so far and some people predict it is likely to hit 75. It has lived through the service periods of B-47, B-58, B-1, and B-2 bombers and will probably be replaced by the B-3, whatever that turns out to be. But then they said that about some of those earlier bombers too.

Throughout the period of service of the system, the contractor will have a role in support of the system in terms of engineering change proposals (ECPs) and the work related to them, spares provisioning, technical data support and possible on-going or recurring training, and in maintaining a good relationship with members of the command responsible for operating the system. Marketing people who may be familiar with the operating organization and system may be able to detect areas where the system could be improved that people in the unit may not yet see but can be encouraged to understand through conversation. The author can recall one field engineer who convinced the customer that they needed a new unmanned aircraft that resulted in a multimillion dollar contract to build this new system and for which the former field engineer became the program manager.

35.5   Logistics support

Logistics support for the system entails several of the IOC criteria listed above stretched over the life of the operational employment of the system. The complete system will pass through the reliability bathtub curve observed in a relatively high failure rate during a period when marginal parts installed in manufacture fail and are replaced followed by a long period of stable failure rates, and a final period of increased failure rates caused by wearout. If the development contract called for the use of burn-in to weed out marginal parts, then the early period of failures should be brief or nonexistent.

The contractor may be asked to provide engineering help to incident investigations at the user's facility to help determine the cause of accidents and recommend changes that would preclude reoccurrence of the cause. The author had this opportunity on several unmanned aircraft accidents that uncovered a failure in the manufacturing process at an altitude sensor supplier, but the primary causes were human error in the maintenance environment. A couple of these human errors were failure to attach a coax connection required for remote control of the vehicle and failure to connect a battery connector needed for recovery. In both cases, the connector had been pushed in to appear connected but the connectors had not been twisted to lock them in place. A firm was found that had the capability to make microscopic analysis of the connector seating surfaces establishing that they had not been connected at impact.

The principal target of long life is the technical data. What starts out a fairly simple process for operation and maintenance grows so as to include warnings and cautions and procedures to ensure that certain conditions are not in effect. It is very hard to think of all of the possible problems that can occur to a product and not possible to conceive of every possible scenario and sequence of events that will take place in the life of a system. During development, we do our best to identify all single-point failure possibilities and discourage related adverse consequences, but it is prohibitively expensive to deal with all dual point failures. It is interesting to note that most accidents are traced to a series of problems involving hardware failure, unintended software sequences, and human errors possibly influenced by unusual environmental phenomena.

In developing systems, we often consciously or by default assign a reliability of 1 to software and humans because we do not know how to deal with them otherwise. These are probably two of the most often cited causes of incidents in the use of systems.

35.6   Fielded system modification

Changes to deployed, or fielded, systems are accomplished through ECPs with DoD or NASA customers or less formal processes in commercial situations, possibly involving product recalls. Each one of these changes involves a mini-program activity that may be very simple or almost as complex as the original complete program itself. There will be fewer of these changes on a program that has expertly accomplished the development and production work in an integrated product development environment. But, even on such a program, the conditions in the customer's world change over time with a potential need for product system changes to compensate.

Unfortunately some companies have in the past based their program profit margin on ECPs by lowballing the contract competition to win even at a loss knowing that it will be made up in ECPs they have determined will be necessary during the run of the contract. This has become more difficult in recent years with DoD because of the depth and thoroughness of their selection process. But, regardless of your customer base's attitudes about, or skills in detecting, this practice, your company is damaged because it is very hard to develop a proper IPPT mentality when you go into contracts with this false position. To the extent that programs can be managed from a position of honesty and integrity it carries over to the way employees approach their work.

We said earlier that at first article the integration emphasis should transition to the production facilities' orientation. If this has already occurred when a change has to be made to the system, a fairly clean situation materializes. You can form an ECP IPPT to develop the change if it is substantial in scope or assign responsibility to the residual PIT if it has only a small impact. As first article acceptance of the change passes, the integration responsibility can move to the facility-oriented teams.

If, on the other hand, the change surfaces before first article of the basic system, the change development work can best be integrated using the existing IPPT structure. If the number of changes is large as the basic product transition to production, it can become difficult to accomplish effective integrated product development from a management and technical perspective. The changes can become interactive resulting in maddeningly complex relationships.

Where the IPPT have dissolved and transitioned to production, the Program PIT must perform the integration work corresponding to logistics factors such as training, spares provisioning, technical data, and support equipment.

35.7 Disposal and possible rebirth

For anyone who has been involved in military aviation for some time, it is painful to view the old warbirds in the Davis Mounthan Air Force Base bone yard in Tuscon, Arizona. Their paint is faded but they still proudly show their last operating unit markings. You know that some of those old B-52s flew some pretty rough operational missions in the past. You can feel the excitement of the crew as that F-4 approached the carrier deck on Yankee Station after a raid on Haiphong or the F-105 as it returned from a Wild Weasel mission to its base in Thailand. But, they have served their country well and are lined up to make one last contribution to their country. They will be stripped of useful material and chopped up for scrap aluminum. This is the end of the line for a system. It is called disposal. The sad end of an old warbird however is mitigated to some extent by the knowledge that from these humble endings a Phoenix will rise from the ashes to fly again.

35.8 Integration during system phaseout

As the product system nears the end of its usefulness to the customer, it will approach a need for phaseout and replacement by another system. The resources of the old system may not all need replacement. Some of them may be useful in a planned replacement system or within the context of other systems under the customer's control.

Integration work must be performed to partition all system resources into those with extended utility in specific applications and those that must be disposed of. Resources in the former category will have to have their final destination negotiated with new users or the contractor that wins the contract for the replacement system as customer-furnished equipment. Other resources must be economically and safely disposed of. Some of these resources may be sold or turned into scrap and sold. Other resources may entail hazards to the public. An obvious case is made for systems with nuclear weapons installed. Less obvious cases include systems with dangerous contained or associated chemicals (propellants, solvents, sealants, and so forth).

Ideally, the customer and contractor for the system would have realized the need during system development to identify disposal concerns and to have mitigated these concerns during the development process. But this kind of concern for disposal during development is a relatively recent phenomenon. Many systems in use today will not, as they approach disposal, have a prepared set of requirements and corresponding plan covering that disposal for this reason. As a result, some of these systems will pose very difficult questions for those who must after the fact devise a disposal process.

In the past, trying to interest a program manager during a proposal or early development work in disposal concerns has not been good for one's career. A program manager at this time only wants to focus on giving birth to his or her creation and the death of this system is very hard to think about. Hopefully, we will become in the future more sensitive to these needs in the development of systems that unavoidably must use potentially dangerous materials.

part seven

Closing

chapter thirty-six

Closing*

36.1 Integration at the beginning

We set out in the beginning to bring order to the system synthesis component of the system development process by identifying a finite number of integration processes and applying them to the several commonly observed integration situations necessary in requirements development, design and integration, production, verification, deployment, operations and maintenance, and phase-out periods of the system life cycle.

We saw that we could set the parameters of the integration spaces at any level of complexity we wished resulting in silly simplicity or maddening complexity at the extremes. We chose a three-dimensional approach that appeared to offer a reasonable balance between these extremes. We chose this organized approach for the identification of the component parts of the integration process motivated by the reasonable demand for completeness.

It is questionable whether this comprehensive, set-theoretic revelation about integration work will result in a direct improvement in the practice of system integration work in industry. But, it is hoped that it will make it possible to more effectively teach the process to all engineers such that a larger population will be more effective in applying the integrated product development, or concurrent engineering, approach to the development of systems to solve complex problems.

We saw that computer tools can be very helpful in aiding our integration efforts and that some particular improvements in these tools (interoperability for traceability) by the tool suppliers would provide even greater utility. Developments in computer communication are having a profound effect on the development environment by enabling the virtual team concept permitting engineers to effectively contribute to team activities from remote points including their homes and from the other side of the world. These networks are acting to knit together the internal mental processes going on in the minds of the many engineers participating on teams and programs.

But, no matter how refined and expensive our computer tools may become, no matter how technically advanced our integration processes may become, we must understand that it will continue to require educated system engineers on the floor armed with great human communication

---

* The material in this chapter is from Grady, J., *System Validation and Verification*, CRC Press, Boca Raton, FL, Chapters 9 and 10, 1994. Used with permission.

skills to make the process work. So, in addition to the contents of this book and the technologies associated with their company's product line, those who would become system engineers focused on integration work should master spoken and written communication skills, interpersonal relations, and organizational dynamics appropriate to small teams.

36.2   Process improvement

This overall process is founded on a need to decompose large problems into related sets of smaller problems that can be successfully attacked by small teams of specialists because we individuals are knowledge limited and because we have found that a group of people who can work well together and share their experiences will conceive more alternative solutions from which to select the preferred alternative that is consistent with the requirements. That is, a coordinated group will expand the solution space beyond that available to a single individual engineer, however talented that engineer may be.

Is there a better way to develop complex systems? We do not know at present but there is good reason to think that we are not at the peak of perfection in our theory and certainly not in our practice. Progress is primarily unidirectional and ever possible. Given that we continue to seek improvements to current methods over time in a continuous stream of advancements, why should we not expect eventually to make substantial improvements in our methods. But, this process requires a plateau from time to time where we can collect our thoughts and take stock of our advance. This book attempts to create such a plateau describing what the author has found to be the best our profession has to offer in the opening decade of the twenty-first century. Incremental changes to this process will continue to improve the process at the margins in terms of reduced cost and schedule demands for given product system performance capability advances. Certainly, computer applications will continue to reduce the number of mind-deadening repetitious tasks we must perform and enable greater opportunities for what we humans do best—creative thought.

Explosive improvements in the development process will probably not come from incremental changes, however. They will come from shocking revelations and changes in approach based on research into how the human mind works and the application of computer technology to solve the fundamental conflict between human knowledge limitation and the need to effectively apply more knowledge than any one person can master. Is it possible to avoid the heavy hand of order? Or is order necessary to provide an affordable framework for problem decomposition and solution integration? Can we replace our current process with its inherent heavy demand for order with an appeal to creativity across the whole process not just in these small islands that we create by decomposition?

Let those among us who have the necessary skills and experience seek out the potentially revolutionary ideas that will toss the structured development method on the ash heap of history and replace it with a method that acts to elevate the human spirit and appeal more effectively to man's creativity. But, in the meantime, let us, the rest of us, seek to use this plateau as a basis for additional incremental improvements in the process that has worked reasonably well for many years. After all, the structured development process described in this book may really be the ultimate approach.

We must also compare and contrast the traditional systems approach, encouraged by the military competitions at work prior to the early 1990s, with the needs of the commercial market place. Is the optimum commercial development process an accelerated version or a subset of the DoD-inspired process or is it fundamentally different in nature? If the traditional process is generally the right approach, we need to speed up its effects, and the techniques and machinery discussed in Part one may be useful to that end. There may be other techniques not exposed in this book to increase the speed and effectiveness of the communication and understanding of ideas. We should actively seek out these potential opportunities.

If the optimum commercial process is radically different from the traditional approach, how may it be characterized? Do ISO 9001, IEEE 1220, or ANSI/EIA 632, all of which appear to deviate very little from the fundamental concepts upon which the DoD process is based, reflect the needs of the commercial market place? If they do not describe commercial needs, what is the answer?

The author's prescription, expressed in this book, for commercial firms is a tailored version of the DoD approach that places the balance point closer to creativity than the order extreme with correspondingly greater risk than DoD would be comfortable with. Documentation is necessary, but can be radically simplified by replacing plans, specifications, test procedures, and many other documents with raw database content observed directly on one or more larger computer screens. The review and decision-making process can be made very responsive through the effective information sharing techniques covered in Chapter seven.

There is one other challenge for system engineering practitioners in industry that need not await any new developments. The author has found in talking to many fellow practitioners in industry that few are content with the process they currently apply in their company. Despite the availability of skilled system engineers on staff, many companies do not make progress in system engineering. They suffer a management staff smart enough to speak supportively of the systems approach who are actually married to the ad hoc approach they preferred as design engineers and managers in the past. Simply awaiting the retirement of these dinosaurs is not a solution to this problem since they are created in the design community as fast as they retire from management and, in engineering organizations, management people are commonly going to be advanced from the design community.

Therefore, system practitioners must find a way to influence the education of design engineers in universities and engineering management people in industry to understand that their success as design engineers and managers is encouraged through cooperative efforts that preserve the maximum solution space for design teams consistent with the minimum order needed to ensure synergism of effort and to protect their efforts from avoidable errors that will require many changes after the principal design work is complete.

It is possible that the worldwide market place, which includes some developing countries that are new suppliers of fairly complex products using a labor force that can live on low wages relative to more advanced countries, will force a reduction in the living standard in more advanced countries to preserve their competitiveness. It is possible that in some countries, the workplace will become very regimented in support of high productivity goals. World War II offered good evidence that armies composed of free men under arms could defeat armies composed of people functioning under a highly authoritarian command structure. But, can free men with a relatively rich lifestyle compete with a society encouraging a highly regimented workforce where humanity takes second seat to production. It would be tragic if the American solution to a need for improved efficiency in enterprises in response to market forces followed the failed totalitarian methods experimented with by some governments in the twentieth century.

36.3   The current cutting edge

In the summer of 2007, when this chapter was being written, some new ideas had been flowing through industry in a chaotic jumble that will very likely overflow into a new development paradigm at some time in the future. If there is to be an explosive process change, it is likely to come from some of these ideas.

36.3.1   Universal modeling set for system architecting

Work on-going on the evolution of SysML, derived from UML 2, will very likely lead to the emergence of a universal modeling capability equally useful for problems that can be solved in hardware, software, people, or facilities as well as at the system level. In Chapter twelve, the door was opened to this merger. We said that today one could form a universal modeling capability using the union of traditional structured analysis (TSA) and UML but would have to deal with some difficulty in establishing

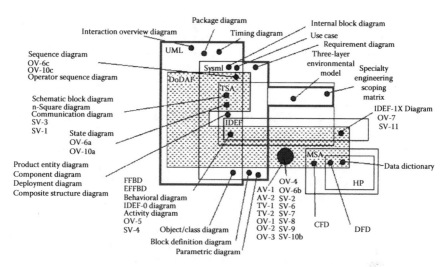

Figure 36.1 Universal architecture modeling set today. (From Grady, J., *System Validation and Verification*, CRC Press, Boca Raton, FL, pp. 231–275. With permission.)

traceability across the HW–SW chasms using clumsy techniques like those discussed in Chapter twelve. Figure 36.1 reveals a Venn diagram illustrating several of the current modeling artifacts available and in use today in an attempt to explore commonalities and differences and expose today's universal model. Early OOA is excluded, but modern structure analysis (MSA) and the Hatley Pirbhai extension as well as UML and Department of Defense Architecture Framework (DoDAF) are included, from the software sphere. Overlapping areas in the diagram imply common applications of particular modeling artifacts and isolated spaces suggest unique applications. The reader will note that all sets of artifacts do include some form of functional modeling. The activity diagram is from UML and SysML, of course. The functional flow block diagram (FFBD), enhanced functional flow block diagram (EFFBD from CORE), and behavioral diagram (from RDD-100) are from TSA and IDEF-0 is from IDEF languages. These different artifacts have unique features with simple functional flow diagramming being the least capable that the others collapse into if the unique features are stripped away.

The heavy boarder bounds a set of artifacts that the author would fold into a universal architecture modeling capability as of the date this was written. Several DoDAF artifacts are illustrated grouped outside this boundary, but at the time this was written, people were working on a UML framework for DoDAF Modeling (UFDM) that will result in that model moving into the universal set.

A system engineer at the time this was being written was obligated, in the author's view, to at least understand and be able to discuss features of all of the models shown in Figure 36.1 with the responsible modeler for a particular system development application. Ideally, the system engineer would be able to apply at least some subset of these artifacts in actually performing the modeling work as a means to gain insight into appropriate requirement for the system and its entities.

Many system engineers probably could not live up to that expectation at this time. To the extent that one wishes to continue working as a system engineer, we should all work toward mastering these modeling languages, however. Figure 36.1 reflects the minimized set at the time, because SysML was not yet ready to completely replace TSA, in the view of the author perhaps only because there are not enough people who can apply it successfully yet. However, enterprises that are involved in the development of systems to solve complex problems should be working toward the replacement of TSA with SysML and seeking to participate in its development with the Object Management Group (OMG) if they have strong feelings about the outcome. Part of the problem with SysML entry in many companies is probably that not enough system engineers have taken the time to understand it. But the author believes that there are several remaining problems with UML and SysML including a questionable ability to deal with specialty engineering and environmental requirements, lack of a common requirements capture mechanism, availability of a common product entity structure depiction, and an apparent aversion to specifications that system engineers have become perhaps overly fond of.

SysML added the artifact <<requirement>> that had not been identified in UML. It could conceivably take care of specialty engineering and environmental requirements in SysML, but it is not backed up with any organized mechanism to encourage completeness through which one gains insight into what might be appropriate content in these two areas. Some organizations attempt to close this gap with quality requirements working groups, but a more organized approach is encouraged.

In TSA, one can apply a specialty engineering scoping matrix, illustrated in Figure 36.2 and described in Section 23.2.3, to coordinate specialty disciplines with product entities installing an orderly process through which specialty requirements may be coordinated with the entities. The specialty engineer thus identified must write one or more requirements from his or her discipline for this indicated specification. There is, perhaps, no restriction in SysML that would preclude a system engineer from applying this construct to help determine what to write specialty engineering/quality requirements about within an application of SysML.

TSA also has available a three-tiered environmental construct useful in helping to identify environmental requirements alluded to in Section 23.3.2 and pictured in Figure 36.3 that is not available in SysML. Given that one can see the need for a particular environmental stress, it could certainly be identified as a SysML requirement but, once again, the analyst should

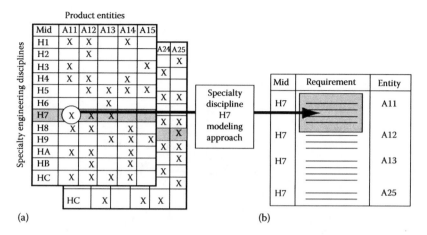

(a)   (b)

Figure 36.2 (a) Specialty engineering scoping matrix and (b) requirements analysis sheet (RAS). (From Grady, J., *System Validation and Verification,* CRC Press, Boca Raton, FL, pp. 231–275. With permission.)

have an organized approach to gain insight into all of them that might be appropriate. The three-tiered environmental approach employs tailored, selected environmental standards mapped to spaces within which the system entities will be employed to identify natural environmental stresses combined with a threat analysis for hostile stresses and a thoughtful effort to identify noncooperative stresses at the system level.

End item environmental requirements will yield to a three-dimensional service use profile through which one coordinates links between system environmental stresses, physical process steps, and product entities plus a thoughtful attack on identifying any internal sources for self-induced environmental stresses. Component-level environmental analysis can be encouraged by a zoning analysis of end items identifying physical zones that can be defined in common environmentally. Components mapped into these zones in a packaging exercise inherit the environmental stresses of the zones. Once again, there does not seem to exist any prohibition in SysML about how the analyst gains insight into the environmental stresses identified as SysML requirements so there may be a work-around in this area as well.

The author recognizes natural, noncooperative, hostile, and self-induced environmental classes that must be identified and included in specifications. He further recognizes that cooperative system environmental relationships are handled during development as system external interfaces because there is always someone with whom you can talk to reach agreements on the relationship. Few of these classes have a useful application in software environmental analysis so it may be necessary to form

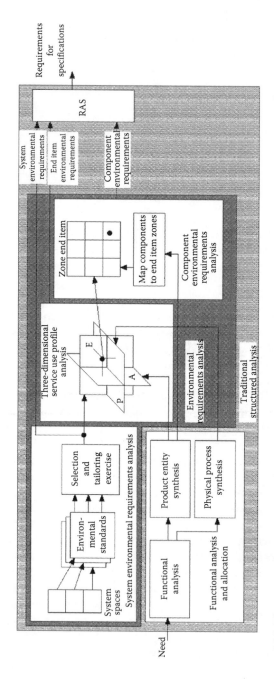

Figure 36.3 Three-tiered environmental analysis. (From Grady, J., *System Validation and Verification*, CRC Press, Boca Raton, FL, pp. 231–275. With permission.)

a new class called software environmental stresses that can include languages, machine structure, compiler, operating system, and other factors.

The author accepts that SysML can be employed to identify performance requirements that deal with what the system/entity must do and how well it must do it. Also, there are adequate constructs to identify interface requirements. With an acceptance of some machinery to help identify environmental and specialty engineering requirements/quality in SysML, the author would accept that SysML was ready to move into prime time with one final possible exception.

A common requirements capture mechanism is encouraged in the form of what the TSA devotes would call a requirements analysis sheet that can be formed as a table in a requirements database tool. This table links the modeling artifact from which the requirement was derived, the requirement, and the product entity to which the requirement is allocated or assigned. UML and SysML use two different ways to model the static structure of the system composed of product entities. UML uses four different constructs objects and classes, components, nodes, and subjects; while SysML uses blocks. A common method is encouraged, and until these two models come to closure, the product entity diagram of TSA is encouraged.

Computer software engineers who apply UML appear to rely almost entirely on the use-case analysis for identifying requirements and reject the idea of capturing these requirements in a format prescribed by DoD in what is called a software requirements specification (SRS) preferring to use a use-case specification that everyone seems to have their own standard for. The author actually would like to see all paper specifications trashed with everyone relying on the content of databases directly, but it is unlikely that DoD among others will move down this road in the near term. Therefore, the industry should either abolish the SRS format replacing it with a proper use-case specification standard or determine how to embrace an SRS format for software developed using UML in common with one developed using SysML. One might ask why this is important to the system engineer. The concern is that SysML also employs use-case analysis and encourages the development of use-case specifications. Before everyone leaps off the dock onto the good ship SysML we should first reach some understanding about the intended effect on how we currently capture requirements. The author has no worry about a leap into model-driven development through this path, but it should be better advertised if it is intended that specifications as we know them today will forever be changed or even eliminated.

The author has devised a universal specification format that can be used no matter the modeling basis for requirements derivation that is taught in courses his company offers that is planned for a future book covering the content of this paragraph. It is likely that the paper specification used in system development for a very long time will become a relic in the not-too-distant future with movement from problem space modeling to solution space modeling without the benefit of a paper specification.

Once any remaining problems with SysML are resolved, the team that can use the universal modeling paradigm composed of the minimized set of modeling capabilities shown in Figure 36.1 (UML and SysML very likely) so as to first gain insight into requirements, rotate the modeling activity interest to design concepts, and then couple the design modeling effort into simulation work will be yards ahead of those racing them to the pot of gold.

36.3.2   Model-driven development

The most exciting cutting edge activity in work at the time this book was being written was the idea of model-driven development. The programs that developed the Lockheed P-38 Lightning, the Boeing B17 Flying Fortress, and the Grumman F6F Hellcat could be fairly said to have been applying a document-driven development approach. Paper contracts and specifications led to the preparation of paper engineering drawings by engineers using slide rules, pencils, T-squares. Paper test plans were prepared to guide the verification process.

The author can remember this development environment stretching into the 1960s where one often saw people collecting around a drafting board within a large room full of drafting boards to discuss a current design problem. There were no computers except an old analog machine upon which flight control system work was performed and a big mainframe for heavy-duty engineering and business-related analysis. Engineers on the floor were still using slide rules but a few brave soles were experimenting with handheld calculators.

As digital computers became available as stand-alone workstations, functional departments in companies began developing ways to accomplish their work faster and better using them resulting in programs consisting of human islands forced by specialization and on these islands were separate computers that could not exchange information any better than the specialists. One could call this a database-driven development process. Everyone had their own database but the exchanges that occurred between them were accomplished between the engineers who owned them using a spoken language that in some cases would result in changes in the data in those databases.

The ultimate development machinery might be formed by humans interacting with databases that are interrelated in useful ways through the network that joins them and servers that have access to their data on which are running applications that can process various combinations of data in those databases to create the equivalent of the thoughts that human engineers now derive from the isolated data. For example, the mass properties engineer finds it necessary to change the weight allocation of an entity that today might lead a control system engineer to conclude that it would be necessary to change a term in a gain equation elsewhere in the system. If these relationships can be made clear to the computer network then the network could be caused to detect this change and either signal the humans that there was a need for a change or simply make the gain equation change.

There will probably first evolve a semiautomatic effect where the humans are notified of a needed change followed by a cross-functional discussion and either a decision to permit the change or to reject the change. There may be hundreds of these kinds of relationships at work at any one time as there are in the minds of the collective program staff presently. There is a question about the extent to which we would ever permit an automatic action in this kind of situation.

Will we ever observe in the work place the results of the problem space modeling evolving into the solution space modeling and the resulting code for an ASIC passed on to the machines that create the chips. That is quite a stretch, but it is more believable than the results of modeling work resulting in a structural design automated in CAD and passed on to the appropriate tooling for hogging out the part from an aluminum billet. But we can all accept that there could easily be a better combination of human and machine participation in the development process than we now experience.

Over time, we will come to recognize that the computers are extending the minds of the humans staffing programs and through networks permitting these computers to carry on cross-functional work on their own. There is clearly a great deal of research to be done in the computer tool world, management science, and the system engineering community to evolve a pattern of human and machine responsibilities and actions that could collectively improve the process of developing solutions to complex problems. In the process of working toward a solution, it will be wise to include some consideration for the utility of some of the ideas floating in the river regarding the topic of complex systems development.

36.3.3 *System development in a broader context*

From the beginning of time, man has worked to master the control of his environment evolving the knowledge domains of mathematics, physics, and engineering to permit man to understand problems and fashion deterministic solutions to those problems using linear methods appealing to an orderly understanding of phenomena. Even where the problem was not entirely linear in nature, we have simplified it to allow us to deal with the problem in a simpler framework. Man has done this while knowing from frequent experiences that he has always lived in a world that includes other possibilities than just the orderly systems that he has tried to create.

Orderly systems generally function in a closed loop repeating some portion of their functionality generally in the interest of economy through reuse of some portion of the system assets. These systems obey the theorems of mathematical logic and laws of science permitting us to determine from a current system condition how it will react to specific stimulus and the next state in which we should expect to find the system. These systems are deterministic meaning that all effects have sufficient causes.

There exist other deterministic systems that are not orderly and these systems follow the laws of chaos. The fact is that the next state of such a system is predictable given that you have a good description of the current state and the action that disturbs it. The weather largely follows this pattern. The emergence of turbulence in the study of wing design can be understood using chaos. System engineers often consider the emergence of chaos as the opposite of a systematic system development process when they really mean random, suggesting that not even the program manager has a clue about what is going on or how to control it.

We apply the laws of chance to many phenomena in system development such as in determining needed and predicted reliability and maintainability figures. In other cases where we find it difficult to determine a fixed value of a parameter that will always be true, we can predict a mean and the amount of variation about that mean in a statistical fashion.

While there are many things of interest in understanding a problem to be solved by a system, there are two things of fundamental importance. First, we wonder about the initial conditions at the time we first become interested in the situation. We also wish to understand how the situation will change over time and given an initial condition, to what extent it is possible to predict a future condition. From this perspective, we can partition all systems that we might show an interest in into the four subsets illustrated in Figure 36.4. Of these, we can predict a future condition given the history of the system for ordered and chaotic systems. This is simply not possible for random and complex systems. In the former case, we may be able to identify a mean and variance that gives us some insight into a future condition, but many a failed gambler will question any approach claiming to offer better than a 50% chance in a binary situation.

Complex systems are composed of many interconnected entities each autonomously acting under its own principles that could be based on order, chaos, randomness, or complexity with no clear means to predict future events from the perspective of an aggregate system viewer. Many complex systems are adaptive in nature meaning that they can autonomously respond to stresses in their environment and change their behavior to stimulus without outside control. At the time this book was written,

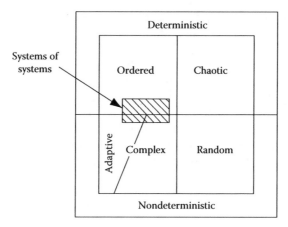

Figure 36.4 System possibilities. (From Grady, J., *System Validation and Verification*, CRC Press, Boca Raton, FL, pp. 231–275. With permission.)

there were no clearly defined principles and techniques proven effective in the development of complex systems though research had begun into discovering some.

Some complex systems may qualify as systems of systems but ordered systems can be formed from collections of ordered systems. Many military systems under development at the time this book was written qualified as systems of systems. Some of these systems may be conceived at one time as systems of systems, each component system possibly under the development of a different acquisition organization or program with different sets of development contractors. An experienced system engineering will instantly see the difficulty of managing interfaces between these evolving systems where there may not even be any contractual basis for the formation of an aggregate ICWG. As a result, progress may require a higher incidence of false starts and iteration than anyone would be comfortable with.

In other cases, systems of systems may simply mature from a collection of existing systems that can be thought of as accomplishing a greater capability than any single system or subset of them. The current situation may include several systems that users come to understand through experience with them that they are actually interacting but in a very inefficient fashion often through humans passing on information from the results of one system to those responsible for the operation of another system. This phenomenon has been at work for a very long time improving the systems that serve us.

The term "family of systems" is also of interest as a way to describe a collection of systems that share certain characteristics of interest. There may be no synergy between the systems in a family as there must be between entities in a system. For example, one may be interested in the family of systems that are operated to accomplish space transport from Cape Canaveral Air Force Station. These systems have little in common other than accomplishing the same function. They may share certain resources once they are in flight, but they probably all have their own unique composition, independent launchpad, and dedicated crew.

The range of the word "system" is thus expanding in time, embracing a more comprehensive and complex collection of entities. As the science of complex systems expands, we will see more commonality exposed between man-made systems and natural systems and perhaps the development of man-made systems will take on some of the characteristics of the evolution of natural systems that we postulate to have happened in earlier times.

36.4   Integration at the end

Many engineers just entering the work force from the university feel that there is little new for them to experience for the first time. All of the great thoughts have been thought. All of the great discoveries have been made. This is simply not so, of course. It is hard to see how the few leading edge topics discussed in a very cursory way in this chapter could fail to ignite in a new hire a fire that may last them a whole career in this fascinating profession called system engineering. Hopefully, the plateau offered by this book will provide some system engineering practitioners with a foundation upon which they may base future initiatives to improve the understanding and practice of system engineering as it applies to system synthesis within their companies and their profession. Those who would improve our processes will hopefully understand the platform upon which they are standing or they will surely repeat many of the errors we made en route to realizing this platform.

Bibliography

American National Standards Institute (ANSI)/Electronic Industries Association (EIA), 1998. ANSI/EIA 632, *Processes for Engineering a System*; New York: ANSI.

Barabasi, A.-L., 2002. *Linked: The New Science of Networks*; New York: Perseus Publishing.

Blanchard, B.S., 2003. *System Engineering Management*; New York: Wiley.

Boehm, B.W., 1988. *A Spiral Model of Software Development, Tutorial: Software Engineering Project Management*; Washington, DC: IEEE Computer Society Press.

Byrne, J.A., 1993. The Virtual Corporation. *Business Week*; February 8, 1993 issue.

Cross, N., 1994. *Engineering Design Methods*; New York: John Wiley.

Davidow, W.H. and Malone, M.S., 1993. *The Virtual Corporation*; New York: Harper-Collins.

Electronic Industries Association (EIA)/Institute for Electrical and Electronics Engineers (IEEE), 1995. EIA/IEEE J-STD-016, *Software Development and Documentation*; New York: EIA/IEEE.

Electronic Industries Association (EIA), 2002. EIA 731, *System Engineering Capability Model*; New York: EIA.

Forsberg, K. and Mooz, H., 1991. The relationship of system engineering to the project cycle. *First Annual Symposium of the National Council on Systems Engineering (NCOSE)*, Chattanooga, TN.

Grady, J., 1993. *System Requirements Analysis*; New York: McGraw-Hill.

Grady, J., 1994. *System Integration*; Boca Raton, FL: CRC Press.

Grady, J., 1994. *System Synthesis Management*, Boca Raton, FL: CRC Press.

Grady, J., 1994. *System Validation and Verification*, Boca Raton, FL: CRC Press.

Grady, J., 1995. *System Engineering Planning and Enterprise Identity*; Boca Raton, FL: CRC Press.

Grady, J., 2000. *System Engineering Deployment*; Boca Raton, FL: CRC Press.

Grady, J., 2005. *System Requirements Analysis*; Burlington, MA: Elsevier Academic Press.

Grady, J., 2007. *System Verification*; Burlington, MA: Elsevier Academic Press.

Grady, J., in press. *System Management: Planning, Enterprise Identity, and Deployment*, 2nd edn.; Boca Raton, FL: CRC Press.

Headquarters, Air Force Systems Command USAF, 1964. Air Force Systems Command Manual 375-5, *System Engineering Management Procedures*, Andrews Air Force Base, MD.

Institute for Electrical and Electronics Engineers (IEEE), 1998. IEEE 1220 *Standard for Application and Management of the Systems Engineering Process*, New York.

Institute for Electrical and Electronics Engineers (IEEE)/Electronic Industries Association (EIA), 1998. IEEE/EIA 12207 *Software Life Cycle Processes*, New York.

International Standards Organization (ISO), 2008. ISO 9001 *International Standard for Quality Management*, Geneva, Switzerland.

Juran, J.M., 1999. *Juran's Quality Handbook*; New York: McGraw Hill.

Machol, R.E., 1965. *System Engineering Handbook*; New York: McGraw-Hill.

Morais, B. and Mar, B.W., June 2004. Bridging the Gulf—Systems engineering and program management. Presentation at the *2004 INCOSE Symposium*, Toulouse, France.

U.S. Department of Defense, Never released. MIL-STD-499B, *Engineering Management*; Washington, DC: Department of Defense.

U.S. Department of Defense, 1975. MIL-STD-881A, *Work Breakdown Structures for Defense Material Items*; Washington, DC: Department of Defense.

U.S. Department of Defense, 1994. MIL-STD-498, *Software Development and Documentation*; Washington, DC: Department of Defense.

U.S. Department of Defense, 1988. DI-CMAN-80008A, *System/Segment Specification*; Washington, DC: Department of Defense.

Index